CREATING
INNOVATIVE PRODUCTS
Using Total Design

CREATING
INNOVATIVE PRODUCTS
Using Total Design

The Living Legacy of Stuart Pugh

Stuart Pugh

Edited by
Don Clausing
Xerox Fellow in Competitive Product Development
Massachusetts Institute of Technology
Cambridge, Massachusetts

Ron Andrade
Professor, Product Development and Quality Management
Federal University of Rio de Janeiro
Rio de Janeiro, Brazil

ADDISON-WESLEY PUBLISHING COMPANY
Reading, Massachusetts   Menlo Park, California   New York
Don Mills, Ontario   Wokingham, England   Amsterdam   Bonn
Sydney   Singapore   Tokyo   Madrid   San Juan
Seoul   Milan   Mexico City   Taipei

Many of the designations used by manufacturers and sellers to distinguish their products are claimed as trademarks. Where those designations appear in this book and Addison-Wesley was aware of a trademark claim, the designations have been printed with initial capital letters.

The publisher offers discounts on this book when ordered in quantity for special sales.

For more information, please contact:
Corporate & Professional Publishing Group
Addison-Wesley Publishing Company
One Jacob Way
Reading, Massachusetts 01867

Library of Congress Cataloging-in-Publication Data
Pugh, Stuart.
    Creating innovative products using total design: the living legacy
    of Stuart Pugh / Stuart Pugh, with contributions by Don Clausing,
    Ron Andrade.
       p.   cm.
    Includes bibliographical references and index.
    ISBN 0-201-63485-6
    1. Design, Industrial.   2. Engineering design.   3. Computer-aided
    design.   I. Title.
    TS171.4.P84   1996   95-26006
    745.2--dc20   CIP

0-201-63485-6

1 2 3 4 5 6 7 8 9–MA–99989796

First printing March 1996

Contents

List of Illustrations   *vii*
List of Tables   *xiii*
About the Author   *xv*
Foreword   *xvii*
Preface   *xix*
Acknowledgments   *xxv*
Introduction   *xxvii*

Part I   **Design in Education and Industry**   **1**

1   Engineering Design: Time for Action   *5*
2   Engineering Design: Teaching Ten Years On   *17*
3   Engineering Design at the Postgraduate Level   *25*
4   Engineering Design Education with Real-Life Problems   *33*
5   Projects Alone Don't Integrate: You Have to Teach Integration   *47*
6   Marathon 2550: A Successful Joint Venture   *55*

Part II   **Design Process and Philosophy**   **77**

7   Total Design, Partial Design: A Reconciliation   *79*
8   Design: The Integrative Enveloping Culture, Not a Third Culture   *87*
9   Systematic Design Procedures and Their Application in the Marine Field: An Outsider's View   *97*
10   Design Activity Models: Worldwide Emergence and Convergence   *115*
11   Integration by Design Is Achievable   *131*

Part III   **Design Techniques and Methods**   **137**

12   The Dangers of Design Methodology   *141*
13   A New Design: The Ability to Compete   *157*
14   Concept Selection: A Method That Works   *167*
15   State of the Art on Optimisation in Great Britain   *177*
16   Enhanced Quality Function Deployment   *183*

Part IV   **CAD and Knowledge-Based Engineering**   **205**

17   CAD in the Context of Engineering Design:
      The Designer's Viewpoint   *209*
18   CAD and Design Education: Should One Be Taught Without the
      Other?   *223*
19   The Application of CAD in Relation to Dynamic/Static Product
      Concepts   *235*
20   CAD/CAM: Hindrance or Help to Design?   *245*
21   CAD/CAM: Its Effect on Design Understanding and Progress   *261*
22   Knowledge-Based Systems in Design Activity   *273*

Part V   **Design Teams, Management, and Creative Work**   **289**

23   Further Development of the Hypothesis of Static and Dynamic
      Concepts in Product Design   *293*
24   Engineering Design in Practice   *305*
25   The Organisation of Design: An Interdisciplinary Approach to the
      Study of People, Process, and Contexts   *325*
26   Engineering Design: Towards a Common Understanding   *343*
27   Balancing Discipline and Innovation   *349*
28   Organising for Design in Relation to Dynamic and Static Product
      Concepts   *359*
29   The Design Audit: How to Use It   *381*

Part VI   **Design for X**   **389**

30   The Engineering Designer: Tasks and Information Needs   *393*
31   Design Is the Biggest Exposure   *403*
32   Manufacturing Cost Data for the Designer   *411*
33   Engineering Out the Cost   *425*
34   Quality Assurance and Design: The Problem of Cost
      Versus Quality   *433*
35   Give the Designer a Chance: Can He or She Contribute to Hazard
      Reduction?   *443*

Part VII   **Design Research**   **449**

36   Research and Development: The Missing Link—Design   *451*
37   Engineering Design: Unscrambling the Research Issues   *461*
38   Long-Term R&D Outcomes: Will They Miss the Market?   *473*

Part VIII   **Total Design: Summary of the Whole**   **485**

39   Towards a Theory of Total Design   *487*
      Appendix: Chronology   *529*
      Index   *533*

List of Illustrations

1.1   The Design Activity Model (1982)   *11*

2.1   The Design Activity Model   *21*
2.2   Subject and Time Allocations for the First Term of the Design Course   *23*

4.1   Layout of the Existing System   *38*
4.2   Alternative Cycle   *39*
4.3   Closed-Loop System: Trays Bumper to Bumper, Smooth Motion, Flow Continuity   *40*
4.4   Trays with Hinged Lids, Overhead Handling System with Suction Cups   *41*
4.5   Transfer Paddle Wheel   *41*
4.6   Trays with Hinged and Restrained Lids, Tray Rotated Through 360° for Quench   *42*
4.7   'Moon Rider' Concept   *43*
4.8   Quench Station   *43*
4.9   Fill Station   *44*
4.10   Kiln Re-entry Station   *44*
4.11   Prototype Equipment   *45*

5.1   Design Core Bounded by Product Design Specification   *51*

6.1   Marathon Concept A   *66*
6.2   Marathon Concept B   *66*
6.3   Marathon Concept C_1 *66*
6.4   Marathon Concept C_2 *66*
6.5   Steering Subassembly Mounted at the Centre Pivot Point of the Chassis   *70*
6.6   Photo of Marathon 2550   *72*
6.7   Photo of Marathon Chassis on Test Rig Undergoing Simulated Conditions of Many Years' Rigorous Site Use   *73*
6.8   The Chassis (Mainly of Folded Sheet Steel Construction)   *73*

7.1   Professional Divisions Encourage Partial Design   *82*
7.2   Total Design Product Model   *84*

8.1   Design Viewed as the Third Culture   *90*
8.2   Basic Factors to Be Considered in Design   *91*
8.3   Design Viewed as the Combination of Art and Science   *92*
8.4   Balance Between Art and Science: (a) Fossil-Fuelled Power Station, (b) Family House, (c) Textile Fabric, (d) Sculpture, (e) Painting, (f) Textile Loom   *93*
8.5   Design in Balance Among the Arts and Sciences   *95*

9.1   The Design Spiral (Atkinson)   *101*
9.2   Simplified Design Spiral (Defence Research Establishment Atlantic, Canada)   *102*
9.3   Design Core Bounded by Product Design Specification: Dynamic Concepts   *104*
9.4   Design Core Bounded by Product Design Specification: Static or Fixed Concepts   *107*
9.5   Elements of Product Design Specification   *108*
9.6   Design Core   *112*

10.1   Design Model After Swinkels (1985)   *118*
10.2   Design Model After Bleker (1985)   *119*
10.3   Design Model After Boekholt (1985)   *120*
10.4   Pugh Design Model   *122*
10.5   Design Model After Cooke   *124*
10.6   Business Design Activity Model: Pugh   *125*
10.7   Business Design Activity Model: Blank   *126*
10.8   Business Design Activity Model: Architecture of Shipbuilding   *127*

11.1   Total Design Product Model   *135*
11.2   Partial Design Characteristics of Higher Education   *135*

12.1   Scope of Design Methods   *146*
12.2   Prediction of Future Available Energy Sources   *154*
12.3   Energy Consumption Trends for Various Distribution Systems   *154*

13.1   The Wheel of Competitive Strategy   *160*
13.2   Price Versus Fuel Consumption: Cars, Various Manufacturers, 1979   *163*
13.3   Retail Price/GVW Versus Acceleration 0–60 mph: Cars, One Manufacturer, 1979   *163*
13.4   Fuel Consumption Versus Road Area Occupied: Cars, Various Manufacturers, 1979   *163*
13.5   Fuel Consumption Versus Engine Capacity: Cars, Various Manufacturers, 1979   *163*
13.6   Maximum Torque Versus Engine Capacity: Cars, Various Manufacturers, 1979   *163*
13.7   Purchase Price Versus Engine Power Rating: Cars, Various Manufacturers, 1979   *163*
13.8   Vehicle Length Versus Vehicle Width: Cars, Various Manufacturers, 1979   *164*
13.9   Motor Power Versus Weight: Food Processors, 1981   *164*
13.10   Price Versus Weight: Food Processors, 1981   *164*
13.11   Height Versus Weight: Food Processors, 1981   *164*
13.12   Selling Price Versus Motor Power: Food Processors, 1981   *164*
13.13   Selling Price Versus Recommended Retail Price: Kitchen Scales, 1981   *164*
13.14   Payload x Lift Height Versus Unladen Weight: Rough Terrain Fork Lift Trucks, 1972   *165*

14.1   Comparable Car Horn Concepts Produced by One Student Group   *174*

16.1   Basic Quality Function Deployment   *185*
16.2   The Basic Process of Enhanced QFD   *188*
16.3   Core of House of Quality Example: Xerox Copier   *189*
16.4   Concept Selection: Total System Architecture Example—Xerographic Copier *190*
16.5   Total System Design Matrix   *190*
16.6   Concept Selection: Subsystem   *191*
16.7   Subsystem Design Matrix   *192*
16.8   Concept Selection: Piece-Part Example—Retard Roll   *193*
16.9   Piece-Part Design Matrix   *193*
16.10   Static/Dynamic Status Examples from the Xerox 1075 Copier   *193*
16.11   Aerial Access Platforms   *196*
16.12   Elements of Product Design Specification   *200*

17.1   Spectrum of Design Activities   *212*
17.2   Design 1: Conventional Open Lattice Structure   *213*
17.3   Design 2: Propped Cantilever Structure   *214*
17.4   Photo of Parabolic Dish Aerial for Tropospheric Scatter Communication   *214*
17.5   Photo of the Marathon 2550 Dump Truck   *215*
17.6   Chassis of the Marathon 2550   *216*
17.7   Photo of the Giraffe Site Placement Vehicle   *218*
17.8   Intensity and Direction of Design Activity   *218*

18.1   Teaching   *226*
18.2   Practice   *227*
18.3   Relationship of Technique and Technology to Design Phases   *228*
18.4   Computing Level: Design Phases   *229*
18.5   Photo of the Marathon 2550   *231*
18.6   Chassis of the Marathon 2550   *232*
18.7   Computing Pattern for the Marathon 2550   *233*

19.1   Spectrum of Design Activities   *237*
19.2   Innovation and Stage of Development   *239*
19.3   Design Core Bounded by Product Design Specification: Dynamic Concept   *241*
19.4   Design Core Bounded by Product Design Specification: Static Concept   *242*

20.1   Spectrum of Design Activities   *248*
20.2   Design Core Bounded by Product Design Specification: Dynamic Concept   *251*
20.3   Design Core Bounded by Product Design Specification: Static Concept   *252*
20.4   CAD Facilities Matched to Design Core: Static   *254*
20.5   CAD Facilities Matched to Design Core: Dynamic   *255*

21.1   Facilities Offered by Turnkey Systems   *264*
21.2   Facilities Offered by Turnkey CAD Systems   *265*
21.3   Degree of Iteration of the Design: Static Concept   *267*
21.4   Degree of Iteration of the Design: Dynamic Concept   *268*
21.5   Software Availability: Static Concept   *271*

21.6   Software Availability: Dynamic Concept   *271*

22.1   Spectrum of Design Activities   *276*
22.2   Spectrum of Design Activity: Boundary A and Boundary B Operations—Dynamic and Static   *276*
22.3   Software: Static Concept   *278*
22.4   Software: Dynamic Concept   *278*
22.5   GRASPIN Approach to Software Design   *280*
22.6   Constraints on Design (Gairola)   *282*
22.7   (a) Relationship Between Diagnostic and Generative Systems. (b) Knowledge Acquisition   *285*

23.1   Water Valve   *297*
23.2   Well-Based Rim   *297*
23.3   New Rim   *298*
23.4   Parameter Relation to Aerial Access Platforms   *300*
23.5   Parameters Related to Robotic Arms   *301*

24.1   Elements of Product Design Specification (PDS)   *309*
24.2   Design Core Bounded by Product Specification   *311*
24.3   People and Design Core Phases   *313*

25.1   Elements of Product Design Specification   *328*
25.2   Design Core Bounded by Product Design Specification   *329*
25.3   Business Design Activity Model   *331*
25.4   Business Design Activity Model   *332*
25.5   Design Core Bounded by Product Design Specification: Static Concept   *334*

26.1   Early PDS Model: 1971   *346*

27.1   Total Design or Product Delivery Process   *354*
27.2   Establishing the Real Voice of the Customer   *356*
27.3   Elements of Product Design Specification   *357*

28.1   Elements of Product Design Specification   *362*
28.2   Business Design Activity Model   *363*
28.3   Business Design Activity Model   *365*
28.4   Design Core Bounded by Product Design Specification: Dynamic Concept   *369*
28.5   Design Core Bounded by Product Design Specification: Static Concept   *371*
28.6   Spectrum of Design Activities: Boundary A—The Dynamic Core Model, Boundary B—The Static Core Model   *376*

29.1   Using the PDS Elements to Establish and Control the Design Audit   *384*
29.2   Suggested Stages of Design Review   *385*
29.3   Activity of Design Audit: Resources Required   *386*

30.1   Design Activity as a Planned and Organised Activity   *395*

30.2   Differentiation Between Design Team Markers Members' Responsibilities and Areas of Interest   *399*

31.1   PDS Boundary   *407*
31.2   Linkage of PDS Elements to the Design Core and Stages of Design   *408*
31.3   Linkage of PDS Elements to the Design Core and Stages of Design   *409*
31.4   Interaction Between the Activity and Resources Required for Design Review   *409*

32.1   Parametric Analysis: Payload Versus Lift Height for Rough Terrain Forklight Trucks   *415*
32.2   Weight Versus Unladen Weight   *416*
32.3   Photo of Giraffe Site Placement Vehicle   *417*
32.4   Polymer Component Costs   *418*
32.5   Polymer Component Costs   *418*
32.6   Polymer Component Costs   *419*
32.7   Approaching the Target Cost   *420*

33.1   The Engineering Business   *428*
33.2   Business out of Balance with the Market   *430*
33.3   Business in Balance with the Market   *431*

36.1   Design Core Bounded by Product Design Specification   *457*
36.2   Design Core Bounded by Product Specification   *458*

37.1   Factors Making Up the Total Design of a Product   *463*
37.2   Effect of Shift of Technology Mix on Product Viability   *465*
37.3   The Bridging Effect of Research on Product Outcome   *466*
37.4   Methods in Common Use Linked to the Design Core Strategy   *468*
37.5   Market Pull and Technology Push Interactions   *469*

38.1   The Sequence of Methods for Successful Product Design   *476*
38.2   QFD Integrated with Total Design   *478*
38.3   Market Pull and Technology Push Interactions   *479*
38.4   System Configuration for Liquid Waste Processing   *481*

39.1   Pugh's Product Design Activity Model: Dynamic Concepts and Static Concepts   *501*
39.2   Elements of Product Design Specification   *502*
39.3   Business Design Activity Model   *504*
39.4   Business Design Activity Blank   *504*

List of Tables

Table 12.1   A Decision Matrix for a Brush-Making Machine   *149*

Table 12.2   A Decision Matrix: Methods of Applying Heat   *152*

Table 14.1   Evaluation Matrix   *170*

Table 14.2   Evaluation Chart for a Motor Horn   *175*

Table 16.1   Matrix Analysis: Microscope Manipulators for Neurosurgery   *198*

Table 16.2   Matrix Analysis: Microscope Manipulators for Ophthalmology   *199*

Table 24.1   Examples of Technique and Technology Information Related to Groups of People in the Engineering Industry   *314*

Table 24.2   Data Sheet for a Variable Delivery Swashplate Pump   *315*

Table 27.1   Principles of Total Design   *352*

Table 30.1   Interaction Between People and Techniques and Technology   *400*

Table 34.1   Product Design Specification Factors   *437*

About the Author

A graduate of London University in mechanical engineering, Stuart Pugh joined the British Aircraft Corporation as a graduate apprentice. In 1956 he became project engineer for the Mach 6 wind tunnel at Warton Aerodrome and in 1963 was invited by the Marconi Company to become chief designer in its Mechanical Products Division. He then moved to the Hydraulic Equipment Division of the English Electric Company as chief designer and progressed to divisional manager with that company.

In 1970 he was appointed Smallpeice Reader in Design for Production at Loughborough University of Technology and then director of the Engineering Design Centre. In 1985 he became Babcock Professor of Engineering Design at Strathclyde University and head of the Design Division. As founder of the discipline of total design, he introduced and taught the subject across the Faculty of Engineering. He also contributed to the MBA program and taught short courses on total design to industrial managers.

Professor Pugh consulted widely in the field of design method implementation for product success and also conducted executive seminars and field research with leading firms throughout Europe and North America.

Foreword

This is a book for the ages. *Creating Innovative Products Using Total Design* is both a collected works and an original contribution to the literature. Stuart's untimely death in October 1993 only increases the value of this volume, and the reader will find the content and organization of the book to be interesting and of lasting value. This work is a significant contribution to the literature primarily because of the general unavailability of Pugh's original published work. Stuart was not the typical academic; in fact, he was not an academic at all! He was an extraordinary design engineer who spent most of his life in industry actually doing design and many of his later years in an academic environment. Stuart had little respect for the "publish or perish" paradigm and regularly gave very significant conference presentations without handouts or other permanent record of his remarks. On many other occasions he wrote and presented very complete papers for conference presentation and proceedings publication. Many of the chapters of this book are based on these conference presentations. Little of Pugh's work appears in journals, but now a wide audience will be able to profit from the thinking of this great design scholar.

I met Stuart in the fall of 1975 at Loughborough University in England. He attended a seminar I gave on engineering optimization. I recall that he asked several difficult but interesting questions during the Q&A period and later introduced himself after my talk. Stuart found it difficult to understand my interest in computers, computing, and things numerical, as he put it. Why wasn't I spending my time and energy on more relevant and significant issues? I soon learned that I had very much more to learn from him than he could possibly learn from me, which was quite humbling to me at the time. We stayed in communication almost continuously after that first meeting, and I developed a genuine respect for his work, especially on concept selection, and for his approach to design education. Later (in the early 1980s) it was my pleasure to introduce Stuart to Don Clausing and Genichi Taguchi in Tokyo at a conference where we were speakers. Don and Stuart developed a close working relationship, and it is appropriate that Don and Stuart's former student, Ron Andrade, have contributed so significantly to this volume. Don and Ron have paid the price to make this book a reality, and we all owe them a debt of gratitude.

Stuart died before his time, but now his work lives on. Read, enjoy, and use. Design engineering and the world is a better place because of the work of Stuart Pugh.

Ken Ragsdell
Professor of Engineering Management
Director of the Design Engineering Center
University of Missouri–Rolla
August 1995

Preface

Stuart Pugh was one of the great leaders of product development (total design) methodology and practice. Soon after we first met in 1984 he told me that when he left industry in 1970 he thought that he knew all there was to know about design: after all, he had been a successful design engineer and manager. When he arrived at Loughborough and was confronted with the need to teach design, however, he quickly concluded that he knew very little about design. It was just this experience that informed Stuart's teaching, writing, and practice. Very few people have duplicated Stuart's experience of spending almost half of his career in successful industrial practice and then the remainder of his career in a university. Through this dual career Stuart developed a comprehension of and insights into total design that went far beyond those supported by the more traditional monolithic career, whether in industry or academia. These profound insights culminated in Stuart's book *Total Design*, published in 1990.

Stuart's published papers have always generated strong interest, but most of them were presented at design conferences and not widely distributed. Soon after the publication of Stuart's book, Ken Ragsdell (of the University of Missouri) and I urged Stuart to publish his collected papers to make them readily available to design practitioners and managers. Stuart agreed and set to work to select the papers, put them into systematic order, and write introductions. Unfortunately, Stuart became ill in the fall of 1991 with the illness that claimed his life in October 1993. During the summer of 1992 his illness was in remission, and he vigorously worked on this book. It was nearly complete at the time of his death and now serves as a memorial to his life and work. As such, we decided to leave the text in Stuart's native British English, which is also the international professional language throughout the European Union, rather than translate it into American English. Stuart organized the book into an introduction and eight parts that include thirty-nine chapters.

READING SUGGESTIONS

Some readers will go directly to favorite topics. For those readers, practitioners, and educators who are already familiar with the basics of product development, who would like help in starting, I suggest that Chapters 24, 14, 23, 16, 25, 39, 28 and 3 be read first.

They are listed in a sequence that many readers will find rewarding to follow. (The reader who is especially interested in the management of product development might wish to read Chapters 25, 39, and 28 first.) These chapters will provide a good foundation for exploring the remainder of the book to discover additional insights and an extended historical perspective.

Chapter 24 introduces the product development specification (PDS), which is the beginning of product development. Chapter 14 is Stuart Pugh's famous paper on concept selection, which is done in response to the PDS. Chapter 23 describes one of Stuart's most important concepts, the difference between conceptually static products and conceptually dynamic products. It is important to become familiar with this distinction before reading more chapters. Chapter 16 is Enhanced Quality Function Deployment, which takes the concepts of Chapters 24, 14, and 23 another step forward by integrating them with QFD, which was developed in Japan. Chapters 25, 39, and 28 integrate all of Stuart's writings in the context of management and teamwork. Chapter 3 describes the most innovative post-graduate curriculum in total design that we have yet seen. This is the logic for the suggested reading to start the book. If it agrees with your interests, you will probably enjoy starting with this sequence.

PART I: DESIGN IN EDUCATION AND INDUSTRY

This part directly reflects Stuart's concerns that the academic teaching of design was aloof from industrial practice, while industrial practice suffered from the lack of reflective structuring that can best be achieved in the university. Chapters 1 through 6 develop the integrated approach that Stuart knew to be essential for design success.

A successful integration of education and industry is described in Chapter 6, Marathon 2550: A Successful Joint Venture. In this joint venture between the academy (Loughborough University of Technology) and industry (Liner Concrete Company Ltd.), a team of graduate students developed a dump truck that was a commercial success.

The master of science curriculum that Stuart developed at Loughborough was unique: nothing like it has been seen before or since. This very successful nontraditional curriculum is described throughout this part, especially in Chapter 3, Engineering Design at the Postgraduate Level. The curriculum should be strongly considered as a model by all academic departments that teach engineering design.

PART II: DESIGN PROCESSES AND PHILOSOPHY

This part emphasizes two key elements of Stuart's approach—total design, rather than partial design, and design activity models. It also includes an interesting discussion of the role of design in society in Chapter 8, Design: The Integrative Enveloping Culture, Not a Third Culture.

Total design was Stuart's consistent goal. He recognized the all too human tendency to abstract from any design situation those elements that we feel most comfortable with, to the exclusion of other important elements. Stuart constantly urged designers to go beyond this partial approach to achieve the total design that more fully satisfies the needs of society.

Without a design activity model, design is ad hoc problem solving and very unlikely to achieve completeness. Stuart's model is probably the most complete and graphic that has been developed. It is complex, but then total design is inherently complex. A great problem of ad hoc design is that it ignores complexity, resulting in a partial design that leaves many important aspects only fuzzily addressed. Stuart's design activity model helps the practitioner to address the full complexity of the design and does not gloss over important elements.

PART III: DESIGN TECHNIQUES AND METHODS

This part contains in Chapter 14 the paper for which Stuart is most famous: Concept Selection: A Method That Works. First presented at the International Conference on Engineering Design (ICED) in Rome in 1981, it had already undergone extensive development based on feedback from designers in academia and industry. Since 1981 it has been widely implemented. I introduced it at the Saturn Corporation (GM cars) in 1986, and when I returned in 1991 it was in widespread use and was credited with much success at Saturn. Academics are repeatedly trying to "improve" the Pugh concept selection process, primarily by reinstalling features that Stuart and his colleagues found dysfunctional and eliminated. As Santayana observed, those who do not know history are doomed to repeat it. One virtue of this book is that it documents an important part of design history.

Enhanced Quality Function Deployment—Chapter 16–is also dear to my heart. Stuart and I wrote it in one week in April 1990. During that week I enjoyed the hospitality of Stuart and his wife Sheila in their home outside Glasgow. Every day Stuart and I went to his office at the University of Strathclyde and wrestled mightily with the concepts and each evening I fear that we bored Sheila with our further discussions. Finally it all fell into place: Stuart and I were exultant.

Part III emphasizes one of Stuart's primary concerns—creative work. On the one hand, methodology is necessary, but care must be taken that it does not stifle creativity. On the other hand, in the absence of methodology creativity is often empty and produces much change with little improvement. It was one of Stuart's constant admonitions that designers must struggle to stay on the narrow path between the strait jacket of rigid methodology and the random outbursts of unstructured creativity.

PART IV: CAD AND KNOWLEDGE-BASED ENGINEERING

This part introduces one of Stuart's most powerful ideas—static and dynamic concepts. The observation that design approaches that worked very well for perfecting static concepts would not work at all for dynamic concepts is a main feature of Stuart's work.

Computer-aided design (CAD) must be very different for static and dynamic concepts. This idea was further developed as a powerful concept beyond CAD (see Part V). His work on static and dynamic concepts inspired my reusability matrix, which I first presented in 1991 and is now undergoing further development at MIT in cooperation with Ford and Xerox.

This part reflects Stuart's caution about CAD. I remember him saying that unless total design is used to develop CAD, CAD will just help us to produce rubbish faster. He felt that specifications from the users of CAD should guide its development but that it has been to a considerable extent a technology-push product. In Chapter 22, Knowledge-Based Systems in Design Activity, Stuart issued a challenge to the providers of computers in design that is still timely today.

PART V: DESIGN TEAMS, MANAGEMENT, AND CREATIVE WORK

This part further explores the appropriate balance between static and dynamic concepts. Critical operational issues of the organization of design and the best process for total design are addressed in terms of the position of the product within the static-dynamic conceptual spectrum. The best organization and process for static concepts will be a complete failure for dynamic concepts and vice versa, a point that has been largely missed by practitioners and writers and resulted in waste of industrial resources and confusion of academic effort.

This part includes a chapter by Stuart Pugh and Ian Morley, a social psychologist at Warwick University. They worked together for several years, a rare (perhaps unique) collaboration between an eminent design leader and a social psychologist. Although much has been written in the last ten years about design teams, almost all of it has been put forth by "team people" who know little about design or by design people who know little about interpersonal interactions. Thus the work by Stuart Pugh and Ian Morley is an especially valuable integration of design and psychological viewpoints.

PART VI: DESIGN FOR X

Chapters 30 through 35 address design for cost, design for assembly, and all of the other "design for's." Stuart put these on a rational basis by relating them to his product design specification (PDS) (see Part II). The PDS has thirty-four elements, such as cost, weight, safety, and testing, and each of the thirty-four could be the source of a "design for" methodology. This part presents the view that these elements can and should be integrated by the PDS instead of being a large number of independent methodologies.

PART VII: DESIGN RESEARCH

Stuart points out in Chapters 36 through 38 that design research can address total design or partial design and static concepts or dynamic concepts—essentially a four-quadrant matrix. Research into partial design of static concepts will reveal very different results than research into the total design of dynamic concepts, yet other than in Stuart's writings these distinctions have not been clearly made. Stuart's insights bring clarity to a mass of apparently conflicting research studies. Stuart emphasized that design research cannot be done independently of design and urged that the objective of design research should be to improve design practice, yet much academic research is done without reference to industrial practice.

In this part Stuart also addresses the critical activity of industrial research that leads to new products and distinguishes between market-pull products and technology-push products. The integration of research and technology development into the total design activity is clearly described.

PART VIII: TOTAL DESIGN—SUMMARY OF THE WHOLE

This part—one marvelous, extended paper by Stuart Pugh and Ian Morley—to a considerable extent is a summation of all of the concepts presented in earlier papers. All design organizations could benefit greatly from the studying and implementing of its concepts.

CLOSING

In closing, Stuart Pugh was a pugnacious Yorkshireman, quick to challenge the many underdeveloped concepts of engineering design. Stuart accepted the challenge of helping others to see the imperatives of total design as clearly as he did, which comes through clearly in the collected papers that comprise this book. Few people have matched Stuart's record of success in industry and in the academy, and the full sweep of his comprehension of design issues was often too great for others to grasp fully. However, as a friend and collaborator he was a wonderful person to know and work with. He was never facile in his discussion and always gave serious consideration to any well-formed idea. He sorted out the chaff and assimilated the wheat. We are fortunate that he left us with the thirty-nine loaves that comprise this book.

Don Clausing

February 18, 1995

In reviewing this book just prior to final production, I was struck by the fact that Stuart Pugh was far ahead of the rest of us in writing about concurrent engineering. Although the term concurrent engineering was not used much until 1988, Stuart emphasized the essential factors in the early 1980s, and to some extent in the late 1970s. As you peruse these papers, look for the integrative factors in process and organization, which are still fresh and valuable today.

Don Clausing

October 9, 1995

Acknowledgments

The author is indebted to Mrs Doreen Sehgal for text correction and typing and for organising the chapters to give what I hope is a logical balance to the book. The author would like to acknowledge the *European Journal of Engineering Education*, the Institution of Mechanical Engineers, the Conference on Design Engineering, and *Quality Assurance*. The author would also like to thank all of those people whose cooperation and contributions helped to make this book possible.

If the book stimulates discussion and debate among professionals in design, it will have achieved its primary objective.

Stuart Pugh

On behalf of my son, daughter, and myself, I would like to express sincere appreciation and gratitude to Professor Don P. Clausing, MIT, Cambridge, MA, USA; Professor Ronaldo Andrade, Universidade Federal de Rio de Janeiro, Brazil; Professor Ken Ragsdell, University of Missouri, Rolla, MO, USA; and Dr. Ian Morley, Warwick University, UK, for all of their help and perseverance in bringing my late husband's book to fruition.

Sheila Pugh

When Stuart Pugh died in October 1993, I undertook to have this book published. In this endeavor I have been greatly helped by Dr. Ronaldo Andrade, former student of Stuart Pugh and professor at the Federal University of Rio de Janeiro, who was a visiting scholar and visiting professor at the Massachusetts Institute of Technology from 1992 until 1995. Dr. Andrade helped Stuart to rewrite the introduction to the book when he was becoming incapacitated by his illness and later on, among other activities, prepared the chronology with full bibliographic references that appears in the Appendix.

Don Clausing

February 21, 1995

Introduction

With this book I have an opportunity to collect some of the work I have published over the past twenty years in my attempts to show practitioners, academics, and enthusiasts that design is an activity to be *practiced* and not just for studying and researching. You notice my emphasis on practice. *Design is for doing*. This fact is not very well understood even today.

In 1970, when I was chief engineer, Mechanical Products Division, with the English Electric Co., I became aware of my lack of understanding of design, notwithstanding my having done and having been responsible for it for many years. When the occasion arose to move to a university, I had the opportunity to study, research, and practice design in academia. The emphasis on practice remained.

I visited the United States in 1975, did a university and college tour, and found stirrings of engineering design in a few isolated pockets, although most of it was based on traditional engineering. There is nothing wrong with that, except that it tends to be applied to what I call partial design and becomes rather narrow and restrictive. If design is to be effective, the traditional department and subject boundaries must be crossed, and the fundamental and important contributions of each one integrated into what I define as *total design*, the subject of this book. That stirring continued apace in the 1980s and 1990s as the United States grasped the nettle of the methods by Taguchi, myself, and others, hopefully in a continuous rational manner, both in education and industry. That this is bearing fruit is manifest by the number of courses, conferences, books, and so forth on the topic and the number of companies adopting the procedures in ever increasing numbers.

Looking back over my years as an industry-based design practitioner and then as an academy-based design practitioner and educator, I still see a lack of understanding about the essence of the design activity. This book is an attempt to rectify that situation. It represents the emergence of a cohesive, structured approach to engineering design developed over a period of more than twenty years and firmly based on practice.

Pondering upon the collection of papers that form the book I could not prevent being at the same time pleased and disappointed when revisiting the first paper in the chronology, Engineering Design: Towards a Common Understanding, published in 1974 (this is Chapter 26 of this book)—pleased with my insight then and the coherence of my work afterwards. Disappointed to realise that if its date were erased and that paper presented in a design conference today, it will cause the same impact as it had nearly twenty years ago! Then I wondered, What is going on? Or rather, what did not go on?

I invite my colleagues in academy and industry to go through this book, reflect upon the propositions, and try the methods that emerged from my work. If you do not agree with them, use your own alternatives, but please, not just in exercises or gaming situations. Put them into practice for real, and make them work as I did. After all, that is what design is all about.

COMMUNICATION DIFFICULTIES

A perennial problem that arises at design conferences and discussions is understanding just what is meant by *design* and *design engineering*. In 1974 I attended a conference, ostensively for designers, at which I was to present a paper on *costing*. It quickly became apparent that the audience was confused since their view of design was variable, random, and mixed. This caused me to change tack and to rewrite and present Chapter 26, which I must admit changed the whole scenario but also enabled us to achieve overall consensus. In it I described design as a highly manipulative activity in which the designer has to continuously and simultaneously pay attention to and balance several factors that impinge upon and influence design. I related this action to the circus act where the performer keeps several plates rotating on top of sticks. The designer performs his act with the primary elements of the product specifications. Since that time I have always clarified matters and defined design this way. One then makes progress and overcomes communication difficulties, even with academics.

A step further was the proposition of the design activity model (Chapter 30), which provided a visible and simple description of the design process and greatly contributed to improved understanding in practice and education. The model depicts design as having a core of basic interconnected activities enveloped by the product specifications. Each activity draws information from the others, knowledge input from science and technology, and operational procedures from techniques related to the core. The model is easily conveyed and imparts sufficient appreciation to raise consensus even among designers who work in different fields or levels of detail. We made significant progress, and this was recognised by Sharing Experiences in Engineering Design (SEED), an organisation based in the U.K. academia at varying levels. This model now forms the basis of design teaching in more than eighty U.K. institutions.

However, the lack of understanding about the essence of design remained. In the beginning of 1982 I prepared a paper (Chapter 8) prompted by the views of some of my academic colleagues that design should be a third culture together with sciences and arts. I refuted that argument and still do. Design is not a body of knowledge. It is the activity that integrates the bodies of knowledge present in the arts and sciences.

The last paper in the chronology of this collection (Chapter 39) summarises my understanding of design and adds to it the concept of disciplined creativity, a process of doing creative work to converge to successful products. This is achieved by giving a good visible structure to the design process, such as the design activity model I put forward, assisted by appropriate methods, such as those presented in this book.

RESULTS

The net outcome of all the work done in the 1970s and 1980s has been the emergence of more refinements of the structure of the design model and, perhaps of much greater significance, the emergence of new technology-independent design methods that have been proven to work in practice (see my book *Total Design: Integrated Methods for Successful Product Engineering*, Addison-Wesley, 1990).

Those new methods that may be attributed to me are summarised in Chapter 16 and consist of the following:

- *Parametric and contextual analysis*: The first method consists of cross-plotting parameters of competing products and looking for revealing relationships. By and large this exercise is usually not expensive, since it can be done with data drawn from published technical catalogues, and the results can be surprising. In my experience with its application I have never failed to find a relevant unknown relationship, even when working with very experienced design teams. Chapter 13 shows how to use it. Contextual analysis is supported by matrix analysis or reverse concept selection to select the best products in the market and develop insights into existing designs. These methods are most useful when preparing the product design specification (PDS).
- *Structured product design specification formulation*: I believe that the crucial point in the design process is that of the product design specification formulation. The PDS envelopes the design process, sets its boundaries, and is its sole reference and its controller. A systematically and thoroughly generated set of specifications is absolutely essential to successful product design. Its formulation evolves from a rigorous interrogation of data using the PDS elements as primary triggers as discussed in Chapter 16.
- *Concept selection*: The proof of the effectiveness of this method is its widespread worldwide usage. It goes beyond conventional concept selection in the sense that it promotes the evolution of concepts towards a strong nonvulnerable solution by means of controlled convergence. It is common in its application that the design team ends up with a concept totally different from any of those it started with and a better one! This method has received much attention throughout the years, and attempts have been made to modify it by introducing numeric values into its matrix of analysis. I do not recommend doing this and sustain a word of caution: do not be fooled by the numbers in this exercise. They can give you an illusive sense of accuracy and take you in the wrong direction. The method as first published is shown in Chapter 14.

Gradually, as the plot unfolds, the relationship of these methods with other technology-independent methods such as Taguchi is made. In Chapter 16, Don Clausing of the

Massachusetts Institute of Technology and I combined these methods into enhanced quality function deployment (EQFD), which provides a wide framework for improved product development.

In the ongoing evolution towards total design it has been recognised that it is essential to apply all technology-independent methods to each different level of the product or concept:

- Total system architecture (TSA): the whole system,
- Subsystem (SS): the subsystems that go to make up the whole, and
- Piece-part (PP): the piece parts that comprise the subsystems.

Some reasons for partitioning products into these levels are to allow better cognition of the problem by approaching it from different orders of abstraction, to allocate responsibilities in accordance with degrees of competence, to afford control and management of the whole design and development process, and to accommodate company functional divisions. No matter the reasons, the level one is working on has to be perceived, otherwise the integration with which the design activity operates will be disturbed.

A further consideration is in regard to the status of the product and process concepts. In 1983 I advanced the hypothesis of the static and dynamic concept status (see Chapter 19, an appreciation of the applicability of CAD software). A static concept is one that has reached a conceptual plateau established in time and has subsequent products developed based on it. Variations are made on rearrangements, new materials, processes, or changes in some sublevel of the system. The classical example is the automobile, which has been evolving on top of the same total system concept since the beginning of the century. Dynamic concepts originate from new approaches to existing problems or are generated by the necessity to find solutions to new problems. These are located towards the innovative end of the design activity spectrum as opposed to static concepts, which are at the conventional practice end. In this regard the design activity carried out in industry should always be structured in a dual fashion in order to accommodate for the different status. Identification of the product status is crucial when dealing with highly competitive and rapidly evolving markets. This is clarified in Chapter 23.

Many companies find that their greatest difficulties are in separating the concept levels and identifying their status. Within my experience many remain scrambled throughout the whole design process leading to confusion about application and less than adequate products for the market. This reinforces the necessity for a logical, structured approach and for formal recognition of how and where you are or are about to enter into the process. It must be systematic for success.

WORLDWIDE STATE OF THE ART

A word on technology-dependent methods. Earlier I alluded to technology-independent methods such as described in this book. Be alerted to the fact that the differences between the two are not necessarily recognised. Methods of circuit analysis and stress analysis are technology-dependent. Concept selection and Taguchi methods clearly are not.

In Europe, generally speaking, the technology-dependent methods dominate, although things are changing at a pace. In the United States the same is true, although the rate of change is much faster in U.S. industry, with universities lagging. The latest technology applied in the wrong context can lead to failure; the converse is also true.

In parallel with the above, Japanese industry was developing approaches that by definition were based on practice. We saw the emergence of Taguchi quality engineering and of quality function deployment, which although they have the same basic aims as total design may be said to lack a certain rigor in front-end operations since by and large they are grounded in existing products.

By linking together all this work this book is more up-to-date and current than others in the same field. This is described in Chapter 16, on EQFD.

When this book was in preparation, Don Clausing of MIT was finalising his book *Total Quality Development: World-Class Concurrent Engineering*, which goes beyond EQFD. Total quality development (TQD) combines the best designing, the best engineering, the best management, the best strategy, and the best teamwork to far exceed traditional product development. In this manner, technology-independent and technology-dependent methods are integrated into a wide framework that allows engineers to create and make things work in the most efficient way. TQD draws support from total design to attain an improved product development process. Clausing's book is a valuable attempt to shift the current partial-design-oriented paradigm of product development.

As stated earlier, the rationale and adoption of technology-independent methods is now widespread in the United Kingdom and United States, although in many ways the practices of Continental Europe are still somewhat narrow, based I believe on the NIH (not invented here) syndrome.

One of the most significant things I have learned is to adopt and adapt approaches and methods (irrespective of origin) that work and provide results and not to try doing things differently for their own sake. The other point is to recognise that in most designs the designer or design team is grappling with intangibles, which unless you practice can escape notice.

EDUCATION AND INDUSTRY

My experience with design both in industry and academia allowed me to develop a deep appreciation of the importance of design understanding and learning. Indeed, the design practice makes one learn its process, but that alone is not sufficient to promote the integration that total design affords. In education as well as in industry the exercise of doing projects has to be complemented by the attitude and practice of advancing into other fields of knowledge and bringing in whatever is necessary to feed the design process. By and large such a task is made difficult by enclosure of specialisations or barriers created between departments or people.

Throughout my academic career, initially at Loughborough University and then at the University of Strathclyde, one of my main efforts has been to demonstrate to my academic colleagues the importance of design as an integrator of knowledge and experience. In the former institution I have been amidst the running of a very successful and pioneering postgraduate course that within a year managed to teach design core and related disciplines; assemble design teams of students with different backgrounds, experiences, and nationalities; confront them with problems in areas they had never even thought about before; and generate final designs of products that succeeded in the marketplace. And how both students and teachers learned in that process! At that time I developed some of the basis of my work and benefited very much from the association with Douglas Smith. The nucleus of our joint work has been adopted by SEED, and now it is seen on a national scale in the United Kingdom, quite simply because design practitioners relate to it. Part I recollects some of those experiences.

At Strathclyde University, as I proceeded with my work I became much more involved with the undergraduate level. As head of the Design Division I led the introduction of total design courses to almost the whole of the first-year intake of the Faculty of Engineering. The response of the students was enthusiastic, which was not unexpected since the approach was to cross the unnatural boundaries of departments. In addition a four-year course leading to a master of engineering degree in product design was started in 1991. A glimpse of those experiences is seen in Chapter 7.

In teaching and learning design it has to be kept in mind that one of the main abilities to be developed is to see it in its totality and simultaneously in its specific partial aspects and act on that. This way the designer will achieve a state of awareness that will allow him or her to work with disciplined creativity and control the outcomes. The required information for the process will become more explicit, and most of all the decision of what methods and techniques to use more clear.

A word of advice: no matter how efficiently we combine practice, understanding of basic knowledge, visible and clear design models, adequate methods, and systematic approaches in order to develop designing abilities, success requires us to transform the defense barriers of knowledge and functional feuds into interfaces of cooperation and integration.

I trust that this book brings this message across and that my colleagues from academia and industry get some inspiration from it and press on in that direction. There is no doubt that the role of the universities as educators of future professionals is critical in that change. However, those in industry can make their moves also. In my view, it was the use of the integrative power of design that opened markets to Japanese industries. There are positive signs of action; the present interest in concurrent engineering is evidence of that. I feel, though, this movement may fade away as many other attempts of the kind have faded for lack of a systematic approach. Total design fills this void.

DIVISION OF THE BOOK

Although one of my reasons for compiling this book was to help those working with design gain easy access to a selection of the work I have published over the years, I felt that adding some structure to this collection would enable readers to immediately focus on the topic of their primary interest.

I have organised the papers into eight parts, making every effort to deal not only with the evolution of design in a total sense but to subdivide the subjects in a logical fashion, hopefully to appeal to education and industry. It was mainly for those committed or involved in teaching and doing design that this retrospective was made. Total design has a much wider application than just hardware, and the technology-independent methods it comprises are being successfully used in nontechnical situations including service industries. In this sense the propositions and successful methods in this book can be extended to areas beyond engineering.

Each part has a brief introduction.[1] The original material of the papers has been preserved and abstracts, notes, or comments included where I felt it was necessary to bring them more up-to-date. Among the papers you will notice some overlapping and the occasional repetition. That is inevitable since they were prepared for different occasions, and sometimes past work is included to sustain the new ideas or provide a framework. I made no attempt to edit them, and I believe this will not be a nuisance to the reader. By and large overlaps reinforce the points I make.

In addition to the table of contents, I have prepared a chronological list of papers to show the historical development of the work. You will note that this list is followed neither between nor within chapters. That is the way design goes! Performing disciplined creativity within known boundaries allows you to generate apparently disconnected and nonsequential results that in the end fit nicely into a more general structure. By no means be constrained by the sequence I suggest. You may pick your own order of reading. The linking between the papers will become self-evident as you proceed.

The book is divided into an introduction and eight parts, as follows:

Part I: *Design in Education and Industry*. This draws on the successful results of the master of technology course that was run in the now-extinct Engineering Design Centre at Loughborough University and the underlying approaches that enabled an effective university-design-industry combination. For those in academia the lessons are valuable.

Part II: *Design Process and Philosophy*. This offers reflections on the essence of design and the integrative action of the activity. Commonalities of the design process in practice and in the literature are also investigated. The design researcher and the practitioner will benefit from a realisation of the common base of the design process.

Part III: *Design Techniques and Methods*. From a note of caution not to move into methods just for the sake of change, it proceeds to the presentation of methods that I devised and are now successfully used worldwide. An integration with other well-known methods is presented. Design practitioners can start experimenting with these methods at once.

---

[1]After Stuart Pugh's death, Don Clausing prepared introductory comments which also appear at the beginning of each part.

Part IV: *CAD and Knowledge-Based Engineering*. This offers a critical view of the computer in the design process. It warns about the risks of letting computer fascination move software applications away from real design needs.

Part V: *Design Teams, Management, and Creative Work*. The doing of the design process is visited here. It shows that a simple and consensually agreed definition of design and its process is the key for successful team work. It provides a road map for design practitioners and managers.

Part VI: *Design for X*. This part examines the information needs for design and advises that designing for a specific aspect can be misleading. Balanced design is achieved when all X's are kept in mind. Practitioners and design planners will understand that despite all the attention given to some aspects of design, resulting products still may be inadequate.

Part VII: *Design Research*. The origination of design research and the applicability of research outcomes are the issues here. It shows that total design can be successfully applied to R&D. Those in academia and concerned with the connections with industry will profit from the clarification and guidelines provided.

Part VIII: *Total Design—Summary of the Whole*. This final part provides a systematic framework for design theory and practice, and for all those in pursuit of successful designs.

I have attempted to describe and illustrate the emergence and evolution to the current day of a whole rationale of total design, which reports the latest thinking and practice in Europe and the United States. The reader should benefit from this experience, particularly in the United States, since as stated earlier the stirrings continue and as with the United Kingdom, reports and studies are now appearing in great numbers (See *Improving Engineering Design: Designing for Competitive Advantage*, ISBN 0-30-904478-2, National Academy Press, 1991). One therefore has to a limited extent the opportunity to minimise abortive repetitive work, which we in the United Kingdom could not benefit from in the early 1970s, and thus avoid at least some of the pitfalls.

TOTAL DESIGN MOVES ALONG

Surely this is not the last stop for total design. In its evolution over twenty years it has been constantly improving. I take pride in the fact that along the way I kept my consistency and coherence. Do I see things the way they are and not the way people like to dress them? I don't think so. I rather think that I always have seen design with a integral view and freedom, not in part or constrained by imaginary delimitation of frontiers. In such a manner I moved in search of a generalisation of the design activity into a single form that enables practitioners to use it and industry to appreciate the contribution of good design to its business success. In Part VIII, the first moves into this generalisation are presented. It is a result of my work with my friend and organisational psychologist Ian Morley, Department of Psychology, University of Warwick with whom I set the grounds for what we call the theory of total design. Along with this current development

I have my interest focused on usable methodologies for accurately linking all PDS elements without omission and for systematically changing concepts from static to dynamic without creating confusion up front in the design process. But those are subjects for the next retrospective to come, twenty years from now!

PART I

Design in Education and Industry

COMMENTS ON PART I

The symbiosis between design education in universities and design practice in industry is the foundation of Stuart Pugh's path to design success. Based on much observation and personal experience, Stuart believed that one could not understand design if one's views were formed primarily by either academic experience or industrial experience. The insights that are formed by a combined career—first in industry and then in the university—lead to an appreciation for the processes and practices that enable success.

The design educational program that Stuart developed at Loughborough is still a model. Part I should be read for the design education concepts that were far ahead of their time. Even as this book is published these models are beacons for design educators. They guide the way to great improvements in design education and can be studied carefully to great benefit by educators and by industrialists who are interested in the design practitioners of the future.

Stuart saw that total design is the great integrator of the engineering curriculum. It brings together the various fundamental courses and applies them to useful purposes. Likewise total design is the integrator between the academy and industry. At the core of the integration is the design activity model. Contrary to much common opinion, projects per se do not achieve the necessary integration. Projects can still be very diffuse and unintegrated. The design activity model and the structured methods that make it operational integrate diverse people and knowledge into effective team decision making on the critical issues.

In these chapters the reader can see the design activity model developing as Stuart had time to consider and test it. The reader has the opportunity to carefully study the evolution of the design activity model—the better to understand and be guided by the final version. Design activity is complex, and a model such as Stuart's is essential to avoid getting lost and to successfully accomplish the necessary integration. By studying the evolution of the design activity model the reader can develop a stronger understanding of design activity than by simply observing the final version.

Stuart recognized that the essential requirement for outstanding design education and practice is to have integrated academic and industrial design projects in which student teams work with an industrial company to develop a product. An outstanding example of such a project is described in Chapter 6, Marathon 2550: A Successful Joint Venture—a student-developed product that actually was put into production by an established company. This can be read for its insight into successful design education. It can also be read as a guide to successful industrial projects. To readers in industry, can your integrated product development teams do this well? To readers in academia, can your student teams do this well? What can you learn from this case to improve the performance of your teams?

Part I provides the foundation for the remaining seven chapters. Parts II through VII are specific pieces of the total design puzzle, which then culminate in the final grand integration in Part VIII. Part I is the machine that could have changed the world (and still might) and that produced the specific methods of Parts II through VII and then loops back on itself in Part VIII, the final realization of the vision that started in Part I. Read this as a preview of the rest of the book. On your journey through these six chapters look for a design activity model, total design as integrator, university and industrial projects as the capstone of design education, and Loughborough postgraduate design education as a model. Then come back and read this part after you have finished the whole book and discover the new insights that have been opened to you.

—Don Clausing

INTRODUCTION TO PART I

Part I details the emerging interaction between the requirements and demands of education at the tertiary level and those of industry.

Chapter 1, Engineering Design: Time for Action, considers the whole spectrum and calls upon various reports to substantiate the necessity for engineering courses to be based firmly on design as the central integrator. This theme is expanded further in Chapter 2, Design Teaching Ten Years On, and Chapter 3, Engineering Design at the Postgraduate Level, where design at the postgraduate level is reconciled with design in practice. The transition between the two is bridged with increasing success, as exemplified in Chapter 4, Engineering Design Education with Real-Life Problems, where the example of a project in a process industry is described in detail.

As Chapter 5, Projects Alone Don't Integrate: You Have to Teach Integration, shows, the assumption, very prevalent in the 1970s (and still made today) that projects automatically induce interaction and integration of traditional taught material is clearly not true. Any structured project does induce some integration, but unless the teaching is properly structured, whatever integration is achieved is done so randomly and is inefficient.

This part concludes in Chapter 6, Marathon 2550: A Successful Joint Venture, with the description of the progress of Marathon 2550, a postgraduate student project which was and is successful since its basic product—a truck—is still on the market some twenty years later—a tribute to the students who did the work. They created an efficient design based on a static concept at the total system level and found a persistent demand for the product albeit in varying forms.

CHAPTER

Engineering Design: Time for Action

ABSTRACT

This chapter considers the question of engineering design education both at present and in the future, but of necessity looks at the past in order to provide a meaningful foundation upon which to build the arguments for the present and the future. It confines itself to tertiary education and beyond since to introduce into the equation design teaching in schools at both primary and secondary levels would need considerably more space than this chapter allows.

The distinction is made between engineering education per se and design education, since it is felt that the lack of the latter in education circles continues to lead to an increasing confusion over the whole question. Indeed, in the discourse is a brief description of the postgraduate course with which the writer is associated and, of particular importance, a discussion of the evolution of the course to the present day and the thinking behind it.

The chapter concludes with a suggested strategy for the future.

From B. Evans, J.A. Powell, and R. Talbot (Eds.), *Changing Design* (Chichester: Wiley, 1982), 85–97. This chapter is published with the permission of David Fulton Publishers (first published in the United Kingdom).

THE PAST

In considering the somewhat emotive question of design education it is essential that this be done against the background of engineering design in practice, since ultimately it is concerned with the provision of systems and artefacts to satisfy humankind. It is not, as would appear in many instances, to be a recyclable topic never to see the light of day outside academic circles.

Some twenty-five to thirty years ago the U.K. engineering industry could and did design and manufacture products which the rest of the world accepted gratefully, since we were at that time one of the few nations capable of such achievements. In fact, we could *engineer* almost anything and the world would purchase it. The markets for our products were well established and relatively stable, and the educational system therefore was geared to producing engineers armed with a purely technological and applied engineering science education which satisfied and was complementary to the needs of industry at that time.

Subjects typical of engineering courses at that time were applied mathematics, theory of structures, thermodynamics, electrical power, and so forth. Particular subject selection implied a degree of specialisation, since generally universities were concerned with specialisation and the production of specialists. Indeed, one emerged from the system as a mechanical engineer, electrical engineer, or structural engineer. Note that there was no design connotation. One then either gained employment in industry in a specialist role, usually analytical, in which case the narrow problems became real ones instead of textbook examples and the transition from education to practice was hardly perceptible, or one entered the Elysian fields of design, where it rapidly became apparent that the transition was abrupt and something like stepping off a precipice. In the latter context, the relationship between education and practice was obscure, and only time bridged the gap.

In fact, at that time, the vast majority of engineering designers took the route of the Higher National Certificate, which they gained through part-time education whilst at the same time holding down a job. Practice was thus the foundation upon which the acquisition of knowledge was based, and in this situation the reconciliation of education with design in practice was an ongoing process and was probably imperceptible to most people who undertook this route to design. In other words, the university system usually produced the specialists. The technical colleges, institutes of technology, and the like produced the generalists or, to use a medical analogy, the general practitioner.

Having set the scene of some twenty-five years ago, let us now consider the intervening years to the current day. Today more and more countries have become industrialised, resulting in competition which has increased to a point where it has become positively painful. Against this background of progressively increasing competition and hence rapidly changing market expectations, the educational system and structure have changed. We have seen the almost total elimination of a Higher National Certificate route to design, in that possession of a university degree now is a prerequisite to professional recognition. In medical parlance, we have almost destroyed our source of GPs. To imagine the medical profession without a general practitioner foundation would be recognised immediately as being an intolerable situation and hardly likely to happen, yet in the engineering profession, for whatever motives, it has happened, and what is most important it has happened against this rapidly changing economic background. Consider such a change in the education of medical practitioners in circumstances under which the design of the human body is changing, and one starts to appreciate the magnitude of the problem!

That something is wrong is at last being recognised, in that, for instance, the Institution of Mechanical Engineers has recently announced the formation of an Institution of Technician Engineers. Whether this will partially rectify the situation remains to be seen. One hopes that it will, but one suspects that it will not, since there is little evidence that an increase in the number and diversity of professional institutions has enhanced our competitiveness in world markets. It is this situation which has brought about the dichotomy in our engineering courses in an effort to respond to what everyone concerned will acknowledge is a changed situation.

This brings us to a consideration of the present situation.

THE PRESENT

Today the vast majority of our young people with aspirations to engineering in all its facets are being educated in the universities and polytechnics. It is granted that first degree courses in engineering have become in a sense more liberal in that they now include subject matter such as management, economics, and the like, which not many years ago would have been considered irrelevant to the education of an engineer—in spite of the fact that in practice he or she would have been involved with the application of such topics. However, might I suggest with due humility that, whilst our educational institutions have been aware of the aforementioned changes in background, the majority of teachers concerned with engineering have no direct experience of the change itself (Pugh, 1977a). Meanwhile, industry repeatedly states that the universities do not produce the type of people they need, and, whilst this is not an unreasonable request (always assuming that industry knows what it wants), it is completely unrealistic to expect young embryonic engineers to glide into industry in a smooth transition. It is unrealistic in the sense that, except to a limited degree, it cannot happen since the majority of engineering teachers are remote from what I would call 'hard practice' and involvement in industry. In cases where industrial links have been established, many are at specific research interfaces which, laudable and necessary though these may be, do not give the individual concerned first-hand knowledge of the competitive design and market interface—the real world of engineering. In fact, whilst it is accepted that research is fundamental to a university's operation, it must be recognised (and by and large it still isn't) that engineering is not research and design is not research: they may utilise research, they may cause it to be carried out, but they in themselves are not research. To quote Eric Walker, President Emeritus of Pennsylvania State University, 'Research is often confused with engineering, but it often has a more limited purpose. Research is the collection and analysis of data which may or may not be pertinent to the solution of an immediate engineering problem. Research by itself is not engineering; it is not even a necessary part of engineering. There are times when sufficient facts are not available for the designer/inventor to provide a solution to an engineering problem, and the research must be done to find the missing facts' (Walker, 1978).

Yet the majority of engineers are being educated against a research background. How then can they be expected to piece together an educational jigsaw comprising specific subject matter of the type mentioned earlier? Subjects are still by and large taught in what I can only describe as 'subject compartments' with little attempt to bring about the integration of such subjects demanded by practice (Pugh and Smith, 1977). The inclusion of project work, usually implanted to bring about the desired integration, helps; sandwich courses also help; but they are of limited value if the subjects themselves are not taught against an integrative background from the beginning. Design in practice is always an amalgam of specialist inputs, and the design of an artefact does not exist that is not based upon such an amalgamation: it is the picture on the box that matters as well as the individual pieces of the jigsaw. Yet how many of our engineering teachers are aware of this and have direct experience of it? Unfortunately very few. Two or three years in industry or last practice twenty years ago may have been adequate when viewed against a stable, slowly evolving market background; it simply is not good enough or appropriate enough in the situation of a rapidly changing background that confronts us today.

However, whilst appearing to be critical of my academic colleagues I must at the same time come to their defence. With almost the entire emphasis as regards career prospects being placed upon research and little, if any, credit in academic terms being given for practice, who can wonder at the almost overwhelming preoccupation with research at the expense of practice? Whilst reiterating that research is a primary function of the universities, it must be balanced by broadly based practice being recognised as having equal merit: the intellectual demands of the latter are at least equal to and often greater than those of the former. It is a restoration of balance that is required, since somebody has changed the weights on the scales, and we are behaving as if we have not noticed it. What chance then of students having any alternative other than to attempt to leap the gap between education and practice? I shall first consider postgraduate level of education and then the undergraduate level.

POSTGRADUATE LEVEL

In recognition of this gap, some fifteen years ago in 1967 a postgraduate course in engineering design was established at Loughborough University of Technology. The fundamental aim of the course was, and still is, to form a bridge between education and practice via the mechanism of a twelve months' course. The first three months of this course would, in effect, reveal to participants not only the picture on the box but also bring home to them with some force the interrelationships between the traditional subject compartments necessary in order to produce effective competitive designs: what one might call the general case. This is then followed by a group project of nine months' duration, during which the students are expected to put together a definitive jigsaw as defined by a particular company and market situation and in so doing to evolve a more competitive, more effective design than existed at that time. Whilst this latter objective was achieved in some instances, it was recognised that we had a problem (Pugh and Smith, 1978). It had been brought home to us over the years by our students, not so much by their emerging from our course as ineffective design practitioners, but by their expression of the difficulties of understanding and putting together the course content into a cohesive and comprehensive whole. We considered that the course content was substantially right for the time allotted and in keeping with a nonspecific discipline or general engineering design type of course (Moulton, 1976). We had come to this conclusion by critical analysis of our actions and those of our students over the years in pursuit and achievement of successful designs. It is not that the approach did not work, but we felt that it could work a lot better.

This reconciliation between design in practice and what we teach took a long time and is by no means completed; however, I think it safe to say that we have established a fundamental format for teaching design which matches practice, uses the same language as practice, and which above all else enables students to understand the complex nature of design and to put this understanding into practice in an increasingly effective manner. This third statement applies to all students, irrespective of their specific engineering disciplines (based upon the traditional view of such disciplines)—that is, electrical, mechanical, civil, and so on.

A generalised model has been formulated for engineering design (or for that matter any other form of design) which I call the 'design activity model' (Figure 1.1). This model is *not* presented to the students in its entirety at the start of the course as being what engineering design is all about; it is unfolded progressively in sympathy with teaching which puts flesh on the skeleton of the model until the whole is revealed. In parallel with this evolution students are given exercises which gradually bring into play more and more components of the model until, at the end of the first three months, they are producing simplistic designs having utilised all components of the model. Effective operation and manipulation of the model is considered an essential prelude to the commencement of the major design project, which no longer is a design exercise but is carried out with real constraints, the objective being as already stated—better designs than exist at the time.

I do not propose in this chapter to discuss the model at great length, but perhaps readers might find a brief description useful. If one considers the design activity as having a central core, the essential phases of which are market, specification, conceptual design, detail design, manufacture, and selling, one has an iterative process which proceeds generally from market needs to selling: it must be of necessity to proceed in this way, otherwise nothing would ever be produced! The outcome, the selling of a product or system,

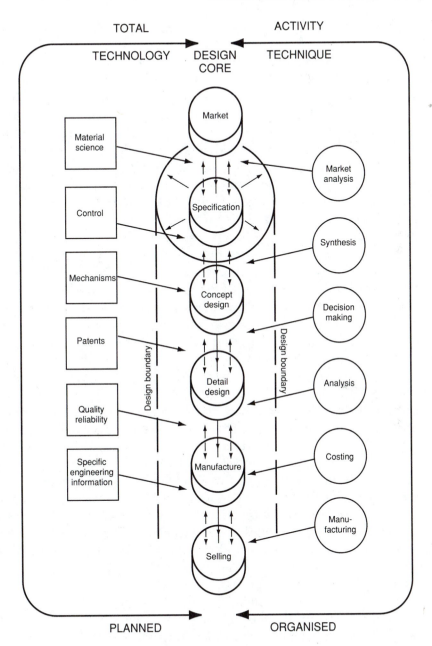

FIGURE 1.1. The Design Activity Model (1982)

itself influences the market and therefore sets us another cycle of design activity which interacts with the market and causes it to change. This flow of activity is, or should be, carried out within the envelope of an evolving specification which embraces many factors (Pugh, 1974) as well as function and which itself evolves as the design proceeds due to the interaction with the core activity. The specification is, therefore, our design boundary.

In order to be able to bring the core to life and make it work, one needs to have not only an understanding of the iterative nature of the core activity but also knowledge and awareness of the techniques available to aid its working—in other words, the tools available in order not only to set it in motion but also to improve its function and performance. These we call 'techniques related to the design core'—typically costing, ergonomics, creative methods, appropriate analysis, and computing. Again, these are inputs to the design core and interact with it. It is interesting to note here that as yet we have made no mention of technology, which in our view is not only unnecessary at this stage of understanding but may cause confusion and mitigate against a clear understanding of design. While considering techniques related to the design core, we have a finite situation in view. However, 'technological support subjects' are also inputs to the core, and these open up infinite possibilities in that the whole of technology is potentially available for use. Under this heading subjects are included such as materials technology, control, and specific technological information from a variety of disciplines. In any given design situation this knowledge and information will be related to the particular design area with which one is concerned. The whole of this activity is carried out under an umbrella of planning, organisation, and control; these we call *management subjects*. Place this total package within a corporate business structure, and one has a picture of how engineering design fits or should fit into a modern business. This is what happens in practice; students understand it and use it to good effect.

I would at this point make a comment about the great variety of 'models of the design process' in existence and still being produced. From a personal viewpoint, I find that many of these models make sense to me having had twenty-five years experience at the design and market interface. However, if I discount this experience and consider such models from a student's viewpoint, then I can honestly say that the majority would be incomprehensible. Therefore, to teach design against an incomprehensible background at best becomes extremely difficult and unrewarding and at worst becomes impossible. It is essential, in my view, that any teaching package must be interrelated, comprehensive, coordinated, and above all else meaningful. This statement applies to the teaching of anything, but is of vital importance in design teaching since it is difficult to understand and has an infinite number of facets.

To return to the model in question, having established it one finds that the teaching material required to build up the model, particularly the design core area, does not exist in textbooks today,[1] which in itself is not surprising, and such material has had to be developed in sympathy with the model. There are also other ramifications stemming from this approach in that much of the conventional material available today is not written in a form compatible with the model, and this is having to be written. In case I create the impression that the task is complete, let me immediately dispel such beliefs; it has only just begun and will grow in magnitude and complexity as our understanding of design itself increases. However, students are now producing more effective designs than in the past, so there must be some value in the model.

It might be argued that not all designers in industry will be called upon to operate in such a comprehensive manner, and it is a point of view with which I would wholeheartedly agree. However, I do feel strongly that all engineers, *with or without aspiration to design*, should be made aware of the total picture. In industry, hierarchies exist in design offices for reasons which I assume are based on levels of competence (Pugh, 1977b). It is essential that the whole spectrum of a design team be made aware of the total activity, otherwise not only are their aspirations likely to wither and die but also the work that they undertake becomes less meaningful and lost in the day-to-day activity of industry.

UNDERGRADUATE LEVEL

Having become increasingly involved with undergraduate teaching over the past few years, I now have serious doubts as to whether design teaching for undergraduates should be attempted at all. Having made this statement, let me now elaborate. (This is very much an early view.)

In theory, I do not agree with my previous statement, since if the whole gamut of traditional subject matter could be taught against a design background then I see no reason for such teaching not being integrated by means of a design teaching package of the type described earlier. It is here that we hit the first snag; such an assumption would require the teachers to have ongoing experience of the relationship which their particular subject material has to design in practice, thus enabling them to relate such material to the realities of the world. But this is not generally the case, and I elaborated on this point earlier in the chapter. This being the situation, if an attempt is made to graft on to the traditional engineering teaching format a design teaching package of the form previously outlined, there is a high risk of developing a form of schizophrenia in the minds of the students, resulting in bewilderment and confusion.[2]

Let me expand on this particular problem. First, the timing of the graft is significant in that, if it is carried out before the end of the final year of the course, the students have a limited understanding of and ability in traditional subject matter. For example, if in a design exercise a particular method of analysis is called for, students may not yet have covered it in their syllabus, the net result being a defrayment of effort and limited success in carrying out the exercise due to a lack of knowledge.

Second, if such a design teaching package succeeds in opening the minds of the students, changing their attitudes, and making them conscious of the fact that in reality all questions are open-ended, they will then encounter difficulty and conflict when posed with closed-ended questions in examinations. This I can foresee causing tremendous difficulties for the staff concerned with specific discipline teaching. Students would not answer closed questions in the appropriate and expected manner. The only way to avoid this conflict would be for specific subjects to be taught in an open-ended fashion —which it is unrealistic to expect for reasons mentioned earlier.

Thus, in order to accommodate this situation and avoid such difficulties, it becomes necessary to teach design in a somewhat artificial manner. The usual approach is for students to undertake a closed-ended project which is necessarily artificial and which, although useful in itself, does not bring about in the students a recognition of the totality of the design activity; it does, however, retain compatibility with the teaching.

At the undergraduate level, therefore, we are in somewhat of a cleft stick in so far as design teaching is concerned. We also run the risk of detracting from the teaching and understanding of fundamental engineering principles upon which all engineering in practice is based. Engineering design teaching is not a substitute for traditional subject matter; it is the integrative mechanism by which such teaching is made meaningful. Basic engineering science teaching should not be neglected.

Thus we have a dichotomy of alarming proportions. These are the facts of the situation. What can be done about it?

THE IMMEDIATE FUTURE

For the immediate future I can see only one sensible approach to this problem. Traditional single-subject discipline teaching should remain as it is currently taught, since it is not easy to suggest a practical alternative under present circumstances. Engineering design teaching should form an essential part of the final year of all undergraduate engineering courses to act as the integrator of the traditional material which is essential to practice. However, it could be argued, and I would support the view, that in order to do this properly more time needs to be made available in already overcrowded syllabi. Is not the answer then to extend the engineering courses by one year in order that design may be taught in a proper fashion? I must stress at this point that I am not in a roundabout way alluding to the 'enriched engineering courses' which are becoming the vogue. With certain exceptions these courses appear to me to be giving the students who partake of them some more pieces of the jigsaw, in terms of additional subjects and enhanced knowledge of traditional subjects, with little attempt, if any, to reveal the picture on the box. The result is almost entirely predictable, in that people will emerge from such courses with more knowledge in greater depth, but more confused than ever when it comes to the bridge between education and practice. I see little evidence that before deciding that one solution to a national problem is the enriched course, a realistic attempt was made to define the problem. The problem is basically that in order to survive in the markets of the world we have to design the right products for the right market at the right time. We therefore have to teach design aspirants to design properly against this backcloth and that of increasing competition. How then does this correlate with the enriched and extended degree courses now being postulated? In my opinion it doesn't, except in relation to the education and training of the specialist engineer, and so yet again we are intent on producing more specialists, which, essential though they are to industrial operation, cannot be expected to operate as general practitioners. One exception seems to be the University of Southampton (Engineering Department, University of Southampton, 1978), which, in its first announcement of a four-year course, stated, 'A major emphasis will be upon Engineering Design as an application of the engineering theory and science covered in earlier years of the course.' What is also highly significant about the Southampton proposal is that it appears to be crossing the traditional and divisive departmental boundaries. Bravo, Southampton! More universities could well take a lead from this initiative.

THE LONG-TERM FUTURE

In the future we must have a progressive shift towards teaching based upon the realities of the situation, and then it might become realistic to attempt design teaching from the first undergraduate year onwards, at least to some extent. This we have done at Strathclyde since 1986. A rationalisation of university and polytechnic structures is required to avoid the divisiveness of departmentalism whilst at the same time maintaining specific research interests. It is salutary to refer to the Galloway Report (Galloway, 1975), which, although it deals specifically with the development of mechanical engineers, reads equally well and is equally meaningful if the word *mechanical* is omitted. In the appendix headed 'Industrial Academic Objectives' the actions necessary in the long term (long-range planning) are stated to be 'to ensure that by individual and joint endeavour the profession is recognised and utilised for the benefit of industry, the community and the country; to ensure that needs are foreseen, that the standards of education and training are in accord with future needs.' The actions required of the academic fraternity are to 'sustain academic training at a high level to ensure development of the country's technology; be aware of and meet effectively the diverse needs of industry; maintain the research role but not in a self-isolating form; develop academic staff so that they are seen to be *top quality, practicing professional engineers.*'

These are strong words indeed. Needless to say, *long term* is not elaborated on. However, the expression *long term* may be entirely hypothetical. Are our industrial competitors going to mark time until we have adjusted to the realities of the situation with traditional British alacrity? I do not think so. Many companies are already experiencing or predicting a void in their design departments. Of the designers who emerged from the system of twenty-five years ago, many are reaching retirement age, with very little evidence that they will be replaced or that others have learned from their experience. We have to move very rapidly if this situation is to be corrected, and, unless we adapt the structures of our university and polytechnic engineering departments to meet this new challenge, we shall fail. We must stop shadow boxing and recognise that we are in a real fight.

ADDENDUM

Since I wrote this paper originally, the Corfield Report on product design has been published (Corfield, 1979). The following quotation, amongst many, is not only revealing but is supportive of the arguments put forward in this chapter: 'In practice we find that our educational establishments produce scientifically and technically qualified people of specific disciplines and there is some, but very modest, increase in interdisciplinary education at graduate level. There is good provision for training in development or design in our university system except in a few isolated cases. The whole burden of developing competent development engineers or design engineers at present falls upon industry itself and consequently industrial management has to overcome both the hurdle of accepting design engineering as an important and costly phase in product innovation and the hurdle of finding staff motivated and trained in the process of design.' After such a quotation, I feel that any further words from me would be entirely superfluous.

NOTES

1. My book *Total Design* (Reading, MA: Addison-Wesley, 1990) does to some degree address these issues, as do the publications of SEED.
2. The establishment of the master of engineering in product design in 1991 overcomes these problems to a large extent.

REFERENCES

Corfield, K.G. (1979). *Product Design.* London: National Economic Development Council.

Engineering Department, University of Southampton. (1978). 'First Announcement: Four-Year Courses in Engineering, Appendix 3.' University of Southampton.

Galloway, D.F. (1975). *Meeting the Needs of Industry.* London: Institution of Mechanical Engineers.

Moulton, A.E. (1976). *Report on Engineering Design Education.* London: Design Council.

Pugh, S. (1974). 'Engineering Design: Towards a Common Understanding.' Paper presented to the Second International Symposium on Information Systems for Designers, University of Southampton, July.

Pugh, S. (1977a). 'Submission to the Committee of Enquiry into the Engineering Profession' (Sir M. Finniston, Chairman).

Pugh, S. (1977b). 'The Engineering Designer: His Task and Information Needs.' Paper presented to the Third International Symposium on Information Systems for Designers, University of Southampton, March.

Pugh, S., and Smith, D.G. (1977). 'CAD and Design Education: Should One Be Taught Without the Other?' Paper presented to the International Conference on Computer-Aided Design Education, Teesside Polytechnic, 13–15 July.

Pugh, S., and Smith, D.G. (1978). 'Design Teaching 10 Years On.' *Engineering Design Education*, Supplement (June), 20–22.

Walker, E.A. (1978). 'Teaching Research Isn't Teaching Engineering.' *Journal of Engineering Education* (January), 303–307.

Engineering Design: Teaching Ten Years On

ABSTRACT

This chapter considers the evolution of the postgraduate course in engineering design at Loughborough. It outlines many of the problems faculty and students encountered during curriculum development and offers guidance for the future. It discusses the course in detail and hopefully will enable others to avoid some unnecessary pitfalls.

With Douglas G. Smith, in *Engineering Design Education* (Spring 1978), 20–22.

ENGINEERING DESIGN AT LOUGHBOROUGH

The Engineering Design Centre at Loughborough emerged as a direct outcome of the Feilden Report of 1963 with the specific dual objectives of educating and training young engineers in the arts and skills of engineering design. Denis Chaddock, the first Professor of Engineering Design as such in the United Kingdom, constructed a course based upon the many interactive facets of engineering which together form a basis for engineering design. The course itself consisted then, as it does today, of a wide range of interrelated topics not normally taught in undergraduate courses in engineering and was not concerned with any specific discipline. This was combined with practice in design through the mechanism of industry-based project work. That his foresight in establishing such a programme in an academic environment was indeed correct is borne out when we see such schemes as Total Technology and more recently the Teaching Company not only gaining academic respectability but also stemming from a sense of responsibility to the nation's needs as a whole.

In being intimately concerned with the structure, operations, assessment, and monitoring of the course almost since its inception, we feel that committing some of our thoughts to paper will not only clarify our own thinking in this area but may also form a useful back cloth for others concerned with engineering education, both now and in the future.

PROBLEMS OF FORMULATION AND UNDERSTANDING

It is an inescapable fact of life that almost everyone possesses 20:20 hindsight vision. In our course this has manifested itself not so much as changes in what we teach, although there have been changes over the years, but rather in the form of the fundamental question, 'Why do we teach the course syllabus in the way that we do, and what is the basis for such an assemblage of course material?' Until very recently we felt intuitively that the course content and balance between lectures, tutorials, and practice were about right, and over the years there is no doubt that we have produced a number of potentially good designers by subjecting them to our programme. We must, however, admit to a sense of unease and, to put it mildly, frustration in not being able to completely reconcile in our own minds that what in fact we were doing was coherent and understandable. Because it is a characteristic of design that if pure logic had to be applied *before* proceeding with a design then little or nothing would be designed in this world, such a situation was therefore not foreign to us.

Simply put and by analogy, the problem was 'How can we teach mathematics if we do not fully understand its nature?' In design terms this is very easy to say but somewhat more difficult to answer; in fact, it is the classic problem in design itself. It is not the definition of *a* problem but the definition of *the* problem that is important, and it is to this end that we concentrated our thinking.

That we had this problem has been brought home to us over the years by our students, not so much by their emerging as ineffective design practitioners, but by their expression of the difficulties of understanding and putting together the course content into a cohesive, comprehensive whole. We considered that the course content was by and large right for the time allotted, and that the subject matter was also right for a nonspecific discipline or general engineering design type of course (Moulton, 1976). By a critical analysis of our actions and those of successful designs, we determined that it is not that the approach did not work but that it could work a lot better.

This reconciliation between design in practice and what we teach has taken a long time and is by no means completed. However, this year we have changed our teaching format; we have not changed what we teach but rather the way in which we teach it. Akin to any design, the component parts are always vitally important, but it is the way in which they are put together which dictates whether or not one has the most successful design. We believe that we have at last found a better way of putting our course together in order to reconcile practice with theory, and there is no doubt that this is reflected in the output, attitudes, quality, motivation, and understanding expressed in the work of our present students.

THE RECONCILIATION

In a nutshell we have reconciled what we do in successful design practice with what we teach in a form which we consider is logical and easy to understand and translate into practice. It is still difficult for some, but we think it easier and more logical than it was in the past. We are certain that we have improved our capability for 'putting it all together' in practice and in so doing have achieved an improved format for teaching. It is, however, essential that students be able to put it all together, the outcome being hardware in some form or other. In fact, teaching and practice are reconciled in a manner which the true practitioners of successful design will recognise as being what they do or should be doing regardless of their specific engineering discipline. We also find that in teaching design, whether to postgraduate civil engineers reading construction management or final-year materials engineers, the teaching format we have adopted is equally appropriate in all specific disciplines.

Engineering design in practice is a complex activity, and probably the factor which mitigates most against understanding is the belief that apparently complex problems demand complex solutions. Nothing could be further from the truth! Such approaches in the past have not led to a better understanding of design except perhaps in the mind of the initiator, but they certainly have added to the confusion surrounding design. We depict design in any sphere as consisting of a number of fundamental phases, procedures, or activities which are always followed, sometimes intentionally, sometimes randomly. If this is not done, then matters have to be put right. However, subsequent analysis of any design activity, no matter what the design, can always be matched to the model shown in Figure 2.1, and retrospective analysis of a project relative to the model will often yield some surprising pointers as to what went wrong during the course of the project.

If one considers the design activity as having a central core—the essential prime phases of which are market, specification, conceptual design, detail design, manufac-

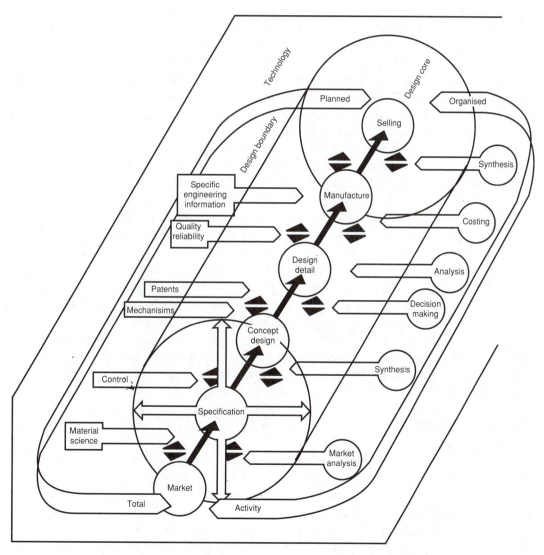

FIGURE 2.1. The Design Activity Model

ture, and selling—one has an iterative process which proceeds generally from market needs to selling. It must of necessity proceed in this way otherwise nothing would ever be produced. The outcome—the selling of a product or system—itself influences the market and therefore sets up another cycle of design activity which interacts with the market and causes it to change. This flow of activity is, or should be, carried out within the mantle or envelope of an evolving specification which embraces many factors (Pugh, 1974) as well as function and which itself changes and evolves as the design proceeds due to the interaction with the core activity. The specification is, therefore, our design boundary.

In order to be able to bring the core to life and make it work, one needs to have not only an understanding of the iterative nature of the core activity but knowledge and awareness of the techniques available to aid its working. In other words, one needs tools not only to set it in motion but also to improve its function and performance. These we call 'techniques related to the design core'—typically, costing, ergonomics, creative methods, and the like. Again, these are inputs to the design core and interact with it. It is interesting to note here that as yet we have made no mention of technology, which in our view is not only unnecessary to this stage of understanding but may cause confusion and mitigate against a clear understanding of design. Still considering techniques related to the design core we have a finite situation in view, however. 'Technological support subjects' are also inputs to the core, and then one has an infinite situation in that the whole of technology is potentially available for use. Under this heading subjects such as control and specific technological information are included from a variety of disciplines. In any given design situation this knowledge and information will be related to the particular design area with which one is concerned. The whole of this activity is carried out under an umbrella of planning, organisation, and control; these we call management subjects. When this total package is placed within a corporate business structure, one has a picture of how engineering design fits or should fit into a modern business.

Thus, the way we see the situation and the previous suggestion of retrofitting specific designs into this model should at least make one more aware of what is or is not happening in a particular organisation. A progressive buildup of this model from the design core to organisation and planning is the background for teaching our postgraduate students in engineering design and structuring our course content.

IMPLEMENTATION

The subject matter covered during the first three months of the course is grouped under the headings referred to previously, and furthermore the timetable itself is structured in sympathy with these areas, details of which are shown in Figure 2.2. It should be noted that each week of this term follows the same basic pattern so that the correct juxtaposition of course material and emphasis can be maintained in concert with the model (see Figure 2.2). Design core subjects take precedence. Thus, for example, costing techniques are begun before students are called upon to do any costing in their core subject

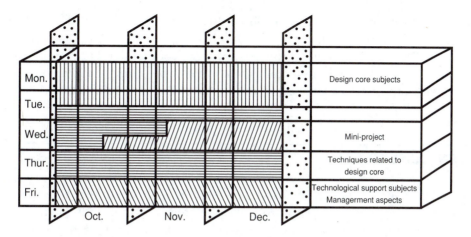

FIGURE 2.2.   Subject and Time Allocations for the First
Term of the Design Course

assignments, but 'technological support subjects' and 'management aspects' proceed steadily through the term.

It is upon this basis that the course unfolds in a structured manner with lectures, tutorials, and coursework on the main topics. The 'design core subjects' are explored phase by phase so that by the end of the term the students have dealt with each phase in some depth and also with the progressive interaction between them. Meanwhile, they will have had specific assignments in each of the aforementioned areas. The first real opportunity to 'put it all together' comes with the 'miniproject' which commences two weeks from the start of term and continues until the end of term. This project has two phases. First, it is carried out as an individual assignment culminating in a presentation to Centre staff. It is interesting to note that during both stages there is a high level of competition; however, the team activity usually results in a vast improvement to the designs, particularly at the conceptual level. In fact, last year one of the group designs was patented.

By Christmas the whole model has been brought into play, and students are by then realising the complex nature of any design activity and are able to practise it with increasing confidence and understanding. It must be said that this year the problems caused by the highly interactive nature of design seem to be fewer than in previous years, a factor which we attribute at least in part to the restructuring of the course material based on the new model.

The first term from October until Christmas is fully taken up with the programme just described, the objective being the acquisition of knowledge of the design activity and its inputs and the acquisition of skill in handling the interactions. Naturally, due to the restricted timescale of the course the technological input side is limited. This we consider an acceptable limitation, as the acquisition of all the knowledge in the world would be of little use if the student did not learn how to handle it in an interactive manner.

From January until September the level of lecturing decreases, and the students, in groups, start work on a real live project taken from industry. We think it highly significant that whilst the increasing cry today is for more cooperation between industry and academic institutions, such cooperation has always been forthcoming on our main projects with some measure of success and we have reason to believe that this will continue.

STUDENT ASSESSMENT

We should perhaps say a few words about our student assessment procedure in order to give a comprehensive view of the course. We operate a system of individual continuous assessment on a three-month basis; this assessment procedure is itself based on some twenty-five criteria against which the students are judged, together with a marking system for all course work, the miniproject, and the main project. You might well ask, Why have we not instituted a traditional examination system? The answer to that question would be, Does engineering design per se lend itself to the traditional approach to this problem? Our view, even today, would be an emphatic no. However, we recognise that our course does impart knowledge to the students, and beginning in 1977–1978 we will institute two examinations to cover those aspects of the course which lend themselves to examination. One paper will examine the breadth and depth of understanding of the design core activity and will be essentially a literate examination as opposed to a numerate one. A second paper will examine the technological support subjects and will essentially be a paper of a more numerate nature. These examinations, together with the procedure described earlier, will then form the new basis for assessment.

THE FUTURE

We shall continue to develop our course along the lines outlined in this chapter, the intention being to improve and raise the level of design competence of our students. We believe that there is still much to be done to enhance our understanding of the design activity. We do however, remain firmly of the opinion that Denis Chaddock started upon the right lines albeit with an embryonic course which has evolved over the years and is still highly relevant to today's needs.

REFERENCES

Moulton, A.E. (1976). *Engineering Design Education.* London: Design Council.
Pugh, S. (1974). 'Engineering Design: Towards a Common Understanding.' *Proceedings of the Second International Symposium on Information Systems for Designers* (pp. D4–D6). Southampton: University of Southampton.

3

Engineering Design at the Postgraduate Level

ABSTRACT

This chapter discusses the need for teaching engineering design at the postgraduate level, to benefit both industry and education and to develop design understanding and methods. Distinction is made between methods applicable to specific engineering products and those applicable to all fields of engineering design. A case is developed to justify both broad-based courses and narrower enhancement courses usually aligned with specific areas of engineering, with the proviso that to pursue the latter without the former may result in better engineers but not necessarily better designers.

The development of the course is discussed from the standpoint of totality of practice, evolution of understanding and methods, performance of students, and acceptability to industry of their performance. The paper continues by outlining some of the problems of implementing courses against a background of traditional engineering educational structures, concluding with suggestions for the improvement in the structure and operation of such courses.

From *Proceedings of Tomorrow's Engineering Design Conference* (Loughborough: Loughborough University of Technology, 1979).

THE NEED

Engineering design courses at the postgraduate level are needed (1) to raise the standards of design and design teaching generally and hence produce better designers and (2) to enable the development of design methods to take place relatively unconstrained by industrial pressures. However, it becomes essential, even at this stage, to differentiate between what may be called broad-based courses and those aimed at specific engineering products (Moulton, 1976).

Both types of course are needed, yet if one accepts and examines the two basic premises outlined above and applies them to the two types of course, then different conclusions must be drawn in respect of the outcome in each case.

I consider the narrower specific courses first, since they are representative of an extension of the traditional approach to engineering teaching and form a necessary part of modern education. Probing this question in more detail and examining the likely outcomes are rather revealing.

NARROW-BASED COURSES

Standards

There is no doubt that standards of design in these specific areas have been raised and that students attending such courses become better at designing the appropriate artefacts, that design competence is improved, and that detail design is improved, and it is hoped that once the students commence practice in earnest, that better quality products will emerge. It is, however, questionable whether student design capabilities are enhanced except upon a narrow front, usually based upon existing practice, and therefore questionable whether such students are not being educated to repeat the sins of their forefathers albeit being better equipped to do so!

Do they have an understanding of design and competence in design to fit them for the rapidly evolving and changing world marketplace? One suspects not. Such courses are provided to meet the needs of industry and are based upon the supposition that industry itself is geared to meet these market demands. From the evidence generally available, however, this supposition is itself questionable.

The question to be posed is this: If the latter statement is wholly or partially true, should not the educational sector be researching and gearing itself to train people to respond to the rapidly changing world situation, whilst at the one and the same time ensuring enhanced and elevated practice in the areas of specialisation rather than concentrating almost entirely on the latter? It is unreasonable to expect industry to unscramble the problem since day-to-day involvement in an industry, with all its attendant pressures, is hardly the environment in which to become introspective and constructively critical. The educational sector, without such pressures, should take the initiative and provide people for industry who have a total view of design (as required today) together with an enhancement in a specialism.

It is suggested, therefore, that whilst to build upon a specialisation is necessary and desirable, to do so with a recipient who does not have an up-to-date understanding and capability in a total design sense is nothing short of inadequate and may be dangerous, misleading, and ultimately damaging to the economy of this country. Whilst design competence and standards will be improved by such courses, their narrow base mitigates against the improvements that are possible.

Development of Methods

The development of methods to improve design practice is a firm responsibility of design education at the postgraduate level. It is appreciated that research, as opposed to courses, is also a likely source for evolving methods. However, one wonders, with the narrow-based courses, whether methods emerge as the result of research in the topic area or whether the courses themselves lead to the evolution of new and better methods of design. One suspects the former rather than the latter, but it must be admitted that the author has only secondhand knowledge of the situation.

This raises the question of whether methods emerging from either of these routes are themselves applicable to design as a whole or just to the area of specialisation defined by the research or the course? It is suggested that the latter is the case, and, without wishing to detract from the necessity and excellence of such work, it should be recognised that whilst such methods will, in design parlance, improve a designer's competence through the possession of better 'techniques,' they will have little effect on his or her understanding and competence in total design terms. To put it more succinctly, they will improve his techniques and hence designs on a narrow front without improving his handling of a total design situation, and by definition will contribute little to methods applicable to design across the whole spectrum of engineering. This point is expanded upon later in this chapter.

BROAD-BASED COURSES

Standards

The master of technology course in engineering design at Loughborough, which has been in existence for twelve years, has in this period evolved as a broad-based course. It was established primarily to produce competent design practitioners without emphasising any particular specification. It must be admitted that whilst the above was the primary objective, considerable difficulty was experienced in establishing a comprehensive, cohesive format for teaching, which in itself would provide a natural lead into design practice, whilst no doubt the recipients of such teaching improved their standards of design teaching per se. Since broad-based design practice formed a major course component, this was considered an ideal situation and environment in which to gain better understanding of design and, stemming from this, the potential for better design teaching. In hindsight, it is difficult to visualise how a better understanding and hence teaching of design could be gained in any other way. This point is particularly important since it is considered with the understanding of a dynamic activity, the data of which are changing rapidly with increasing pressures of market competition, technology, world resources, and so on. My considered view is that to attempt to teach design *without associated practice* is extremely shortsighted, since the design of yesterday may no longer be appropriate today.

The reconciliation between design in practice and what is taught has taken a long time and is by no means complete; however, it is safe to say that a fundamental format for teaching design has been established which matches successful practice, uses the same language as practice, and which, above all else, enables students to understand the complex nature of design and to put this understanding into practice with increasing effectiveness. This latter statement applies to all students irrespective of their specific engineering disciplines (based upon the traditional view of such disciplines).

A generalised model has been formulated for engineering design (or for that matter any other form of design) called the 'design activity model' (Figure 1.1). This model is not presented to the students in its entirety at the start of the course as being what engineering is all about; it is evolved and unfolded progressively, putting flesh on the skeleton of the model until the whole is revealed. Students are given exercises which gradually bring into play more and more components of the model, until at the end of the first three months they are producing simple designs using all components of the model. Effective operation and manipulation of the model are considered essential preludes to the commencement of the major design project, which no longer is a design exercise but is carried out with real constraints, the objective being better designs than exist at the time. A full description of the course is given in (Pugh and Smith, 1978). There is no doubt that better standards of design have resulted in the competitive products, processes, and systems that have emanated from the course (Pugh and Smith, 1974; Pugh, 1978).

Development of Methods

An approach to design teaching which avoids specialisation like the plague forces designers to search for and establish techniques and methods applicable to any field of design. Since the essential mechanism of the course is broad-based practice, the opportunity arises for such methods to be critically evaluated in real situations. That such methods work in practice is evidenced by the quality of the designs produced by students on the course. Not only have new methods been evolved, particularly in the area of concept selection, which is undoubtedly the most difficult and uncertain area of design, but also many existing design methods have been evaluated (Pugh and Smith,

1976). Experimentation with design methods is an essential requirement of any postgraduate operation with the overriding proviso that such methods must be evaluated in real design situations. Evaluation of methods in nonrealistic, artificial situations can give misleading results and create a false impression of success or lack of it.

It is difficult to carry out this sort of work at the undergraduate level since there is neither the time nor a real project available for effective evaluation. Moreover, there is no substitute for the acquisition of fundamental knowledge. The normal three-year courses should not be diluted by spurious attempts to broaden them.

It is the considered opinion of Centre staff that if methods are to be taught, one should at least have an inkling of whether or not they work. Strangely, many of the methods purporting to be useful in design have been established by nondesigners; it is doubtful whether mathematical methods evolved by persons not versed in mathematics would attract the same credibility.

Methods apart, probably the greatest attribute of a broad-based operation is the creation of an environment which has led to the establishment of a 'discipline of design' to which the traditional subject matter of engineering can be logically related. This has led to a greater understanding of design per se by students and staff, and no doubt will evolve further with the passage of time.

RESEARCH

Actively involving the course with the entire field of engineering has given rise to research opportunities in the design area leading to the discovery of techniques useful not only in design teaching but also to design practice at large, based on the premise that if a technique is useful in teaching it should also be useful in practice. This has already resulted in methods being established for the rapid evaluation of the cost of turned components at the design stage (Mahmoud and Pugh, 1979), and further advances in methods and techniques of use to the designer and others are to be expected in the future.

STUDENT PERFORMANCE AND ASSESSMENT

The performance of students has improved over the years as the faculty has better understood design and, what is just as important, better translated this enhanced understanding into a comprehensible teaching package.

The performance of students is measured in two ways:

- Part I of the course (three months) is examined by traditional methods: coursework and two three-hour examination papers at the end of the first term.
- Part II (nine months) is assessed on a quarterly basis against some twenty-five criteria which measure a designer's total competence.

Whilst the Part I method has been in operation for two years, there is already some evidence that a traditional approach to a nontraditional situation is not indicative of a student's true design competence. In the first year, for example, the order of merit resulting from the examination was reversed in the longer period of continuous assessment. If this trend continues, then it is indeed questionable whether our current approach to Part I assessment has any validity, and an approach based on a much longer 'practice paper' may prove to be a more rewarding and useful measure of a student's abilities (Gabe and Menzies, private communication).

THE REAL AND APPARENT NEEDS OF INDUSTRY

In having extensive contact with industry through the mechanism of student projects and also through consultancy work, one becomes aware of a distinctive difference between the real and apparent needs of industry. Any attempt at constructive criticism requires that such a distinction must be made. Many design departments are still operating in almost watertight compartments divorced from the other essential elements of an engineering operation such as marketing, production, and finance. Whilst this may have been acceptable and appropriate twenty years ago, the same cannot be said today. Design is central to any such activity, and it is a management task to rectify this anomaly.

It is of little use educating and training broad-based engineering designers, geared to a rapidly evolving and changing world situation, if the environment in which they gain employment maintains the status quo. Such a reorganisation of industry will require inspired management if designers are not to become disgruntled and dissatisfied; this is the real requirement of industry. Action and implementation is required, *not more reports and recycled verbiage*.

ACCEPTABILITY OF STUDENTS TO INDUSTRY

That our students are acceptable to industry is beyond question, and several companies are populating their design offices with our postgraduates in order to maintain and spread the gospel of a modern approach to design. Many companies indeed come back for more, and we do our best to provide them; however, the resistance of the traditional approach to engineering design is tremendous and will take many years to overcome. This in itself gives added spur to our activities, since someone has to tackle the problem. At least it can be proved from the sound, competitive, and innovative designs emanating from our course that if such designs can be achieved by a random group of postgraduates, then industry itself ought to be able to achieve far more with its (relatively speaking) larger resources and backing. It really comes back to whether industry has the organisation and the will to utilise such graduates in an effective and multidisciplinary manner.

SUGGESTIONS FOR THE FUTURE

The greatest stumbling block to genuine progress in the field of engineering design teaching and training is the traditionalist compartmentalised approach to engineering teaching generally. Engineering design in practice is multidisciplinary, not only in terms of engineering and science disciplines but also in terms of the other activities present in any artefact-producing business, and it should therefore be taught this way. To do otherwise will sustain the compartmentalised approach to engineering teaching which in itself will sustain a mirror image compartmentalisation in industry. To break this vicious circle will require the enlightenment of industrial management and also a corresponding response from engineering faculties. With few exceptions there is very little evidence of such movement in industry or education.

To achieve broad-based postgraduate engineering design, a major step forward will be to establish multidisciplinary design teams engaged on project work of the type carried out by the Engineering Design Centre at Loughborough. If the programme is properly organised, then a strong case can be made for having admixtures not only of postgraduates of different disciplines but also undergraduates who, for example, could gain firsthand experience of a real design project. We have taken a step upon this road at Loughborough in that some library and information postgraduates carry out the project part of their course alongside our embryonic designers. It is intended in the future to involve final-year (extended-course) materials engineering undergraduates in a similar manner. If more mixing of disciplines can be achieved in degree programmes, then industry can only benefit, especially when students so educated slide into industry with a better understanding of the interactions necessary in a business to enable it to successfully compete in the market conditions of both today and the future.

REFERENCES

Gabe, D.R., and Menzies, I.A. Private communication.

Mahmoud, M.A.M., and Pugh, S. (1979). 'The Costing of Turned Components at the Design Stage.' *Proceedings of the Fourth International Symposium on Information for Designers* (pp. 37–42). Southampton.

Moulton, A.E. (1976). *Engineering Design Education.* London: Design Council.

Pugh, S. (1978). 'Engineering Design Education: With Real Life Problems.' *European Journal of Engineering Education* 3, 135–147.

Pugh, S., and Smith, D.G. (1974). 'Dumper Truck Design Highlights Industrial/Academic Co-operation.' *Design Engineering* (August), 27–29.

Pugh, S., and Smith, D.G. (1976). 'The Dangers of Design Methodology.' Paper presented to the First European Design Research Conference on Changing Design (unpublished).

Pugh, S., and Smith, D.G. (1978). 'Design Teaching 10 Years On.' *Engineering Design Education* (Supplement) (June), 20–22.

CHAPTER

Engineering Design Education with Real-Life Problems

ABSTRACT

This chapter expands on Chapter 1, 2, and 3's detailed explanation of the postgraduate course at Loughborough University of Technology, with a full description of a typical assessment system and its ramifications. A description is given of a typical project carried out for an industrial company, which is representative of the type of project used to implement the methods learned in the lecture phase of the project.

From *European Journal of Engineering Education,* Elsevier Science Publishers, BV, 3(2) (Summer 1978), 135–147.

COURSE STRUCTURE

The course structure has changed significantly during the last two years while the lecture content has remained virtually static. In other words, we have not so much changed what we teach but how we teach it. For those not familiar with the course, it is divided roughly into two phases with the major part of the lecture commitments taking place between October and the end of December. The main projects (which are of a group nature) commence in January, and formal lectures decrease to zero by March. The emphasis of the ongoing tutorial commitments changes from the initial tutoring against the background of formal lecture and seminar material to the tutoring of the main group projects.

The lecture content and pattern have evolved and developed so that engineering design is taught as a fundamental integrating activity, which demonstrates to the student that engineering design per se in professional practice is basically a roleplaying multidisciplinary activity. The designer has to bring together the techniques and technology peculiar to his or her own particular product or process design situation and in so doing perform an integrating role within the framework of an appropriate specification and method of working. We therefore teach technology and techniques for the application of technology in parallel with a basic core design activity. This core activity commences with market (or need definition) and proceeds through specification, conceptual and detail design, manufacture, and test to the market. Hence, the design activity loop is closed.

During the teaching of the basic design activity or design core stages, students are assigned tasks at each individual stage. As the core activity teaching evolves, these tasks grow in magnitude to encompass the stages covered previously. Thus full coverage of the core activity is achieved in a highly interactive manner that must not under any circumstances omit any stage. This miniproject is usually concerned with a simple but nevertheless realistic product design.

It cannot be overemphasised that it is essential for students to be thinking and working interactively by this stage in the course, as it is but a prelude to the main group project. This is a very brief description of course. A fuller description may be found in Pugh and Smith (1978).

Student Performance During the First Phase

We teach engineering design in the manner described for two fundamental reasons. First, it is an expression of the manner in which the staff of the Engineering Design Centre carry out design projects in professional practice and has been proven to be very effective. Second, it is a response to a sad reflection upon the state of a major part of higher education today that students by and large have little idea of how to use their specifically taught engineering and scientific subject matter in an integrative multidisciplinary manner.

Since adopting this format of teaching, which now applies not only to postgraduates in the Design Centre but also to undergraduates in other engineering disciplines (materials engineering, mathematical engineering, polymer engineering, and shortly civil engineering), we had been pleasantly surprised and gratified by the number of young people who respond positively to such an approach. They do so with varying degrees of effectiveness and in many instances with great difficulty, since to work in this manner is in direct conflict with the traditional compartmentalised specific discipline teaching still unfortunately prevalent today. However, on balance, the majority of students who have a capacity for multidisciplinary working respond well and never cease to surprise us. I consider it essential for all engineering students to be exposed to this form of teaching since they must be given the opportunity to at least understand the true nature of engineering design, even though some of them as individuals will never have the capacity for effective multidisciplinary working (Pugh, 1977). Postgraduate students, since they are taking a design course, should be assumed to possess aspirations and leanings towards engineering design and therefore have a capacity for working in this way. By Christmas then, the students are performing reasonably effectively as exponents of the total design activity.

Student Performance During Second Phase

Moving to the second phase—group project working—the situation initially is reversed. At the commencement of group projects it could be reasonably expected that with students having performed well and interactively in the previous phase projects should then glide away from rest, take off and gain altitude, proceed smoothly, and land with a thoroughly researched and effectively designed product or process. Unfortunately, in many instances the converse is true. The students seem to freeze up and attempt to reduce the projects to quantifiable packages which they then proceed to resolve as a series of staccato, specific problems, the relationship between the problem packages being more apparent than real. Harry Hermanns (Hermanns and Tkocz, 1976) has commented upon this very problem in teaching involving the use of projects. Whilst in phase one freeze is apparent at the beginning of the phase and, by and large, is eliminated at the end, the miniproject (being realistic but nonetheless artificial in the sense that it will not be manufactured, tested, and marketed) is treated as an exercise by the students and is assessed on the basis of written and oral presentations. Therefore, the full effectiveness of any design can only be *assessed* and not proven. The main project has a different connotation in that ultimately the designs will not only be *assessed* on the basis of written and oral presentations but also in many cases will be *proven* (or otherwise) when they come to be built. I therefore attribute the freeze primarily to this single factor—namely, the fact that the designs will be proven. Engineers from industry and external academics will be involved in this proving process, and thus by the end of the course both student and designs will be fully exposed.

Unfreezing

So begins a period of reconciliation between the attitudes, actions, and performance in phase one with the realities of the project. It is at this point that the skill of the project tutor becomes of paramount importance since it becomes the tutor's task, by logical argument and discussion, to prove to students that the almost random 'problem package' approach will be certain to lead at best to a mediocre design solution. In other words, his main task is to restore confidence to the group and to individuals and to remove the fear of having to perform effectively with the real problem. Once this hurdle has been overcome, the groups almost without exception settle down to interactive multidisciplinary working and ultimately produce design solutions which withstand the scrutiny of the outside world. In order to highlight and discuss some of the aforementioned points, it is perhaps best to describe, albeit in a foreshortened manner, one of our recent student projects.

Student Project: Materials Handling System for Pigment Processing

The initial project brief given to the students was 'to design, manufacture, and commission a replacement materials handling system as part of a pigment processing plant, the system to be as simple and maintenance free as possible.' This is a statement of the overall objective and is not, as many students interpret such statements, a specification for the basic design approach and is not a trigger to commence evolving solutions. The project constraints, limitations, performance parameters, and so on all appear in a product specification which forms the boundaries of the design. To accept the above statement as the sole precursor to commencing design would have proved disastrous and I think indicates a major distinction between a real-life project and many projects at the undergraduate level.

So commenced a period of familiarisation and assimilation of the project and the technologies associated with it. In the existing plant, pigment passes through a kiln on a counter-conveyor system and is fired to 500°C. The pigment is contained in metal trays with removable lids, and an important part of the process is that the pigment should be fired in an atmosphere of its own creation which is contained by the lid. So we have a counter-conveyor system at one end of which removal of lids, emptying and filling of trays, and lid replacement take place. This involves removal of trays from the conveyor, removal of lids, emptying of pigment from trays, refilling of trays, and replacement of lids prior to returning to the conveyor systems. This is shown diagrammatically in Figure 4.1. The handling at the filling and discharge end is carried out by a Versatran robot, and at the return end of the kiln, trays are transferred to the return conveyor by a Simpletran robot which picks up trays from the outgoing conveyor and places them onto the return conveyor, operating in a 'goal post' mode.

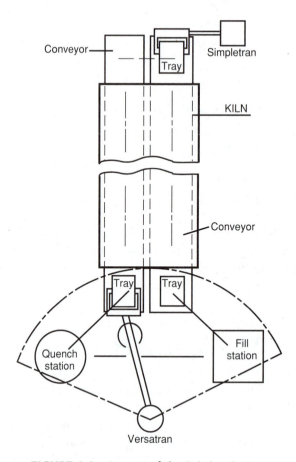

FIGURE 4.1. Layout of the Existing System

The initial problem facing the students was not so much 'How could the problem be solved?' but rather 'What was the real problem?' Was it a question of designing a different kind of machine to handle existing trays and lids, with the very real constraints of retaining the existing conveyor and kiln system such that any approach to handling systems would become heavily constrained? These were typical of the questions asked by students at this stage of the project. It is interesting to note that the group as a whole wished to concentrate on designing a machine combination that would handle the existing trays and lids which, whilst it may have been a safe approach, would doubtless have resulted in the design of a handling system differing from Simpletrans and Versatrans but not necessarily better.

Students thus began evolving engineering solutions to the *apparent* problem and lapsed into producing schemes involving gears, linkages, and so forth without much further thought as to the true nature of the problem. Their thinking was almost entirely directly to solutions rather than to the more difficult and elusive area of problem definition and hence specification, a tendency with all student projects. An instruction was given to the students to *stop engineering* and to concentrate on problem definition. Indeed, drawing boards were literally turned to the wall and not used again for some time. I find that drawing boards and other aids to solution definition are counterproductive at this stage of all projects, since they tend to inhibit thinking and to lure students into elegant solutions to the wrong problem—which may result in *sound engineering* but certainly not *good design*.

A flow chart was drawn, Figure 4.2, which shows immediately that trays with removable lids, whether lids and trays are identical or not, require a considerable amount of handling as demonstrated by the existing system. However, if trays with hinged lids are considered, then the amount of handling required is drastically reduced, at least in theory. This very distinction surprised everyone and threw a different light upon the whole problem. Students became aware that any handling system design considered systems utilising trays with hinged lids, as well as separate trays and lids. They needed to consider the advantages and disadvantages of the different systems in order to gain a better insight into the problem, which in itself may suggest better ways of solving the problem.

A common feature emerging from student projects is that a thorough investigation into the true nature of the problem increases the likelihood that embryonic approaches to viable solutions will come to light, which almost certainly would not happen using a pure technological or engineering approach. It therefore becomes an easier and more satisfying task to attempt to match the appropriate technologies to these embryonic solutions which, having emanated from problem definition, are more likely to lead to a better design approach and hence solution. So it was with the project under discussion.

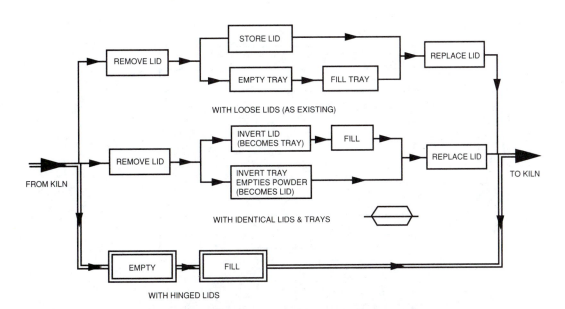

FIGURE 4.2. Alternative Cycle

Problem Conceptualisation

The realisation that a hinged lid and tray approach in principle reduced the amount of handling required in the system led students to consider the total system from the same viewpoint. They approached it with the idea firmly in their minds that effort should initially be concentrated upon consideration of systems where the amount of handling required would be absolutely minimal. The ideal would be a closed-loop system, where the trays recirculate via the conveyor system without the introduction of third-party equipment (Versatrans and Simpletrans) to enable the loop to operate satisfactorily (see Figure 4.3).

Effort was concentrated primarily upon the emptying and filling end of the system, since this was considered likely to prove more difficult to resolve than the transfer end and the solution of one may yield the solution to the other. Needless to say, it was found impossible to achieve the ideal of zero handling, but from this simple premise ideas were evolved which became more complex as they migrated towards a workable solution. The students also knew why they were complex, since they had evolved from a simplistic base. This is an important point, since evolution of concepts from the simple to complex is likely to lead not only to a better understanding of the design but also to reduced conceptual vulnerability. In other words, the students are more likely to know the answers to questions concerned with approaches leading to the final design choice and respond well in these situations.

Figures 4.4, 4.5, and 4.6 show but a few of several systems considered. Attempts were being made to utilise the energy present in the conveyor system to power both the

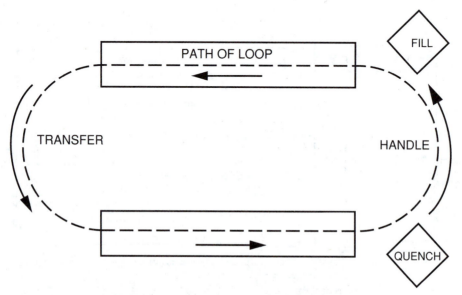

FIGURE 4.3. Closed-Loop System: Trays Bumper to Bumper, Smooth Motion, Flow Continuity

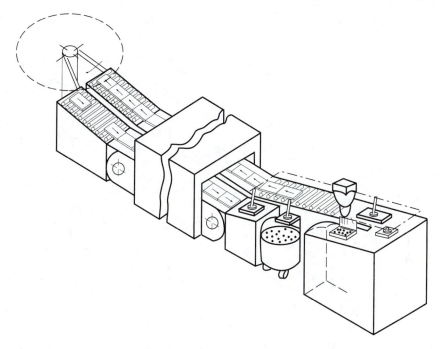

FIGURE 4.4.   Trays with Hinged Lids, Overhead
Handling System with Suction Cups

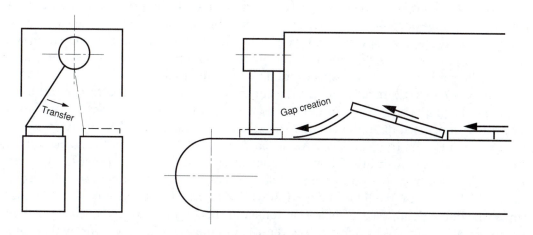

FIGURE 4.5.   Transfer Paddle Wheel

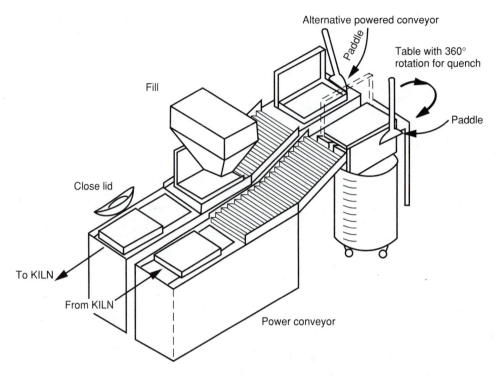

FIGURE 4.6.   Trays with Hinged and Restrained Lids,
Tray Rotated Through 360° for Quench

handling and transfer systems and Figure 4.5 shows an inclined plane up which the trays were to be pushed by the continuous line of trays on the conveyor. If the trays could be close packed on the conveyor, a greater throughput of material would be achieved. However, a continuous line of trays gives rise to the problem of tray separation for handling purposes; this can be overcome by creating a gap in the tray line in order to transfer trays from one conveyor to the other. The concept of gap creation was also to have important ramifications at the emptying and filling end of the system, since any handling system must have space and time in which to operate.

Figure 4.7 shows a paddle wheel–type rotary arm which accepts a tray with hinged lid, a gap having been created in the line of trays in the manner previously described. The tray is supported upon the circular track, and the lid is held in the rotating arm. As the arm rotates, the tray support is removed at the discontinuity in the track causing the tray to swing downwards above the quenching container, depositing the pigment into it. The idea from then on is to provide the radial arm with some form of mechanism which will raise the lid to the vertical whilst the tray is returned to the horizontal, under the filling point, the tray being stationary at this point in the cycle. The cycle then continues, the lid is closed, and the tray returned to the outgoing conveyor by that same mechanism. This system was developed with the aid of the model shown in Figures 4.8,

Hinged lid type tray

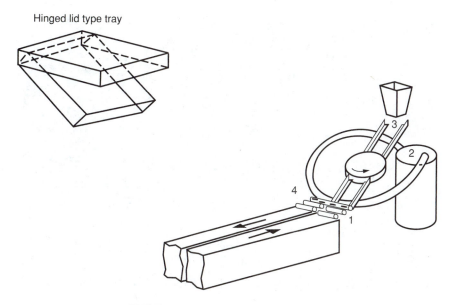

FIGURE 4.7. 'Moon Rider' Concept

4.9, and 4.10, which demonstrated to the students and the sponsoring company that as a concept, it not only worked but was extremely simple in nature and therefore worthy of further consideration.

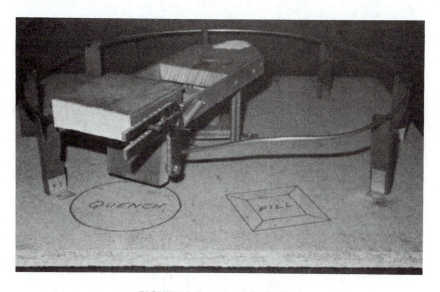

FIGURE 4.8. Quench Station

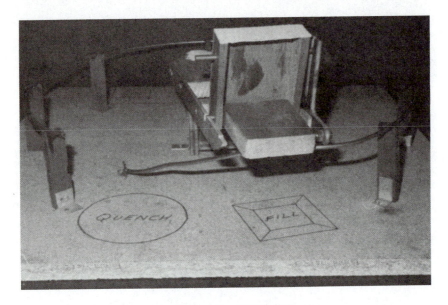

FIGURE 4.9. Fill Station

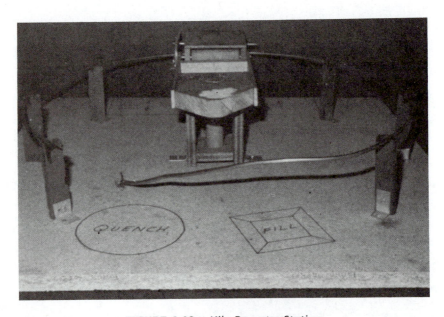

FIGURE 4.10. Kiln Re-entry Station

Detail Design

Effort was then concentrated on the engineering implications of such a system, and detail designs based on this concept were produced which covered the associated control and interlock system to ensure that the correct interface matched with the conveyor equipment.

The sponsoring company manufactured the equipment to drawings produced by the students, and a prototype equipment was built in the company's works and commissioned by the students (Figure 4.11). The prototype worked satisfactorily and is now under development by the company for application to its plant. It is salutary to recognise that the prime cost of a complete system to this design will be of the order of £3,000 to £4,000, whilst the equivalent cost of the robot system is probably between £40,000 and £50,000.

This project has been described here in an extremely foreshortened manner that omits many variants with trays and lids, tests on inclined planes, conveyor and tray friction levels, tray corrosion tests, thermal distortion, and a whole list of other lines of approach which all seemed to underwrite the design approach which finally evolved.

CONCLUSIONS

The project described in this chapter is typical of those carried out in the Engineering Design Centre and stems from the approach to the teaching of engineering design described earlier. There is absolutely no doubt in my mind that the development of an approach which initially treats technology in the abstract leads to solutions of the type that emerged in the pigment handling system. It is the matching of the appropriate technology to embryonic designs (which have evolved from extremely simplistic consid-

FIGURE 4.11. Prototype Equipment

erations) which gives rise to first-class engineering design and thus first-class engineering: the converse is not always true. I am also firmly convinced that students should be exposed to this type of thinking in order that the designs of the future have a chance of being better than those of the past. Finally, we achieved excellent cooperation with industry both from a project viewpoint and also in terms of student sponsorship.

REFERENCES

Hermanns, H., and Tkocz, C. (1976). 'Project Orientation Within the Framework of Integrated Engineering Curricula at the Gesamthochschule Kassel.' *Proceedings of a Seminar on Project Orientation in Education for Science and Science-Based Professions.* (pp. 40–45).

Pugh, S. (1977). 'Creativity in Engineering Design: Method, Myth or Magic.' *Proceedings of the SEFI Conference on Essential Elements in Engineering Education.* Copenhagen: SEFI.

Pugh, S., and Smith, D.G. (1978). 'Engineering Design Teaching: Ten Years On.' *Engineering Design Education Supplement* 2 (Spring).

Projects Alone Don't Integrate: You Have to Teach Integration

ABSTRACT

This chapter considers the question that integration is not achievable to any great degree unless the distinction is made in the teaching between structure and methods. To grasp these differences leads to better performance and grasp of modern design method.

From *Engineering Design Education* (Autumn 1982), 14–16.

Over four years ago, in an article published in the predecessor to *Engineering Design Education* (Pugh and Smith, 1978), an account was given of our experiences in design teaching at postgraduate level over a period of some ten years. Whilst at that time recognising the importance (at least for us) of the model we had formulated of the design activity itself, it was not until more recently, as others have published or discussed their approaches to the same problem, that I truly appreciated the significance and criticality of such models.

Two recent examples are Marinissen and Roozanburg (1982) in Delft, who recently produced a model, and Hubka (1981), and yet I would ask the same question of the modellers as we ask of ourselves: How effective are they in integrating technical subjects and technology in the educational field and in contributing to improved and better designs in practice? Fortunately, Engineering Design Centre students, through the mechanism of their main group projects, which last for a full nine months, are concerned with design in practice, and the results of their work can be compared with the designs produced in industry. In fact, existing designs form an essential element of the Centre's design approach. Student designs are getting better and better in sympathy with refinements to the model and associated design methods, and this I consider to be significant.

Setting aside the postgraduate scene, where one is or should be concerned with design in practice, I would like to consider in greater detail and depth the undergraduate or first-degree level of teaching of engineering design. In various capacities in this field, ranging from teacher, moderator, to contributing to course approval through the CNAA Engineering Board, I have noticed that the number of courses having a design content has greatly increased, that, as a consequence, the number of people teaching and writing about their experiences has also increased, and what I think is the most significant point of all, that one can see and assess student performance and understanding of design. To me, this is a sign of progress in the engineering schools of universities, polytechnics, and colleges, and yet I sense that the improvement is not as significant as it might be. For example, it would appear from the range of models of the design activity currently available, either that teachers have never been concerned with effective design practice and intend that their students should do likewise or that some have obviously been concerned with comprehensive design practice but are still trying to bridge the enormous gap between *doing* it effectively and *teaching* it. Such models range from the naive to the incomprehensible. In saying this, I am not advocating any single model of the activity but encouraging teachers to test the effectiveness of their models by any means available and improve both it and the results emanating from its usage.

The various models currently used may be summarised and categorised as follows:

- Teaching as before, usually by engineering drawing variants but calling it design: still common.
- Using the problem-solving model (Lewin, 1981) but equating it to design in practice: prevalent.
- Implementing the design activity model: still relatively rare.

Many people are teaching design from excellent motives, and all use projects as the manifestation of the design activity—a practice to be recommended. But, as the title of this chapter suggests, in the first two models above, the projects are expected to bring about the desired degree of integration of the taught subjects and technology in a comprehensive manner. In my experience, they clearly *do no such thing*. Unless integration is taught against the background of a model of the design activity which inspires and fosters integration, relates the stages in design whilst remaining comprehensible, and above all else is something to which the recipient can relate, then I fear that we are preparing people for shadow boxing and not for the true ring fighting.

It is our experience (now being painstakingly repeated by others) that one must develop a design activity model. The nearest thing to it, the problem-solving model, does not represent the reality of practice. The fundamental weakness of the problem-solving model is that it inextricably mixes structure and method. Intellectuals may understand it, but they don't have to put it into practice competitively; experienced designers may reflect upon it and possibly relate to it; most young people gain some sort of understanding from it and, on putting it into practice, illustrate their understanding of the model and general inability to design thoroughly.

This distinction between structure and method cannot be overemphasised. To my mind, the structure of the design activity model should provide students with the understanding, relationships, and confidence that they need to apply the appropriate methods in a particular design situation. I regard the model structure as being analogous to a child's climbing frame: it provides the framework on which to climb, it imparts confidence and safety, yet it doesn't prescribe or predetermine the methods by which the child gets to the top of the frame or indeed around inside it. The climbing techniques or methods relate to the framework and enable it to be used effectively.

We discarded the problem-solving approach in the mid-1970s and devised an activity model for design which does work in practice at the undergraduate level, although it is recognised that at that level one can do little more than imbue effective design appreciation because of the constraints of time and the necessary extent of the academic curriculum. This model was 'graphicallised' for the 1978 article and consequently appears at best to be unclear and difficult to read and at worst upside-down. Figure 5.1, unadulterated, gives a clearer depiction of the model, although the description of 1978 still holds.

I would make the following observations on our experiences in teaching design at first-degree level without at this stage relating these comments to the project, which is the integrative vehicle:

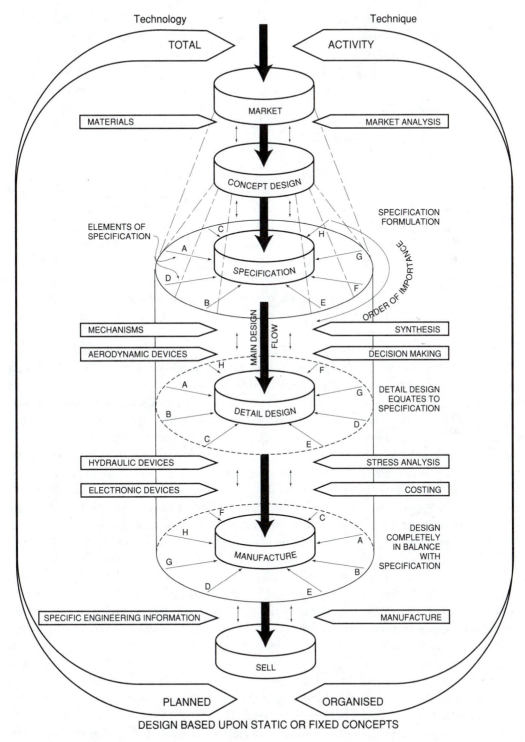

FIGURE 5.1.   Design Core Bounded by Product Design Specification

1. Effective understanding and evaluation of the customer or market requirement or need can be taught irrespective of discipline, engineering or otherwise.

2. From this logical progression into the formulation of a product specification is readily understood, provided that one teaches the appropriate methods and techniques to activate the model (Pugh, 1978). This can be done very thoroughly at this level, and we have had splendid results, even in professional engineering terms, from aspiring civil engineers, materials engineers, and mathematical engineers (in practice, many designs go awry at this stage due to inadequate specifications).

3. Conceptual design can be taught, to a degree, but requires a great deal of time devoted to effective understanding and practice. At this level it probably should be no more than touched upon.

4. Detail design, which used to be taught very effectively years ago, now gives us difficulty, since it should emanate from and be part of the traditional academic subject teaching. We don't do this very well, due to a shortage of teaching material and time. This is being rectified.

5. We regard manufacturing and selling as being essential components of the design core. However, we have not as yet succeeded in having them taught as essential components of the core.

To summarise the above, we have time to cover the whole of the structure of the model in all cases, have little time to teach techniques and methods to activate the model, and, under the heading of technology, do little teaching but rely on the student's having a thorough understanding of the model and then integrating the other subject material accordingly. It would appear that the model, allied to the distinctions made between structure and methods, has improved understanding, enthusiasm, attitude, quality, and quantity of work output. You might well ask how we can draw these conclusions at the first-degree level, since only at postgraduate level do we have the product to prove it. The answer, of course, is that we utilise a thematic project from very early on in the design activity teaching, and students gain understanding of the model, allied to experiencing its effectiveness in a *minor* project exercise, exemplifying the overall effectiveness of the approach. I deliberately emphasise *minor* project exercise for a number of reasons. If, with an individual or group, projects of considerable magnitude are used, such as a replacement for Concorde, then students, in attempting to cope with a large project, gain the false impression in completing them in their terms that this is what design is all about and therefore gain false confidence. Alternatively, the sheer size of the project overwhelms them, and they are deterred from design forever. Large projects can be used effectively in exercises related to specific techniques and methods—say, writing product specifications. One should not in my opinion proceed with such projects into the conceptual or detail design stages, however, as they not only become meaningless due to the shortage of time, but they distract students from gaining a full understanding of the relationships between theory and practice so essential to sound detail design (see Chapter 7).

It is our experience in using a thematic project as the integrator that students gain most from working on some relatively simple, easily assimilated task, like a tin opener, door stop, combined tile cutter and breaker, burglar alarm, and so forth. If the task is kept simple and of a magnitude where from the beginning they immediately recognise that it might be achievable within the time allowed and with 10 to 15 percent thoroughness, then this I regard as progress. Encountering the difficulties of a simple artefact design also provides students with sound data. Our experience on such projects indicates that as a matter of course they start to utilise their specific disciplines, information, and skills effectively, and the design activity model (structure) holds it all together for them logically. Hang-ups are few and far between, and they rapidly gain in confidence and professionalism. Methods are allied to the structure not separately but together in a logical, comprehensive manner. It is dangerous to teach methods without an integrating structure. This has been happening for years, and we see around us designs to prove the point. Therefore, in my opinion, at all levels of design activity, from schools through to postgraduate level and beyond, the efficiency and effectiveness of integration are closely related to the structural model. There is no point in adopting a model or changing it since, as in all things, progress requires change but change is not necessarily progress. In design teaching terms, this to me means that what is needed is enhanced understanding and ability to practice at the appropriate level. The structure is thus *enabling*, and the methods allow the structure to be utilised effectively, provided that they are related to the structure and are taught in sympathy with it. As if destined to give additional weight to the arguments I have put forward, yet another method has appeared since I wrote this article. The Bertoncelj (1982) model of innovation, which I consider to be a design model, is no doubt understandable to the experienced, but I would question its usefulness and effectiveness in teaching for the reasons already outlined.

Finally, in a series of 'Design Teaching Aids for Schools,' shortly to be issued by the Design Council for general school usage in the twelve- to fourteen-year-old age group, a much-simplified version of the model discussed here has been used and called the 'design loop.' Children take to it naturally, and teachers using it report a better understanding and practice for themselves and their students.

REFERENCES

Bertoncelj, J. (1982). 'Programmieren des Erfolges eines Erzeugnisses.' *Schweizer Maschinenmarkt* 18 (August), 38–41.

Hubka, V. (1981). 'Allgemeines Vorgehensmodell des Konstruierens.' *Proceedings of the International Conference on Engineering Design* (pp. 165–176a). Rome: ICED.

Lewin, D. (1981). 'Engineering Philosophy: The Third Culture.' *Journal of the Royal Society of Arts*. 129(5302) (September), 653–666.

Marinissen, A.H., and Roozanburg, N. (1982). 'Design Projects in the Curriculum of the Department of Industrial Design Engineering of Delft University of Technology.' *Proceedings of the 1982 Annual Conference of the European Society for Engineering Education* (pp. 44–50). Delft: ESEE.

Pugh, S. (1978). 'Quality Assurance and Design: The Problem of Cost Versus Quality.' *Quality Assurance* 4(1) (March), 3–6.

Pugh, S., and Smith, D.G. (1978). 'Design Teaching: Ten Years On.' *Design Education Supplement* (June).

CHAPTER

6

Marathon 2550:
A Successful Joint Venture

ABSTRACT

There has been considerable discussion in educational circles in recent years on whether designers can be educated and trained in an academic environment and how it can be done. However, it is recognised in some circles that part of any such training should include work on realistic projects from industry.

This is an account of one project carried out as a joint venture between a university and a manufacturing company. The academic contributor is Loughborough University of Technology through the Engineering Design Centre; the company is Liner Concrete Company Ltd., Gateshead, Co. Durham. The outcome of the project has been named the Marathon 2550, a four-wheel-drive dumper truck with a 2550 kg skip capacity. The project started with a requirement and finished with a preproduction machine, the work having been carried out by a team of postgraduate students for an engineering design course over a period of eight months. The success of the project is vindicated in that the vehicle has been in production for eighteen months, with the design virtually as it left the Engineering Design Centre and with a continually rising rate of sales. To the authors' knowledge, this is the first occasion on which a product has entered production as a direct outcome of a student project.

With Douglas G. Smith, Engineering Design Centre, Loughborough University of Technology, 1976.

BACKGROUND

The Engineering Design Centre was set up in 1966 and offers, as one of its main activities, a postgraduate course in engineering design leading to the award of a master of technology or (ALUT). The course prepares engineers to assume responsible positions in the engineering design field. This is achieved by giving a broadly based training in two senses: students are shown how to approach engineering design problems in general and how to manage and carry out projects from inception to testing of hardware. The course runs for a full twelve months from October until the following September. The first ten weeks are taken up with a full lecture programme on subjects related to design, together with individual assignments. This is followed by a miniproject which gives students the opportunity to integrate the first ten weeks of work before undertaking the major project. The remainder of the course is predominantly concerned with a major design project carried out in cooperation with an industrial company. For the purposes of the projects, students are divided into groups, each under tutorial guidance. All members of the Centre's staff have considerable design experience in industry and maintain their professional expertise through projects and consultancy work with industry.

Among other products, the Liner Concrete Machinery Company manufactures vehicles for material handling on rough building sites—namely, fork lift trucks and a range of dumper trucks. The project described here is concerned with the largest-size dumper truck manufactured by the company.

All projects carried out by the Engineering Design Centre must be current requirements of significance for the company concerned. The Centre does not specialise in one field of engineering but handles a wide range of projects which may be one-off special machines or machines for quantity production as in the case of the dumper. There are many ways in which projects come to the attention of the Centre—the visit of a staff member to a company, the visit of company personnel to the Centre, recommendation, return of a satisfied sponsor, or a direct approach to the Centre. Following the initial contact, a meeting is held between the potential sponsor and the staff of the Centre. If there is mutual agreement that the project is likely to satisfy the requirements of both the sponsor and the Centre, a proposal is made to the company, setting out such matters as proposed method of working, timescale, costs, and patent agreements. If the proposal is accepted by the company, it becomes one of the projects carried out during the period of the course from January to September. As far as the Centre is concerned, the projects must satisfy a number of requirements. There must be a reasonable probability of reaching the required objective during the eight months available to the team of students. Furthermore, the project must give the students experience in dealing with the many aspects of design such as the market, the specification, project planning, unearthing the real problems, ideas, decisions, mathematical analysis, costing, materials, ergonomic and aesthetic aspects, design layouts, detail drawings, production procedures, production problems, and preparing for and supervising test programmes. It is seldom possible to satisfy all the requirements in the most desirable way but the dumper truck came very near to doing so.

In 1970 the company, having manufactured and marketed a successful range of dumpers for a number of years, were considering a replacement for their $2\frac{1}{4}$-ton Roughrider Super 2250 to maintain their competitive position in the market. The company's design department was, at that time, fully committed for some two years ahead, and therefore, if a new design of dumper was to be undertaken, an expansion of the company's own facility would be necessary or an alternative source of design would be required. In fact, the company decided on the latter course of action, and this intention was outlined to one of the authors during a visit to the company. Following a series of joint meetings, it was agreed that the Centre would undertake the design of the replacement vehicle. However, before this could be done, it was essential to assess the market and draw up a specification.

THE MARKET

The market for diesel dumpers is one that has been established over a number of years, and in the United Kingdom alone there are some ten manufacturers meeting the home demand, together with imports, particularly from Germany and France. As such, it is a highly competitive market, and therefore, if we were to succeed with a new design, it would have to be balanced in terms of performance, capacity, maintenance, reliability, and above all, manufacturing cost.

One of the initial tasks was to gather information in two spheres in order that an initial specification could be produced. First, it was necessary to gain an insight and understanding of needs and requirements of the end user and of the plant hire companies, which form a major outlet for such machines. It is interesting to note that during this phase, the requirements of the two were not always compatible, since the plant hirer's main aim is to capitalise on its investment in plant and machinery, whilst the end user's main interest is in site effectiveness with a minimum of maintenance and maximum reliability. To this end, visits were paid to building sites, contractors, and plant hire companies. Second, an analysis of available competitive equipment was carried out, not only to gain a further insight into the market but also to enhance the degree of understanding of the problems in this field, an essential prerequisite to any design programme. This is particularly necessary for student project work, where the combined team is unlikely to have prior knowledge of the market or design experience in the field.

The analysis of the competition also included an investigation of selling price patterns, which were useful as a comparative cross-reference in order to guide the project along.

Perhaps the point should be made here that, in some educational institutions concerned with design teaching, to consider competitive equipment is frowned upon as being an inhibitor of creativity. We feel that not only is this an erroneous viewpoint, but it is also extremely dangerous and likely, in the long run, to suppress the best designs. One has only to look at Japanese products to support this statement. Our view is simply this: the competition exists, and it is therefore a factor which must be taken into account along with all the other inputs within a given design situation. To ignore it, is to do so at one's peril.

In parallel with the investigation of the market, a start was made on preparing the initial product specification, part of which would relate to the market itself.

THE PRODUCT SPECIFICATION

It is essential that prior to the commencement of the design of anything a specification be prepared, no matter how rudimentary. This is an unequivocal statement. In our view no specification—no design! To omit this step is to court disaster and enhances the chances of evolving a product which has a mismatch with the market.

A designer always has a difficult task and therefore must utilise all the information available within a framework of discipline dictated by the specification. This is not to say that the specification is inviolate and cannot be changed; it must be treated as an evolutionary document, so that at the end of the day the product matches the specification. This point often causes difficulties with students who, having prepared a specification, feel that to veer from it is tantamount to committing a crime. This rigidity of thinking often is instilled into designers during early technical education, mainly by the specific problem and specific solution approach to design, which has only one correct answer. Of all the problems associated with design teaching this is without doubt one of the most difficult to overcome as it requires a major shift of mental attitude.

One can state as a matter of fact that students in general, postgraduate students being no exception, are unlikely to have been introduced to or have experience in writing specifications. Indeed, in terms of academic design exercises, specifications are considered to be inhibitors of creative design. Fortunately, this is not so in practice, although many companies do design and market products without the formulation of a proper specification at the outset.

In order to introduce students to specification writing, we produced a skeletal specification suitable for subsequent expansion by students in cooperation with the company, which may be summarised as follows:

1.0. Performance
    1.1. Overall vehicle
        Speeds
        Maximum inclines
        Turning circle
        Articulation (vertical)

1.2. Skip
   Capacity: overall
   Level capacity
   Heaped capacity
   Loading height
   Discharge height
   Skip angles
   Minimum and maximum widths
   Rate of tipping

1.3. Chassis
   Type
   Maximum ground clearance
   Loadings

1.4. Transmission
   Engine
   Clutch
   Gearbox
   Transfer box   } If mechanical transmission is adopted
   Propeller shafts
   Axles
   Wheels and tyres

1.5. Braking

1.6. Steering

1.7. Auxiliary power system
   For steering, tipping, and so on

1.8. Range
   Fuel tank capacity

2.0. Environment
   Temperature range
   Maximum altitude
   Type of ground
   Type of labour
   Liability of sabotage

3.0. Life Expectancy
   Service life of machine
   Periods between service

4.0. Product Cost
   Target cost in production

5.0. Quantity and Production Life Span
   Production rate
   Production facility
   Expected market life
   Batch size

Tooling policy
6.0. Size and Weight
Track
Wheel base
Overall length
Overall width
Overall height
Ground clearance
All up weight
7.0. Appearance and Styling
Overall appearance
Control and access
Driver position
Visibility
Seating
Pedals
Overall finish
Driver safety
8.0. Constraints and Areas of Flexibility
Areas open to design flexibility, ie not constrained
Engine position
Transmission type
Hydraulic system
Chassis
Fuel and oil tanks
Skip
Driving position
Protection must be provided for:
Engine
Transmission components
Front and rear of chassis
Brakes
9.0. Service and Maintenance
9.1. On-site maintenance
Minimum lubrication
Maximum accessibility
9.2. Off-site maintenance
Engine and transmission components easily removable
Periods between services
Working instructions
10.0. Storage, Shipping, and Packing
Shelf life of components
Precautions during storage

Method of shipping: home and export
Lifting facilities
11.0. Auxiliary Attachments
Provision should be made for fitting (without prejudice to overall design):
Fork lift
Digger
12.0. Special Features
Starting: alternatives
Towing
Optional skips: swivel
High discharge
Limited slip: differential or equivalent

All these topics were explored in depth by the students, answers were given by the company, preferences were stated, and thus the initial specification evolved. Whilst all the factors considered are important ingredients of the whole, perhaps three stood out as being of prime importance:

- Production was to be carried out in the company plant. Any design therefore had to be matched to the company facility but not if to do so would prejudice the design.
- Target production costs must be achieved.
- The vehicle's performance and life in service must be equal to or better than the company's then current model.

These three factors are worthy of further comment, for different reasons. First, one so often hears expressions such as 'the correct design,' the 'right design,' 'the best design,' 'the only design,' and so on. These statements in themselves do not mean anything unless viewed against the background of the production facility; what is best for one production facility is unlikely, in entirety, to be so for another. This point is often overlooked in design and production teaching. Second, to achieve target production costs without recourse to long stops, such as value analysis, is difficult to carry out in industry, let alone in a university. Third, the statement 'vehicle performance and life in service' was easy to say but difficult to quantify. Nevertheless, we had our initial specification, and against the background of this specification a work programme was formulated.

QUALITIES OF STUDENT PROJECT TEAMS

Before we embark upon the next phase of the project—namely, the work programme—it would be useful to comment on the vagaries of student project teams. Students registered for the postgraduate course usually hold an undergraduate degree, or equivalent, in mechanical or electrical engineering. They come from a wide age range, have varied backgrounds and experiences, and present different strengths and weaknesses. For the purposes of the projects, students are divided into teams, the selection of which is made after an assessment of the performance of each student following the miniproject. An attempt is made to create balanced teams. For example, in this project the ages ranged from twenty-three to forty-two years with men from various fields of industry, college lecturers, and a student from overseas.

Whilst at first sight this variety may appear to be an advantage, the students themselves are usually limited in two ways. First, generally speaking, they have never had responsibility for, or even been involved with, a project from inception to testing. Second, their breadth of experience is usually very limited, and they are reticent about tackling new fields.

In order to overcome these limitations, students are allocated to projects in fields where they have little or no experience. Although most of the students find this unnerving at first, it is a familiar comment at the completion of the course that the experience has given them greater confidence for the future. It has been found that where a student is involved in a project from his own company, he usually brings with him the current attitudes of the company, which are not always advantageous to the project.

It has been found that a team of three to five students is most satisfactory for a project of eight months' duration carried out under Centre conditions. Such a group usually becomes well integrated, maintains good communications, and finds no place for passengers.

In selecting project teams every care is taken to ensure that the teams are likely to be well integrated as far as personalities are concerned. This factor, and student attitudes, are often much more important for the smooth running and success of a project than technical brilliance. Despite the care that is exercised, personality mismatches do sometimes occur, not only between student and student but also between students and staff! If such situations do occur, no effort is spared by staff in analysing and seeking to resolve them. However, it has been generally observed that it is the students with the least ability who cause the most problems and not, as is generally thought, the high flyers.

FORMULATION OF A WORK PROGRAMME

Eight months can appear a considerable length of time at the commencement of a project. The various stages can be allowed disproportionately long or short periods of time and critical decisions unnecessarily delayed unless some control is exercised. In order to allow for these factors, a work plan was drawn up.

However, it was not possible to do this until the team had built up their background knowledge of the company, the market, and the specification which have been discussed earlier. Further, it had to be ascertained that the basic requirement for the design was still the same. Each point of the specification was probed to ensure that there were no mismatches. For example, the relation between the target cost and the dumper payload was carefully checked. It may have been that the payload required was greater than could have been achieved for a given cost. Had this been so, it would have been necessary to resolve this inconsistency, or else the subsequent design work may have met neither of these criteria. It was also necessary to differentiate between those features of the specification which had to be adhered to and those having flexibility. For example, two optional diesel engines were specified as these units were already in use by the company and widely used in the industry. On the other hand, the method of steering was open to investigation.

In order to satisfy the three factors stated as being the most important criteria of the specification, it became increasingly clear that the chassis was a key feature. How could we be sure that the new chassis would be equal to or better than the existing one? Clearly, the new chassis had to be cheaper to build but at least equal in strength! It was this latter feature, that of 'equal in strength' which now had a significant influence on the project strategy and use of resources. We concluded that the only satisfactory way of ensuring the new chassis satisfied this requirement was to test both. It was considered a better policy to have the chassis right from the start rather than to incur the expense of field rectification and a 'tarnished image' in the event of problems arising. This philosophy has been amply justified. It was one thing to decide to test the chassis but another to implement the tests, as the company did not have a suitable test facility available. It became necessary therefore to plan for the design of not only the dumper but also a test rig.

In order to set preliminary time periods and target dates for the main phases of the project, a bar chart was drawn up for discussion with the company. Thereafter, a more detailed PERT chart was prepared by the students for planning purposes and for use by both parties to monitor and control the progress of the project. Although perhaps an example of the use of a sledgehammer to crack a walnut, the ICL 1900 PERT package was run on the university computer for both time analysis and loading. This enabled students to assess the potential and limitations of computers for planning and to compare them with manual methods.

Students working on the project were expected not only to plan and carry out the actual design work but also to manage the project to a certain extent. This was achieved by the students each taking a turn as project leader. Whilst the leader does not, of course, replace the tutor or give orders to other students, he can handle many day-to-day tasks to ensure the smooth running of the project. For example, he keeps a careful check on the project progress, liaises with the company on day-to-day matters, prepares correspondence, and sets up meetings. All of this is done under the close supervision of the tutor who has daily contact with the team. Indeed, the offices of the staff were only a few paces from the teams' rooms.

IMPLEMENTATION OF THE PROGRAMME

In order to carry out the programme it became necessary to divide the work into two areas: (1) the design of the test rig and implementation of the test programme and (2) the design of the vehicle itself. For the sake of clarity we shall deal first with the vehicle.

Conceptual Design of Vehicle

Within the constraints imposed by the specification, we had in effect the opportunity to start the design with a blank sheet of paper. It was observed, however, that two factors in particular were likely to exert a major influence on any conceptual design; the first being the choice of steering arrangement and the second the type of transmission to be used.

It had become almost a tradition over the years for vehicles of this type to be based on a 'split in the middle' type of chassis, whereby steering is accomplished by articulating the chassis halves about a central vertical axis. This type of steering offers one important advantage over steering axles and the like, in that it enables the front and rear wheels to follow in each other's tracks, a factor of vital importance on soft, muddy ground. In such situations only two furrows are cut by the wheels. If one compares vehicles of the same weight, power, and transmission ratios, the one having a split in the middle chassis, the other having steering axles, the former will get into and out of these situations with alacrity, whilst the other is likely to become bogged down. To retain this feature or its equivalent was therefore of vital importance. The students spent a considerable time, with the aid of creative methods, trying to evolve different and better ways of achieving this facility. They failed to come up with a more acceptable method, probably because they were trying to better a technique which in itself offers so much but has a basic simplicity. This is a factor which manifests itself time and time again on our projects. At this point, it was decided to retain the split in the middle chassis. This was a landmark in the conceptual design thinking.

The transmissions available to us ranged from the pure mechanical, partial hydrostatic and mechanical, through to the pure hydrostatic system. These types of transmissions were explored in great detail, particularly the hydrostatic and partial hydrostatic types. The hydrostatic transmission, apart from its well-known control features, has one thing in its favour over the others in that it allows a greater flexibility in the disposition of the engine within the vehicle confines. However, this asset proved to be more apparent than real in a vehicle where the engine and driver occupy a significant volume of the whole. Add to this a potential cost penalty, at the power level we were considering of some 40 percent over a suitable mechanical transmission and its virtues become a vice. Hydrostatic transmission was discarded.

Partial hydrostatic transmission was considered to be a likely satisfactory compromise, but it was found impossible to match the high and low speed requirements of the specification with hydraulic motor and axle combinations available in this country. This too was discarded.

Our conceptual deliberations were therefore to be based on the criteria of split in the middle chassis and mechanical transmission.

A selection of four concepts A, B, C_1, and C_2 were offered to the company in February 1971 in the form of schematic layouts (see Figures 6.1, 6.2, 6.3, and 6.4).

Scheme A

This is a very notional scheme, having a D-shaped front and rear chassis, with the rear chassis being particularly wide in order to accommodate engine and driver, side by side. The reinforced fuel tank is mounted behind the driver and above the engine. The battery box is beneath the seat, and the hydraulic tank located below and to the rear of

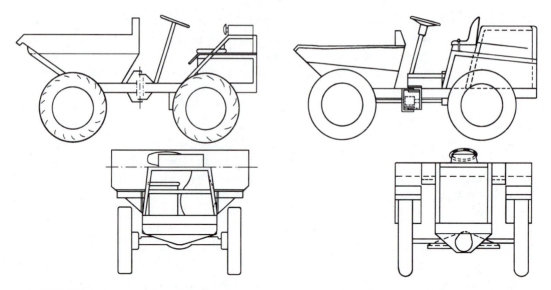

FIGURE 6.1.   Marathon Concept A   **FIGURE 6.2.**   Marathon Concept B

the driver. The rear superstructure protects the driver, tanks, and engine from collision damage. Also proposed were rubber mudguards.

Comment

The concept of tanks enveloping the engine is a good one as it dispenses with guards and engine covers. However, the side driver position gives poor driver visibility. Tanks

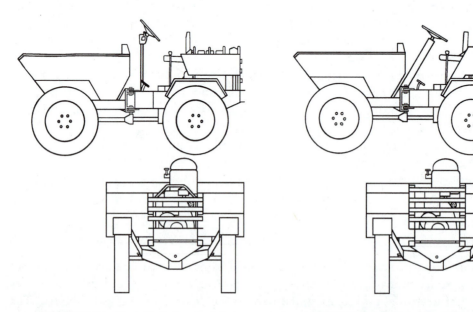

FIGURE 6.3.   Marathon Concept C_1 **FIGURE 6.4.**   Marathon Concept C_2

will be of varying sizes and contribute little to vehicle strength and stiffness. Basically, this is a collection of unique components without consideration of their interaction. Aesthetically, the design is not quite there; angular differences between steering column and superstructure are particularly noticeable, as is the assymetry resulting from the side driver position.

Scheme B

This scheme has a space frame integral with the chassis. The A-shaped front chassis and D-shaped rear chassis afford engine protection and additional torsional strength. The central driver position is over the engine. Fuel and hydraulic tanks are fixed to the upper frame astride the engine. Mudguards are bolted to articulating rear frame. The rear superstructure protects the engine and driver from collision.

Comment

This is a rather better approach. Again, the concept of tanks enveloping the engine is good. The design is also a collection of connected pieces, each performing specific functions. This has much better driver visibility and is more aesthetically pleasing than scheme A. Some attempt has been made to achieve commonality of line, with the symmetry in plan view certainly helping.

Scheme C_1

The saddle tanks atop a D-shaped front and rear chassis and central engine give symmetry in plan view. Tanks act as engine shrouds being carried on U-shaped tubular bars bolted to the rear chassis and are removable for engine access. Mudguards are bolted to rear articulating frame.

Comment

We have the concept of the tanks enveloping the engine but avoiding the issue of integrating the various necessary components in order to make them multifunctional. Load transfer resulting from rear collision is not satisfactory. This is aesthetically jumbled: the vertical steering column and angular mismatches are particularly noticeable.

Scheme C_2

This develops scheme C_1 but has a central driver position. Saddle tanks are designed to carry load and assist the rear chassis whilst providing seat mountings. Symmetry exists in plan view.

Comment

This is a much better attempt at an integrated design in that the tanks can contribute to chassis stiffness and strength. It is aesthetically pleasing in that the angular mismatches of previous schemes are starting to disappear. Good engine protection is provided with potentially a minimum of unique superstructure.

After due deliberation, it was agreed to develop further schemes B, C_1, and C_2 on the basis of lower centre of gravity, central engine and driver position, and enhanced aesthetic appeal. One month later, at the end of March, four more conceptual designs were presented to the company, together with cost estimates for each chassis design. These potential manufacturing costs were based on the historical cost patterns of the company.

It is interesting to record that at this stage in the conceptual design it was becoming apparent that the best choice of concept was beginning to emerge from amidst the various schemes, almost of its own volition. Unfortunately, and this is a problem common to many design situations, as far as we were aware at the time the students did not recognise such a situation developing. This is a basic problem, certainly of any design course and possibly also in industry. It is in our opinion due to a combination of lack of experience, and hence lack of confidence, coupled with an attitude of mind that tends to reject the obvious without thoroughly checking through all the reasons and factors which would underwrite the correct choice of concept. Recognising the concept is one thing; recognising and understanding this type of situation are quite another. So the process began of persuading students to compare the four concepts with the specification. Consequently, at the March meeting it was agreed with the company to carry forward the development of concept C_2. The basis for this decision was that potentially this scheme not only met the specification, but it also offered the simplest form of construction, minimum cost, maximum utilisation of component parts, and, what is more, considerable aesthetic appeal.

There is no easy way, no unique method, which by application enables students to make the correct decision in this type of situation. It is only by a process of examining and reexamining all the factors and their interactions in a relentless manner, with tutors acting as catalysts, does one get through to students, so that in the end the best decision is taken by the students themselves. It is at the conceptual stage of a live project that this must be brought home to students. To proceed to detail design without having been through this process, which to some students is quite a traumatic experience, is to reduce the chance of a successful project.

Suffice to say that the chosen chassis design had been monitored, using a computer program which indicated that all was well in terms of stress levels. In fact, without the program, the chassis would not have attained the high degree of efficiency present today.

Detail Design of the Vehicle

Having made our choice of concept and within its confines, we had to embark upon the detail design of the vehicle. It should be pointed out that the selection of a concept is not an irrevocable decision; one can, at the later stages in any design, face extreme detail design difficulties which on reflection may be due to an error in the concept. We so often find this sort of situation developing and, upon examining the reasons for the difficulty, have to change the concept. This is another feature which repeats itself time and time again and is very difficult to get across to students. One must always be prepared to change!

The chosen concept was basically very simple and therefore offered great opportunities for some really first-class detail design. Naturally, during the conceptual stages, we had been exploring possible transmission arrangements in some detail, together with some preliminary chassis designs. Thus there was, at this point in time, a fair degree of confidence in our selection.

Our philosophy throughout this stage was one of simplicity and economy, whilst ensuring that detail components were matched to the company's manufacturing facility. Great emphasis was placed on establishing the correct pattern of subassembly breakdowns consistent with the required flow-line production.

We had certain constraints in terms of bought-out parts—in particular, axles, gearboxes, and engines—since certain manufacturers produce in quantity for this type of market. The transmission proved to be a particularly interesting challenge. In order to keep the centre of gravity of the vehicle as low as possible, we wished to place the engine as close as possible to the top of the rear differential housing. To achieve this situation meant turning the axles through 180° about the input axis, thereby lowering the centre of gravity of the engine by some 2 inches. The appropriate standard transfer gearbox and change gearbox were then selected, and the transmission system was established. On checking through the complete drive line from the engine to front and rear axles, it was discovered that we had in fact, four reverse and one forward gear, the exact opposite of the specification! What had not been realised was that in rotating the axles we had changed the direction of rotation required at the input to the axles. It transpired in discussion with the gearbox and axle manufacturers that no one previously had wished to turn the axles over, and thus it would appear that engines had been mounted higher than necessary, thereby maintaining the centre of gravity of vehicles at an artificially high level. The outcome of this debacle was that the transfer box manufacturer instigated a crash programme of redesign and produced a box with one less gear in it, which of course, reduced the cost of the transfer box. This transfer box is now a standard in the company's catalogue. The final transmission design is one of extreme simplicity with the minimum number of components to satisfy the requirements.

In this chapter, one can only highlight a few aspects of a full design. We have mentioned one of the bought-out component predicaments, but there were others.

Three areas of particular significance are illustrative of the type of problem faced in detail design, not only with students but in design offices generally. The first concerns the overall vehicle control system—clutch, brake, accelerator, and steering wheel. It has been the practice in the industry to design such controls as appendages to whichever part of the chassis happens to be in the vicinity: the motor car is a good example of this. We persuaded students, albeit with some difficulty, to consider the possibility of an integrated unit, particularly as all the functions were related. We say 'with some difficulty' as a noticeable 'set' had developed which drew them all the time to the conventional approach. In the end analysis, a control column was designed which incorporated all the component parts needed for each function and which could be manufactured as a subassembly. This subassembly, shown in Figure 6.5, sits atop the centre pivot and in itself is easy to assemble and to fix to the vehicle without crawling underneath the chassis.

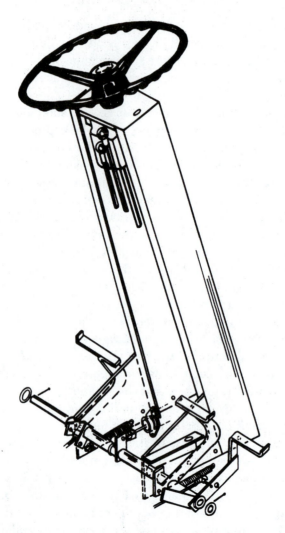

FIGURE 6.5.   Steering Subassembly Mounted at the
Centre Pivot Point of the Chassis

This approach not only simplifies assembly generally, but it also brings in train a number of other benefits in that it enables the floor of the vehicle to be a simple flat tread plate, quickly removable for ease of maintenance and likewise, as so often happens, ease of assembly.

The second concerns the use of computer programs in design. One of the students was given the task of optimising the skip tipping arrangement, preferably using a single hydraulic ram. He spent considerable time and effort trying to write a program which would, when fed with the appropriate data, result in not only the optimum ram size but also the optimum position of the main skip frontal pivot and the ram pivots. After a period of time, the program was very little nearer completion, and, what is more, we had still no idea of the best pivotal positions. Another student was given the task and in *one day* on the drawing board with the aid of a two-dimensional cardboard model had solved the problem. This situation is illustrative of the sort of problem that can arise with increasing application and usage of the computer; the old adage 'There is a time and place for everything' still applies.

The third area of interest concerns the use of ergonomic data for the design of the control and driver position layout. The students sought such information from an authoritative source and used it for the overall layout of the vehicle. The detail design was by now almost complete, and the detail drawings were in course of preparation. It was decided to build a model of the control column, pedals, floor, and seating arrangement to confirm the final layout. This led to an immediate calamity: the vehicle would be capable of being driven with comfort only by extremely short-legged people. At a stroke, the driver had to be repositioned on the vehicle. It transpired that no one had considered checking the ergonomic data, which happened to be concerned with the design of automobiles. We would strongly recommend the use of simple models as an aid to detail design and, indeed, conceptual design.

Each detail of the vehicle was costed by a student who was given special responsibility for the costing work. Thus, the total cost of the vehicle in production emerged. Many more details than represent the total vehicle were costed during the course of detail design. These details were rejected in favour of the lower-cost alternatives which were incorporated into the final design. This is now an essential part of our course at Loughborough.

The detail design and drawings were carried out to timescale, being completed at the end of June with all drawings done by the students themselves. During the detailing phase, regular meetings were held with the company, and the students had, in fact, to sell their ideas and solutions to the company personnel. The students also prepared a full production plan for the company, showing the proposed method and order of manufacture and assembly.

It should be noted that the target cost as defined in the product specification was predicted as having been met to an accuracy of +5 percent, and this has subsequently been confirmed in production.

The Test Rig

It was recognised at the outset that early completion of the test rig was essential so that the existing chassis could be tested and the data obtained used for the new design, and the work was planned accordingly. One of the first steps was to ascertain the in-service loading and loading patterns, which were required not only for the design of the rig but also for the new chassis (Figures 6.6 and 6.7). As a result, it was decided that the rig must be capable of applying both static and cyclic loads to the chassis for the bending, torsion, and combined modes. The purpose of the cyclic loads was, of course, to have the facility to simulate in a few days many years of rigorous site use. The rig was also required to have maximum flexibility as the company proposed to use it for testing other equipment

FIGURE 6.6. Photo of Marathon 2550

in the future. The design and detail drawings for the rig were prepared by students, and the rig was manufactured and erected at the company's works where it was commissioned in May 1971 (see Figure 6.7). It will be seen that the rig was essentially simple, having a steel grillage for the base with a portal frame and hydraulic system for the application of loads, the rams being linked by flexible hoses to allow maximum flexibility in positioning. In Figure 6.7 the chassis is set up to simulate a bending condition which it would experience in service when carrying a full load in the skip over rough but fairly level ground.

Whilst the design of the dumper and the test rig proceeded in parallel, other important features of the project were not neglected. An operating manual for the rig was prepared, and a comprehensive test programme was planned. Further, it was realised at the outset that a computer program would be desirable for determining the stresses and deflections of proposed chassis designs and hence of working towards an optimised solution. A suitable program for the analysis of automobile type chassis was located at Cranfield Institute of Technology (Wardill, n.d.), which is based on the Livesly (1964) method of analysis. The program was in the form where it handled the basic logic of the method but required some development for our own particular requirements and for compatibility with the university computer installation. The program was used in its developed form to predict the performance of the existing chassis under estimated service loading conditions and to predict the modes of failure when subjected to loadings

FIGURE 6.7. Photo of Marathon Chassis on Test Rig
Undergoing Simulated Conditions of Many Years'
Rigorous Site Use

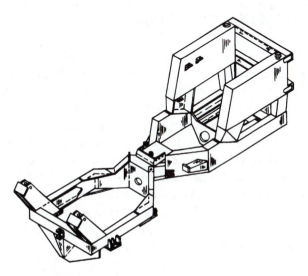

FIGURE 6.8. The Chassis (Mainly of Folded Sheet Steel
Construction)

in excess of service conditions. Good correlation was obtained between the predicted computer results and those obtained from the rig tests which validated the program for use in the design of the new chassis.

A chassis of the new design was built ahead of the preproduction machine itself and was subject to loading patterns similar to those for the existing dumper. But trouble was in store. The new chassis failed very shortly after testing had commenced. Where had we gone wrong? The computer program had worked excellently for the existing chassis, the stress levels in the new one were compatible, so what could the problem be? The answer was simple, but like so many design problems was not immediately apparent.

The new chassis consisted of mainly folded sheet construction—the old one being of heavy plate and hot rolled sections. The difference lay in that there were points of instability in the former and not in the latter. The computer program did not take this factor into account, and we had overlooked the point—a lesson indeed, in the use of computer programs for design purposes.

Despite these problems, we were able to fulfil the criteria as good, if not better, than the existing vehicle, and the appropriate design changes were made for incorporation in the preproduction dumper. The test rig had indeed proved its worth as the areas in question would undoubtedly have given trouble in service.

It perhaps should be added that the testing was supervised by members of the student team acting on a rota basis.

By the time the comprehensive test programme had been completed, the production of the prototype was well under way, and the results of all the previous design work would soon be evident.

PRODUCTION AND TESTING OF THE PROTOTYPE

The detail design and drawing phases were dovetailed into the manufacturing stage so that manufacture of the preproduction dumper could be started as early as possible. Such was the confidence of the company in the design that even at this stage some of the more important jigs and fixtures were manufactured in readiness for the production models.

During the production phase, students visited the company works to assist with the initial assembly work of the preproduction machine and to resolve any problems which arose. Unlike many design courses which finish with only a paper design, any errors were only too plain to see. The students also helped to progress the work through the factory and at the same time acted as a link between the Centre and the company during this critical stage, so that any problems requiring further advice could be reported back and immediate action taken to overcome the difficulty. Such was the cooperation between the company and the Centre that the prototype was completed in its orange livery by mid-August.

A test programme previously drawn up by the students was now put into operation under their supervision. Initial tests were carried out both at the company's works and under the arduous conditions of a local quarry to check the performance of the dumper vehicle against requirements of the specification. All team members attended these final trials, which went without a hitch—a very satisfactory culmination of the project for the student team.

MARKET LAUNCH AND POSITION TODAY

The Marathon 2550 was introduced to the market in November 1972, at the Public Works exhibition held in London. This followed a year of field trials with contractors and plant hire companies with whom it performed well. It attracted a great deal of interest at the exhibition, and many sales were made as a result.

Coincidental with the launch, a film outlining the project was shown on the BBC television programme 'Tomorrows World' as being illustrative of the cooperation that can be achieved between industry and a university.

Over the past eighteen months, sales have been made in ever increasing numbers, both to the home and overseas markets. It is particularly satisfying in view of the country's current balance-of-payment problems that overseas sales are increasing. At the time of writing, vehicles have been delivered to the Middle East, Asia, many African countries, North America, and Europe.

The company now markets two versions of the Marathon—a 2550 kg capacity and a 2000 kg capacity, together with a version fitted with a swivel skip.

This chapter has presented a very brief summary of some of the situations and events leading to the production of the Marathon 2550. We have attempted to balance some of the factual occurrences which took place in respect of the vehicle design with a modicum of our philosophy in relation to design teaching.

This picture would not be complete without some references to features of the vehicle itself, typical of which are

- Pleasant appearance with clean lines,
- Simple but rugged chassis,
- Simple transmission,
- Large-capacity fuel and hydraulic tanks,
- Good driver position with safety and good visibility,
- Engine and transmission protected but easily accessible,
- The minimum number of components so that less can go wrong,
- Low centre of gravity, giving good stability,
- Extra carrying capacity,
- Hydrostatic steering unit with hydraulic ram,
- Disc brakes on all four wheels protected by the big wheels,
- Transmission parking brake.

In all, a successful joint venture between industry and a university.*

*The vehicle now called the Marathon 2550 is still in production in various forms in 1992.

REFERENCES

Livesly, R.K. (1964). *Matrix Methods of Structural Analysis*. London: Pergamon Press.

Wardill, G.A. (n.d.). 'Small Computer Procedures as Tools for Structural Designers.' Advanced School of Automobile Engineering, Cranfield Institute of Technology.

PART II

Design Process and Philosopy

The papers in this part introduce one of Stuart Pugh's most important concepts—the difference between partial design and total design. An interesting experiment would be to survey a group of participants at a typical design conference and ask them the difference. I suspect that the replies would be a bit slow in coming, yet one cannot be a successful practitioner or researcher in design without a clear understanding of the difference. Stuart saw clearly that most academics did not understand the distinction and concentrated their efforts on partial design. Design engineers in industry had a somewhat better understanding of total design but had little successful methodology for practicing it. Read these chapters to get a clear understanding that design engineers can faithfully implement all of the basic engineering courses, successfully apply Newton's law, Maxwell's equations, Hooke's law, and all of the other basic relationships of engineering science, and still end up with a product that is a commercial failure. These staples of the undergraduate engineering education must be integrated and expanded to produce a product that will satisfy customers. The result is total design, made clear in these five chapters.

This part further develops the design activity model, building on the evolution presented in Part I. The reader has a further opportunity to attain insights into the mature design activity model by seeing its development. Also, there is an application to an industry that might at first be considered rather far afield from more traditional industrial practice—marine engineering. Read this to see how the generic design activity model is applicable to even such a unique field and how the transformation from generic to specific can guide the application of the design activity model to any industry.

Part II emphasizes design process—total design guided by the design activity model—and also in Chapter 8 provides a thought-provoking look at the role of design in society, especially its relationship to the arts and sciences. This is presented in the context of C. P. Snow's famous book on the two cultures. Stuart rejected the perspective of some authors that design is a third culture, essentially disparate from science and the arts. Rather he saw design as the integrator of science and the arts. Also, note the interesting mention of teaching design to twelve- to fourteen-year olds. This chapter can be used by the reader as a launching pad into further probing of the role of design in society.

In Part II, then, look for total design more than partial design, the design activity model as the guide to total design, generic product design specification, the role of design as an integrator of cultures, and the innovative freshman design course at the University of Strathclyde.

—Don Clausing

INTRODUCTION TO PART II

In Part II it has been recognised that there is a difference between partial and total design and that lack of recognition of this fact has and does lead to confusion and misunderstanding both in industry and education. Much design teaching is of necessity based on partial design since it is logical and natural that usually the whole is a sum of its parts. This case is argued in Chapter 7, Total Design, Partial Design: A Reconciliation, and such recognition has contributed to a fuller understanding of the design process by both academics and industrialists.

Chapter 8, Design: The Integrative Enveloping Culture, Not a Third Culture, discusses the design process as the natural integrator of the arts and sciences and not as a third culture, as has been advocated by some academics over the past few years.

From the realms of theory and philosophy I have moved into an investigation of the marine field to try to discover whether practitioners in this field fit in with a generally emerging design process as exemplified in Chapter 10. Chapter 9, Systematic Design Procedures and Their Application in the Marine Field: An Outsider's View, confirms the view that by and large they do fit into the pattern and that there is more than a semblance of an emerging common design process. Perhaps this slow recognition is the root cause of the decline of old established traditional industries.

Chapter 10, Design Activity Models: Worldwide Emergence and Convergence, illustrates the worldwide emergence of design activity models which supports the maritime arguments and shows even wider compliance over a wide product range.

Our experience at attempting to teach total design as opposed to partial design is described in detail in Chapter 11, Integration by Design Is Achievable, with specific reference to SEED activity, which underwrites the basic assumption of total design. The fitting of multiple partial design within a framework of total design is shown, and the response of students and practitioners is in a sense the same. Students see the logic of the arguments, whilst practitioners acknowledge it as being obvious in hindsight, although it had not been obvious to them previously.

CHAPTER

7

Total Design, Partial Design: a Reconciliation

ABSTRACT

This chapter considers and contrasts the design requirements of industry with those of education. By considering design teaching in education as being traditionally and necessarily elemental and therefore partial, it demonstrates how products are composed of a summation of many partial design inputs—to give total design. It concludes by describing briefly a new course at the University of Strathclyde designed to cross traditional departmental boundaries and reconcile partial design teaching with the totality of industrial requirements.

From *Proceedings of the International Conference on Engineering Design* (Boston: ICED, ASME, 1987), 1005–1011.

Traditionally, design, at least in British universities and polytechnics, has been taught within a departmental system which has as its origins the very narrow research base essential to progress in the fundamental sciences. Design teaching, methods, techniques, practice, and content have thus usually extended only throughout the subject area—take mechanical engineering, for example. It has taken more and more frequent excursions into the territory of others by the ad hoc addition of subjects such as finance, business management, social studies, and electronics so that its traditional base gradually is being eroded by the addition of subject material, which is doubtless necessary to the education and training of the engineer. In the outside world the nature and requirements demanded of that traditional base have also been changing rapidly. Add yet another subject called design to this dynamic cocktail, and one is likely to destroy whatever cohesion there was in any course without a cogent rationale.

Until recently, we have also avoided definitions and explanations of just what design is, which is in my view a recipe for disaster. This is the situation pertaining within the departmental structures in many institutions, and 'breakout' appears at best difficult. Yet all is not as bad as it may at first appear, since if one does define design, however crudely and inadequately, then the picture becomes much more rational and understandable.

DESIGN: EDUCATIONAL REQUIREMENTS

Let me consider design teaching within the department a little further and state what to many will be obvious truisms (Pugh, 1986b):

- There is very little wrong with the technical content of traditional engineering courses, provided they are kept up to date. It is the way in which they are taught that is the problem.
- Ad hoc design and projects are not substitutes for traditional subjects.
- The major educational problem is to harness traditional subject material to produce an effective design whole.
- The lack of a means of making design visible to students has been a major inhibitor to achieving progress in integration (now a requirement of all U.K. engineering courses). Exhortation has little effect without understanding.

Yet against this background, very necessary analytical techniques and technology must be taught to students but not in isolation from other disciplines. Could it be that the departmental structures are of themselves anti-integrative in a total design sense? I use the expression 'total design' to mean the whole of the process required to produce successful products for the marketplace. Others may call it the new product development process or product delivery process. Taking the subject structures of mechanical, electrical, and civil engineering as examples, it may be safely said that they all teach, research, and carry out partial design to various degrees (Figure 7.1).

The design of a machine element such as a cam or a shaft is very much partial design—essential to the success of the whole but nevertheless partial. Thus systems of partial design teaching have evolved with little cross-talk or diffusion across departmental or subject boundaries. Yet such partial design teaching and the acquisition of practice skills are fundamental to engineering and engineering courses.

It is through the teaching and practice of partial design that our students acquire engineering rigour that ensures sound elemental design—mechanical, electrical, or any other. There is no substitute for this rigour; it is and always has been the foundation of all engineering courses. To achieve such rigour, it may be that to a greater or lesser degree elemental teaching has to be carried out to some extent in isolation from the total design context in order that due focus be given to the elemental material in which, after all, it is not easy to acquire skill and understanding.

But, in my view, what has been and is missing from the British scene (although it is now changing rapidly) is that one must teach the traditional subject material as partial, as contributions to the whole. This should be made quite clear to students right from the outset, placing the partial design into the context of a total design framework which of itself must be demonstrated to become contextual when related to specific product areas.

So we have a natural rationale for partial design, and the teaching should make this quite clear. We require a rationale for total design (Pugh, 1986a), and to place it in context, we require a suitable project to bring it all to life and to achieve effective and efficient integration.

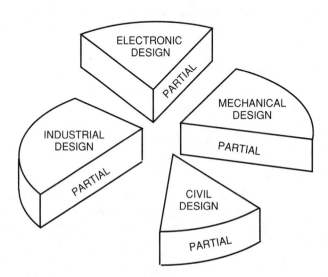

FIGURE 7.1. Professional Divisions Encourage Partial Design

In parallel with partial design teaching, a structured approach to total design should be presented to all students at the earliest opportunity, and teachers should recognise that students cannot nor should they be required to practise effectively over the total spectrum until well into their final year or, indeed, postgraduate studies. This will enable them to progressively relate their traditionally taught material to total design and minimise confusion on the meaning of design. The partial design teaching itself should be disciplined and structured; it should not be allowed to become piled together in an ad hoc manner under a subject banner called *design*.

The structure and framework of total design should be of a form which automatically leads not only to integration but to a natural feeling of integration within the participant irrespective of discipline. Each contributor should be able to see how partial contributions fit into the whole. Total design should be taught and practised in a progressive manner with enhanced information, knowledge, and techniques leading to increased rigour in a total design sense; the diverse question of rigour will be returned to later.

DESIGN: INDUSTRY REQUIREMENT

Design in industry is or should be a total process. Because industry is concerned with all aspects of design in the context of products, processes, systems, or services, the products emanating from industry should be complete in a total design sense. Analysis of any product will reveal that it contains contributions from a mix of traditional design areas, some of which have already been mentioned. The typical total product may therefore be said to have required inputs from many areas of a business including the technological in order to produce the final whole, so total design is the metier of industry.

It is useful to define the product in total design terms typically as in Figure 7.2, where the product is made up of many technological components and a host of non-technical factors which impinge upon the product design and hence the product—typical of which are human-machine interfaces (ergonomics), shape, form, texture, colour (aesthetics). Without the whole being in balance the project may fail in the marketplace. In industrial terms the necessary integration comes about only as a result of all the partial design inputs. Immediately, therefore one has a view of the gap between the two systems which requires to be bridged and reconciled. How, in the educational environment, do we handle partial design inputs, which in practice need to be multidisciplinary? This point will be returned to later.

It is recognised that the goals of industry and education are somewhat different, yet, accepting this, it is perfectly feasible to achieve comprehensive bridging. But in order to do so we must have an understanding of the differences and *also* of the commonalities of the two systems. It is hoped Figures 7.1 and 7.2 go some way to bridging this gap.

If one accepts the gap, then further discussion is required to reorientate the educational end of the bridge, which experience tells me will also allow for some efficiency adjustment at the industrial end.

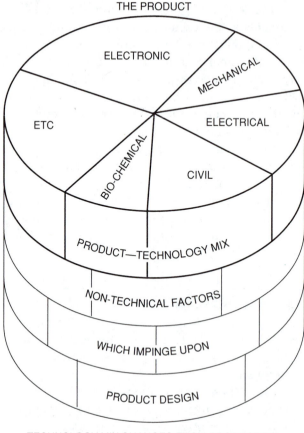

THE PRODUCT

FIGURE 7.2. Total Design Product Model

ENGINEERING AND TOTAL DESIGN RIGOUR: THEY'RE DIFFERENT

It would be accepted without question in educational circles that engineering sciences are based on the acquisition of analytical techniques and skills together with their application. Through a study of engineering based on physics and mathematics, applied through elemental studies, a student acquires an all-round engineering competence. He or she can calculate inertia forces, pressure and voltage drops, surge current, rate of ion transfer, fatigue life, stress a shaft, carry out vibration calculations: the list is limitless if one considers all the engineering disciplines and therefore almost unquestionably acceptable as the basis for any engineering course. However, as has already been demonstrated, the application of such skills and knowledge to engineering elements is partial design. To include the highly optimised, best material, shaft in any design when it is not essential to the design may involve engineering rigour of the highest order, but if the component is not really necessary, wherein lies its design efficiency within the total product picture?

Consider a product such as the Sinclair C5 car, which involved some considerably detailed engineering in terms of chassis design and construction, drives and control, materials, and so on—in other words, engineering rigour of a high order—yet the product failed on the market. Did it fail because, in spite of engineering rigour, its total design rigour was zero, because the market was nonexistent, because the aesthetics and ergonomics left a lot to be desired, because it didn't suit people's aspirations and needs?

We have to get these differences across to our students, otherwise misdirected engineering rigour will *always give rise* to bad total design rigour. This implies that design teams as such should always include nonengineers. Perhaps we should teach and implement this at university, through the inclusion of nontechnologists on project teams—but in a disciplined structured manner, where everyone has a common view of what total design is and therefore subscribes to a common objective with a minimum of misconceptions. Participants should be able to see how their differing partial design contributions fit into the whole scene.

So engineering rigour is essential to product success, but it is different from the rigour required of total design. Indeed, might it not be argued that it may be better to attempt to achieve an understanding of total design at university at the expense of certain engineering topics particularly if engineering and its implied rigour is the only thing being taught under the heading of design.

Success in the marketplace requires total design rigour and engineering rigour of the highest order—never the one without the other. Engineering rigour on its own is not enough.

Perhaps the Rolls Royce story epitomises the difference outlined above—'that technical excellence is not enough without commercial competitiveness' (Hodgson, 1986). We should teach engineering this way.

DESIGN AT STRATHCLYDE

The Faculty of Engineering at Strathclyde has embarked upon a strategy of maintaining traditional engineering excellence whilst at one and the same time blurring the boundaries between the departments. As a major component of this activity, a total design course is being given to almost all of the first-year students (seventeen years of age in the Scottish system). Additionally, we now have a four-year master of engineering in product design which started in October 1991.

The course has been designed to cross boundaries. Commencing with Figures 7.1 and 7.2 and their ramifications, it proceeds through models of the total design (the substance of which is well known) to inculcate in students a picture of design that from the very beginning allows them to relate to the increasing partial design acquisition which occurs as the course proceeds.

In this first year, some 350 students have taken this course covering mechanical, electrical, electronic, civil, building, architecture, naval architecture, chemical, orthotics and prosthetics, production, and the like. We teach a common design course and recommend to the departmental specialists how they can best adjust their material to fit in with our systematic procedures. We teach total design understanding in twenty-five hours including programmed project time. As a supportive adjunct to this we also teach a fifty-hour graphical communications package which is designed to fit in with the total design approach and which is deliberately interdisciplinary. Our own partial design teaching is also being adjusted to fit to this pattern.

We are currently running some seventy projects of varying nature to suit departmental product contexts. For architecture and civil engineering, for example, we are working on a design of a canopy for a shopping precinct. Initially we are seeking limited engineering rigour but expect a comprehensive total design rigour to be adhered to and to emerge as the engineering rigour is enhanced as the course proceeds. Whilst engineering expectations will rise, so will the expectations in total design as more of the elements are progressively taught.

The attitude of students to this approach is enthusiastic—not unexpected since they have not previously experienced the unnatural boundaries of departments. It is essential to involve staff of the different departments in the project work to assist with the partial design teaching adjustments I mentioned earlier.

This work is now proceeding to the second year, and we are now attempting to get to grips with the art and design college ethos. We have a lot to learn from this area in skills and techniques, but, since they also appear to lack a rationale for total design, the acquisition of recognition of such skills will take time. It is in my view essential that we proceed to a situation whereby through a common view of design, we achieve the spirit and practise of the multidisciplinary team within the educational systems.

REFERENCES

Hodgson, G. (1986). 'Roller Coaster.' *Sunday Times Magazine* (October).

Pugh, S. (1986a). 'Design Activity Model: World Wide Emergence and Convergence.' *Design Studies* 7(3), 167–193.

Pugh, S. (1986b). 'Integration by Design Is Achievable.' *Engineering Design Education* (Autumn), 30–31.

8

Design: The Integrative Enveloping Culture, Not a Third Culture

---

ABSTRACT

This paper examines the three-cultures approach—art, science, and design—and rejects it in favour of an integrative approach. The author suggests that this is more likely to allow design practitioners in different fields to communicate with each other.

From *Design Studies* 3(2) (April 1982), 93–96.

The International Conference on Engineering Design held in Rome in March 1981 brought together what turned out to be predominantly academics to present and listen to papers on engineering design and its fringes. Design practitioners, if they were present, kept their heads low, and of the few that were there, discussion with them revealed a condition of bewilderment and disbelief in that they were failing to recognise and bring about the correlation between what they do in practice and the prognostications of the academics, myself included. With some notable exceptions, however, the design philosophers were rampant to such an extent that I allowed myself the genuine pleasure of departing from the pragmatic and indulging in the philosophical.

First, I stated that having heard a great deal about design philosophy, we must not confuse the philosophy of design with the doing of design: they are not one and the same thing, they are not interchangeable. Second, having admitted from time to time that I too am guilty of the substitution, I suggested that, in the context of C. P. Snow's book about (1964) the two cultures, if design were considered as the third integrative culture, then possibly this would resolve many of the philosophical problems. Design is the integrator of the arts and sciences, exemplified by Leonardo da Vinci's painting of the Last Supper, being design at the art end of the spectrum, and a power station, being design at the science end of the spectrum. I concluded my philosophical diversion by stating that all design is an admixture of art and science. I then rapidly returned to the safety of the pragmatic and gave my paper. Since that time, the philosophical side of my design deliberations has been confined to the subconscious—until I attended the Symposium on Design Research held at the Open University in November 1981. Nigel Cross, in the concluding submission, reiterated his allegiance to the philosophy expounded by Bruce Archer (1979), in which he advocates and articulates quite clearly design as the third separatist culture (see Figure 8.1) with its own language, syntax, and so on.

In an instant, everything became crystal clear. Figure 8.1 says it all and is symptomatic of the cultural, specialist, specific discipline worship and adulation which are still endemic in British society today. That design must become something special—on its own, different, separatist—is clearly nonsense, and pursuance of this line of thinking will not clarify matters one iota. Separatism, the cult of the specialist, is the curse of our society, and we are doing little to counter this effectively, although I suppose talking and writing about it does help. Design as the third separatist will be equally as divisive as the two-cultures philosophy expounded by Snow. Philosophers will nod, designer philosophers will be grateful for the new specialism, and in the meantime the majority of humankind will continue not to understand. Pinning my own colours firmly to the mast and at the risk of overemphasis, I am firmly against any consideration whatsoever of design as the third culture considered alongside the arts and sciences.

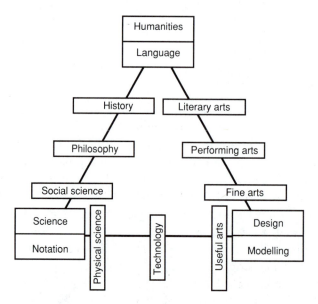

FIGURE 8.1.   Design Viewed as the Third Culture

The problem of the two cultures has recently been occupying other eminent minds. In a recent paper given to the Royal Society of Arts, entitled 'Engineering Philosophy: The Third Culture,' Professor Douglas Lewin (1981) also built a case for a third separatist culture called engineering. He stated: 'Many and varied accounts of the engineering design process have been expounded and it is essential that these confused ideas be resolved and that engineering and design be established as a unified discipline in its own right with a coherent and well-defined philosophy.'

I cannot concede that the adoption of a third separatist culture called *engineering*, or for that matter one called *design*, will lead to the resolution of 'these confused ideas.' Lewin is aware that something is wrong and is searching for the truth, but the suggestion of having another culture with its own language, syntax, and so on will, in my opinion, give rise to another incomprehensible pile of gobbledygook which will not be understood by the occupants of the arts and sciences (if only for the reason that they will not bother to attempt the necessary bridging thought to achieve understanding) or for that matter by design practitioners.

French (1975), in talking of engineering design, sees it as the unifying ingredient of professional engineering education. I couldn't agree more. He then disappoints by saying 'Integrate it: don't add it on'—a separatist view possibly. In a nonmathematical sense, how do you integrate the integrator? The separatist view is confirmed by the statement that 'Design must be infused into the entire syllabus.' Is there another approach to the whole matter which is clearer, more comprehensible, more logical, and less likely to be divisive and separatist and which allows students to achieve not only an understanding of the arts by the sciences (and vice versa) but also an enhanced understanding of and ability to practice design? I think there is!

DESIGN, THE INTEGRATIVE CULTURE: A HYPOTHESIS

In July 1974, I presented a paper at a Conference on Information Systems for Designers at Southampton University. Having sat through the first day of the conference with participants from many disciplines—engineers, scientists, architects, and so on—I began to realise that we were experiencing an enormous communication gap. I therefore changed my approach to the presentation of my paper, the introduction (Pugh, 1974) to which reads as follows: 'The theme of this paper was used to introduce the presentation of my paper "Manufacturing Cost Information: The Needs of the Design Engineer." My main reason for so doing was not in support of the paper but rather an attempt to overcome the communications gap which always seems to be present in any discussion relating to design.' I then went on to outline and articulate a view of design which attempted to cross boundaries not only in terms of traditional science and arts disciplines but also from one sphere of endeavour to another—that is, engineering design, architecture, and so on. Figure 8.2 shows the basic factors to be considered in the design of anything—from a building to a toothbrush.

This was a genuine attempt to bridge the communications gap between design practitioners. (At that time I was not fully aware of the different and wider gaps in communication and understanding between design philosophers.) As I recall, it had a salutary and unifying effect on the audience, and on the day I think it achieved its objectives. Yet today, seven years later and wiser, I consider the situation has not greatly improved in re-

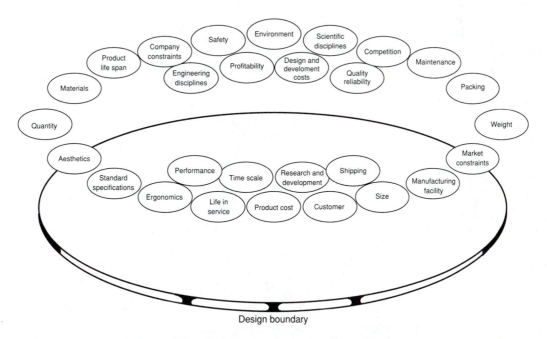

FIGURE 8.2.   Basic Factors to be Considered in Design

spect of a common understanding of design amongst practitioners and, what is probably more important, the unrecognised tremendous commonality of practice between them. We now have the exacerbated dual problem of, first, the lack of common understanding amongst practitioners and, second, with the growth of design philosophy divorced from practice, the subsequent failure to provide a satisfactory framework of understanding for the former—that is, the practitioners (Pugh, 1981).

Might then a humble practitioner, having momentarily descended from the rugged mountains of design practice to the calmer plains of design philosophy, attempt to provide the link not only between practice and philosophy (restoration of communications) but also at one and the same time take another step along the road to resolve the 'two cultures' debate. To formulate a satisfactory hypothesis to resolve the above, two conditions are deemed essential: first, the British tendency to columnar specialisation must be avoided to prevent divisiveness and separatism and, second, the formation of a separate language, grammar, and syntax must be avoided at all costs, otherwise we achieve the same end result as the former. Intellectual piles must be avoided like the plague, especially in this particular instance.

Design, therefore, must become recognisably not only the integrative mechanism for the arts and sciences but also the culture which envelopes *both*. It is not a separate, third culture as suggested by Archer. It is the integrative activity which brings together and embodies (in the broadest sense) the arts and the sciences; they, as component parts, make up the whole, which is design (see Figure 8.3).

Design is not like mathematics or physics: it does not represent a body of knowledge; it is the activity that integrates the bodies of knowledge present in the arts and sciences. Before the engineers start leaping up and down, let me say that I regard engineering as the application of the sciences in the service of humanity, which, through the activity of design, manifests itself in artefacts. However, I would also point out that books, paintings, sculptures, buildings, textile fabrics, and so on are also artefacts. The

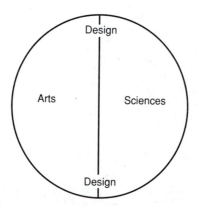

FIGURE 8.3.   Design Viewed as the Combination of Art
and Science

existence of the new artefacts perforce adds to the relevant bodies of knowledge. Making is always part of the design activity: one person can produce a painting or sculpture, several people can fashion a building, many people are needed to build a power station. It is the scale of things that detaches the original designer from the making activity.

Design, when considered in this manner, can then truly be seen to be the integrator of the arts and sciences, and, precisely because it is considered in this way, it must use the language, grammar, and syntax of both cultures to become comprehensive and logically understood. The language, grammar, and syntax exist; we need to use them properly. I submit that there is no case for the development of yet another specialty. We have the opportunity, with this hypothesis, to bring about a better understanding between the specialties. Design, as the one true integrator, can be a powerful aid to understanding and the elimination of society's artificial boundaries.

Let me try to put some flesh on the bones of the hypothesis and bring it to life by considering some of the artefacts evolved and produced by man. As stated earlier, in order to be effective and understood, the hypothesis must fit design in all its forms and fashions. Design, in the context of this hypothesis, is that activity which, from an amalgam of the arts and sciences and their derivatives manifests itself in artefact form of every description from books to aircraft. It is only the balance and distribution of the arts and sciences contents which distinguishes the one from the other (for example, see Figure 8.4).

Engineering, as stated, is considered to be primarily the application of the sciences. However, form, shape, colour, and aesthetics fit into the arts socket, and thus the whole can only logically be explained as design—not engineering design. Do we then allow the common misconception that what is predominantly science application with a heavy engineering content should be called *engineering design*; or, conversely, do we allow the

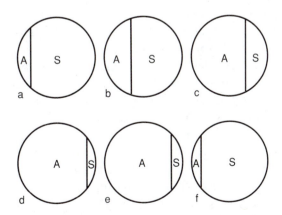

FIGURE 8.4. The Balance Between Art and Science: (a) Fossil-Fuelled Power Station, (b) Family House, (c) Textile Fabric, (d) Sculpture, (e) Painting, (f) Textile Loom

label *industrial design* when we come to the middle ground of artefacts, where beauty and the visual effect is predominant (although far removed from a painting)? Yes, I would consider the above labels acceptable, provided that people really understand what they mean and provided that they are not contributory and supportive of divisiveness and separation.

CONCLUSIONS

This is my hypothesis. In support of this view of the integration of specialties under the banner of design, see many of the recent articles on Japanese quality circles: 'Japan's secret is the way it manages design and manufacturing processes, where its cooperative principles are more effective than the individualistic departmental approach of many of its Western competitors' (Lorenz, 1981).

In repetitive, continuing philosophising about design, one tends to forget that there are many practitioners in differing fields of endeavour who, as yet, lack a common communicative framework. I would humbly suggest that the above hypothesis forms the basis for such a framework and would be the first to admit having worked backwards to this base from analysis of design practice in the broadest sense.

As design educators, we have a duty to enhance design practice and understanding at all levels. Lewin (1981) suggests a systems approach to engineering and goes on to outline the basic steps to evolving solutions to problems, based on a problem-solving approach. Between 1967 and 1975, in the Engineering Design Centre of Loughborough University of Technology, the master's course in engineering design was conducted on this basis. Interestingly and revealingly, we found that students had great difficulty in translating teaching into practice. We have now broken design teaching into three specific linked areas—design core, techniques to work the core, and, in our case, technology, to which the former two are applied (Pugh and Smith, 1978). Students have a better understanding of design and have become more competent in carrying out design in different fields and across disciplines, not only in theory but also in practice.

Finally, in producing a series of design teaching aids for schools, under the auspices of the Design Council and the Department of Industry, for children in the twelve to fourteen age group, a unified, nonspecific discipline integrated approach has been adopted using teachers and pupils to make the aids. The approach, in my opinion, transcends the arts and sciences. Excellent motivation, excitement, and understanding have been generated in the pupils. Interestingly, the topics considered so far, as design exemplars, are a nesting box for birds, a jelly mould, a midday meal, and a form of protective head gear. One finds a commonality of activity present in each one.

As a prelude to the commencement of the work, a document (Design Council, 1978) was circulated to school teachers suggesting that the design of anything was analogous to a guyed mast, where the mast depends upon the guy ropes for support (Figure 8.5). The guys represent the knowledge and experience acquired in the individual traditional school subjects, and without their support the mast will become unstable and may even

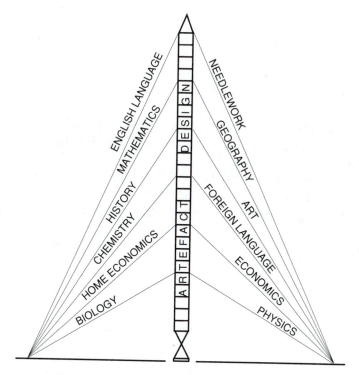

FIGURE 8.5.   Design in Balance Among the Arts and Sciences

collapse. This was an attempt to involve specialists in the design activity and to make them aware of their essential contribution to design and the necessity to integrate the arts and sciences. Design is the integrative culture not a third culture in the context of the arts and sciences.

I can do no better than to conclude with a quotation from Snow (Snow, 1964): 'There seems to be no place where the cultures meet. I am not going to waste time saying that this is a pity. It is much more than that. Soon I shall come to some practical consequences. But at the heart of thought and creation we are letting some of our best chances go by default. The clashing point of two subjects, two disciplines, two cultures—of two galaxies, so far as that goes—*ought to produce creative chances.'*

Is it not the integrative activity of design which creates the chances? Now, back to the mountains and the rarefied air!

REFERENCES

Archer, B. (1979). 'Design as a Discipline.' *Design Studies* 1(1), 17–20.
Design Council. (1978). *The Development of Design Teaching Aids for School (Design Council Notes to School Teachers)*. London: Design Council.

French, M.J. (1975). 'Integrate It: Don't Add It On.' *Engineering* 215(10) (October), 826–827.

Lewin, D. (1981). 'Engineering Philosophy: The Third Culture.' *Journal of the Royal Society of Arts* 129(5302) (September), 653–666.

Lorenz, C. (Ed.). (1981). 'Why Japan Wins the Quality Race.' *Financial Times* (1 September).

Pugh, S. (1974). 'Engineering Design: Towards a Common Understanding.' *Proceedings of Information Systems for Designers* (pp. D4–D6). Southampton: University of Southampton.

Pugh, S. (1981). 'Design Philosophy Is Fine—But Practice Is Revealing.' *Newsletter of the Design Research Society* 14 (October 1981).

Pugh, S., and Smith, D.G. (1978). 'Design Teaching Ten Years On.' *Design Education Supplement* (June).

Snow, C.P. (1964). *The Two Cultures and a Second Look*. Cambridge: Cambridge University Press.

9

Systematic Design Procedures and Their Application in the Marine Field: An Outsider's View

---

ABSTRACT

This chapter explores the establishment of a correlation between systematic design procedures, both developed and developing, in engineering and artefact production and parallel developments in the marine field. In order to establish a rationale for such correlation, the emergence of design activity models will be discussed briefly, illustrating a convergence from diverse fields of practice worldwide, including those attributable to the marine industry and thus indicative of the existence of common ground.

All the major elements of design activity are described and related to marine design, albeit in a nonmarine manner—from competition analysis to the initial product design specification for a vessel, as a subsystem within the overall transport system. The absence of systematic approaches to design specification as opposed to artefact specification (the drawings) will be considered as a major problem affecting all industries.

The conceptual stage is also dissected from the viewpoint of whether in the marine field this is predetermined or not at the specification stage, together with the vexed question of evaluation or optimisation. Computers are considered not so much in their traditional analytical role but more in terms of effectiveness in information management and judicious usage, whilst trying to avoid the effects of 'data-base freeze' upon the designer.

The chapter concludes with the author's view of the common ground between marine design and design in other areas.

From *Proceedings of the Second International Marine Systems Design Conference on the Theory and Practise of Marine Design*, Lyngby (Copenhagen: IMSDC, 1985), 1–10.

INTRODUCTION

Professor Erichsen's kind invitation to deliver a paper at the 1985 Second International Marine Systems Design Conference initially filled me with fear and trepidation, since my knowledge of the marine industry is scant to say the least. An excursion many years ago with a paddle-wheeled submersible and designing hydraulic winches and capstans for sea-borne operation set the narrow limits of this experience. Yet on reflection, memories even within these limited confines, and with, of course, 20-20 hindsight vision, may make this contribution worthwhile.

Attempting to understand the marine design industry and its operation has meant a lot of reading, and I was amazed at the design progression that has been evolving over the years and its correlation with the design progression that is taking place worldwide across industry generally. An attempt will be made to describe the latter whilst retaining close correlation with the former, and what better starting point than with some statements common to all industries which will set the scene for this chapter.

SETTING THE SCENE

It is universal cause for complaint that designers do not properly take into account the needs of the users (the customers), the requirements of production, and the like—that they tend to operate in an almost isolationist mode. Heirung (1982), in his opening address to the last conference, discussed many situations and factors which mitigate against an integrated approach to design, and made the following statement: 'There are many countries where universities and research institutes are more distanced from the industry than in my home countries!'

It is not the intention of this chapter to dwell on the educational sector but briefly, since it is my firm belief that successful product development in any sphere is possible only through systematic design, which, by definition, means integration. Integration is of itself rendered an elusive goal by the separation referred to above, both physically and mentally.

Over the years, reference has been made to this problem (Pugh, 1974), and discussions of engineering design education (Pugh, 1982b) have much in agreement with Heirung. In developing an unscrambling rationale, I attempted to resolve matters a little by discussing and defining research, development, and design as being mutually exclusive but interactive activities, present in any business but each having completely different characteristics (Pugh, 1983b).

Research, it was concluded, is a necessary and vital part of design, and so is development. A piece of research can be developed with no artefactal outcome; a design can be developed with an artefactal outcome. To reiterate, research in any sphere of artefact production is vital. Yet it must be said that applied research is not design in modern parlance, since modern design encompasses many things, from the qualitative, almost artistic, to the quantitative scientific (Pugh, 1982a), and for far too long design has really been considered almost solely in the latter sense.

It has been recognised, in the quest for a rationale and approach which reconciled these two facets and all that they contain, that a similar quest in the marine industry has been articulated by many people. One common major objective seems to have been to establish a suitable model for design.

Mandell and Chryssostomidis (1972), in discussing a design methodology for ships and in particular for examining what they call 'the exploratory phase of the proposed methodology,' resort to a problem-solving model in an attempt to make visible the nature of the design activity. Perhaps almost unwittingly, they also make reference to the computer in the following terms: 'Unfortunately, the direct contribution of the computer to design methodology is small because the capabilities provided by the computer do not augment the user's abilities as a designer but rather as an "analyst."' This point we shall return to later. They also propose developing their methodology by research. This has also been the case within the Engineering Design Centre, but by research in this context, this has meant dissecting the Centre's extensive practice. Thus, the model issue has manifested itself in the marine industry as it has in the general engineering industry, since for efficient use of computers some structure is necessary. Atkinson (1972), in discussing an integrated approach to design and production, albeit implying a concentration on the quantitative end of design, exhibits the model called 'the design spiral.' It is considered highly significant that in describing the implementation of computer programs, he says, 'A useful way to represent this collection of programs is in the form of a spiral, superimposed upon a pattern of radial spokes [as shown in Figure 9.1]. Several spokes represent parameters necessary to define the ship not just as a floating body, but as a unit within a transportation system, ie to fulfil a specific purpose in a defined manner *under agreed restraints*' (author's emphasis). He does not actually say so, but it must be assumed that such restraints come from the 'specification' for the ship, which is not defined. Since he also talks of a single traverse of the spiral yielding a hundred different designs, he is creating a selection of problems of enormous proportions.

Eames and Drummond (1976) also refer to a simplified design spiral; since they may also be experiencing the difficulties mentioned earlier, they have reversed the direction of the spiral Figure 9.2. Whilst Atkinson starts in the centre and works outwards, Eames and Drummond start on the outside and work inwards. Does this suggest that the model has possibilities but is not working efficiently and that its representation is still suspect?

Gallin (1976, 1977), in asking questions such as 'What does ship design mean?' looks into computer utilisation and resorts to may different models, commencing with the design spiral similar to Figure 9.2. He too works from the outside to the inside; ref-

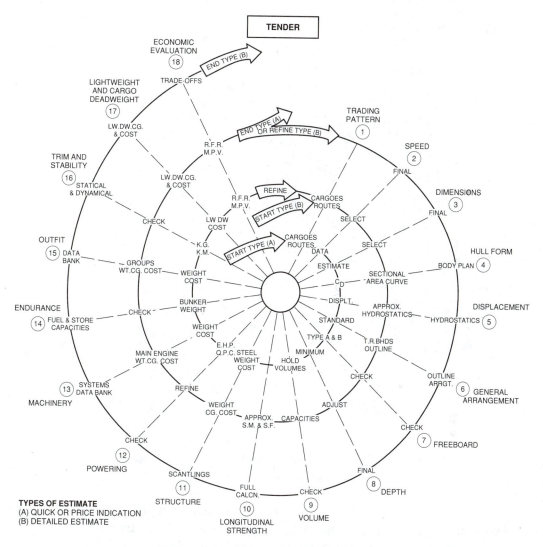

FIGURE 9.1. The Design Spiral (Atkinson)

erence is also made to conventional and nonconventional ship design methods. It is likely that the urge to harness the computer is masking the fundamental design issue—another issue to which we shall return later.

So the design model issue persists from the marine viewpoint as it has persisted in other areas of industrial endeavour. The author considers that not only is the acquisition of systematic design procedures related to the acquisition of a suitable model, but the efficiency of those procedures are then related in some way to the efficiency of the model. This opinion has been formed from the background of extensive practice over many industries, many products, and many models of the design activity.

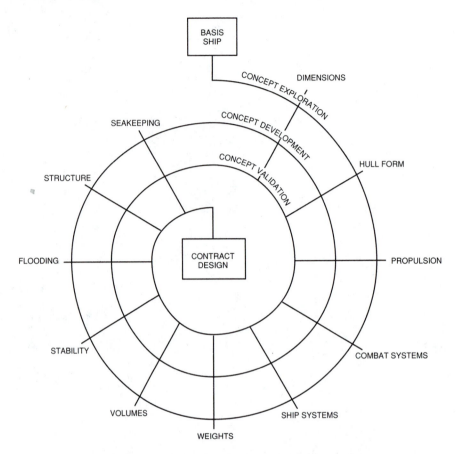

FIGURE 9.2.   Simplified Design Spiral (Defence Research
Establishment Atlantic, Canada)

Pugh (1977) describes a design activity model which attempts to make such activity visible to all but particularly to the designer, since he is the operative. He also attempts to bring together the quantitative and qualitative aspect of design in a rational visible manner. The question of model visibility cannot be overemphasised. It is considered that an effective model must have a number of characteristics:

- All must relate to it.
- All must understand it.
- All must be able to practise effectively and efficiently as a result of it.
- It must be comprehensive.
- It preferably should have universal application owing allegiance to neither discipline nor product.

It is interesting to note that MacCallum (1982), in discussing ship design by computer, finds it necessary and rewarding to refer to the nature of the design process through modelling: 'Especially, design should be thought of as a process of modelling.

In every situation, the designer creates some kind of abstract model which simulates some aspect of the behaviour of the thing being designed. The model may be graphical, numerical, operational or even physical. *The creation of the model, which is a process of synthesis is difficult to formalise'* (author's emphasis). This must be one of the most concrete truisms of the twentieth century. Again, it is interesting to note that the computer is forcing the issue.

It is considered that the emergence of the model issue from the marine industry viewpoint to the current day has been amply demonstrated. The story is much the same from the engineering industry viewpoint, although the criticality of the model issue has not been so articulately aired as in your industry. This lack of airing and discussion, however, has been replaced by a wide variety of models, most of which, in the author's opinion, have inhibited the emergence of a sound framework and structured model of the design activity, having the aforementioned attributes which will resolve and coordinate many of the facets discussed by Erichsen (1982): 'Design may be defined as a task of co-ordinating the use of resources to reach a specific goal.' It is suggested that the goal is the production of a cost-effective, usable vessel. How can it all be linked together?

THE DESIGN ACTIVITY MODEL: A RATIONALE FOR INTEGRATION

Rather late in the day the author discovered in hindsight that, as with anything, a design activity model has to be designed—just as one has to design a ship, albeit with slightly less complexity. In so doing, the opportunity has been taken, at least in theory, to remove the artificial barriers between the sections of any business producing artefacts—marine or otherwise. Thus, everyone in a business must, and does, contribute towards design. They make an input and therefore make that input visible, rational, and not contrived.

The design activity model shown in Figure 9.3 and described fully in the appendix to this chapter is, in content, almost identical to the one mentioned earlier except for one important difference: the graphics are better. As with ship design and graphic design, design matters. However one tries to destroy the model, or whatever the situation to which it is related, it always seems to fit. A device now exists for fitting it to any situation, and this will be demonstrated at your conference. The model is described in detail in the appendix. Its structure is obvious and, no doubt by pure chance, it relates to the marine industry not only in content, which many would recognise, but also in structure. Rawson (1976), in discussing the art of ship designing, having talked about being creative and requiring an objective, at the end says, 'This leads to the third feature of design, important enough to distinguish from the objective although it is often contained within it. *It must be bounded.* A painting must be contained within a physical space, an historical survey within some dates or ethnic influences, an engineering design within cost or behaviour constraints. Such boundaries introduce into the design process the ideas of discipline. Wide as the imagination may wander, it must be disci-

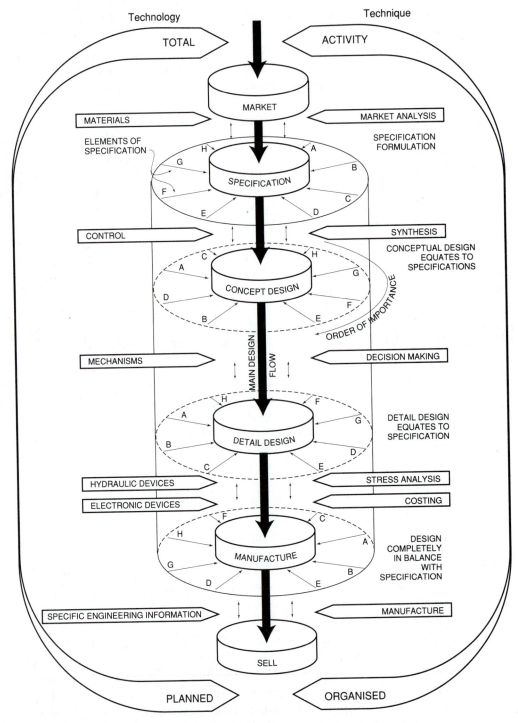

FIGURE 9.3.   Design Core Bounded by Product Design
Specification: Dynamic Concepts

plined to remain within a boundary that has been judged most appealing to society or those who are sponsoring the design activity' (author's emphasis). He then goes on to talk of iteration and so on. Little did he know that he was possibly describing the above design activity model. Erichsen (1982) also refers repeatedly to 'design boundaries.'

So the parallel between marine design and design in other fields is affirmed yet again. Technology mixes may be different according to the type of artefact under consideration—ships, for example, are different from cars—but there should be little difference between the basic design procedures adopted. This convergence onto a common model underwrites this statement. This examination of the design core, from market need to selling, for the sake of brevity concentrates upon the iterative stages to conceptual choice.

THE DESIGN CORE STAGES

Market or User Need

Nordenstrøm (1976), advocating market research within the shipping community, refers to 'bad design and layout of navigational instruments; the bridge of a ship the design is often equally bad.' In fact, his reiterative theme is bad design, but he does not offer any palliatives or cures beyond saying, 'It must be common sense' or 'Research could be your most profitable investment.'

In my opinion, it is neither one, the other, or both; it is a question of disciplined, systematic design. For example, in designing a construction vehicle or an instrument for brain surgery, we carry out massive literature, patent, and legislation searches and, above all else, competition search and analysis to compare what exists with its neighbours (Pugh, 1982c). This is what Nordenstrøm refers to as *trend analysis*. Large parametric plots are carried out, many of which are not initially logical, and when patterns of apparent relationships are discovered, successful attempts are then made to retrofit the logic. This can be carried out for complete equipment, for subassemblies, or at any preferred level very early on in a project and *without exception* much more is known about the product than about the designers and manufacturers—the experts in the field. So detailed understanding of the market or user need is required. Searching for and analysing massive amounts of information need to be carried out with two primary objectives in mind:

- To understand the market or user need situation and also the nature and status of competing products and
- To prepare a comprehensive product design specification (PDS) as a prelude to concept generation.

A PDS does not specify the ultimate artefact; the final drawings do that. A PDS defines, in greater detail, the constraints upon the artefact to be designed: it forms our basic design boundary. It is the control mechanism for systematic design and the document against which emergent designs must be measured and assessed, hence the reference earlier to 'bounded': the specification bounds the design core.

It would appear that there is a distinct lack of systematic front-end activity in the marine field and certainly lack of reference to systematic specification formulation. Langenberg (1982), in his excellent paper given to the last conference, highlights this very point when he speaks of 'inadequate definition of the objectives and inadequate problem analysis.' However, perhaps it should be pointed out that such activity is also in its infancy across engineering industries generally.

Product Design Specification (PDS)

Many other authoritative marine writers refer to specifications, yet none articulate them to any degree and those that do appear to lack structure. Such references to specifications and constraints that are made are usually mixed in with design method statements. Systematic PDS generation is therefore to be recommended in the strongest possible terms. In fact, it is considered as absolutely essential to successful product design both now and in the future. A procedure for this is outlined and gives the primary elements or triggers for the formulation of any PDS shown as radial inputs to the design core (see Figure 9.4)—shades of the design spiral. Inadequacy and inefficiency at the front end of design has led and is leading to the misdirection of much excellent engineering work across whole industries. Is the marine industry in this category?

Meek (1972), discussing sources of constraints relating to ship design, suggests that this may be the case: 'Whereas limitations have, in the past, been mainly physical or technical and reasonably easily comprehended (e.g. past imposed limitation on draft, or structurally imposed limitations on highly stressed hull members), there are now the more esoteric constraints of environmental considerations.' These statements are indications of the much wider considerations to be given attention to at the commencement of the design of anything, so we return yet again to the primary elements of the PDS.

Conceptual Design

Much has been written about the topic of conceptual design, and it probably incites more emotion, at least in engineering circles, than any other phrase in the English language. Systematic concept generation, the generation of alternatives to meet the PDS, is essential to successful product evolution. However, it may well be that in some industries true conceptual design in totality is no longer necessary or, in fact, possible. In fact, certain artefacts —such as cars—may be said to be conceptually static (Pugh, 1983a). In twenty years time, cars will still probably have four wheels, one at each corner, four seats, front engine, and transmission and disc brakes. In other words, it is suggested that car designers have no conceptual choice whatsoever: the concept is assumed, and specifications are written upon that assumption. The design activity model then appears as in Figure 9.4.

In Pugh (1983), ships were also referred to as possibly being in this category. Yet having made an extensive study of the marine literature, I am not now so certain. In fact, this is an understatement: it is reasonably certain that ships are conceptually dy-

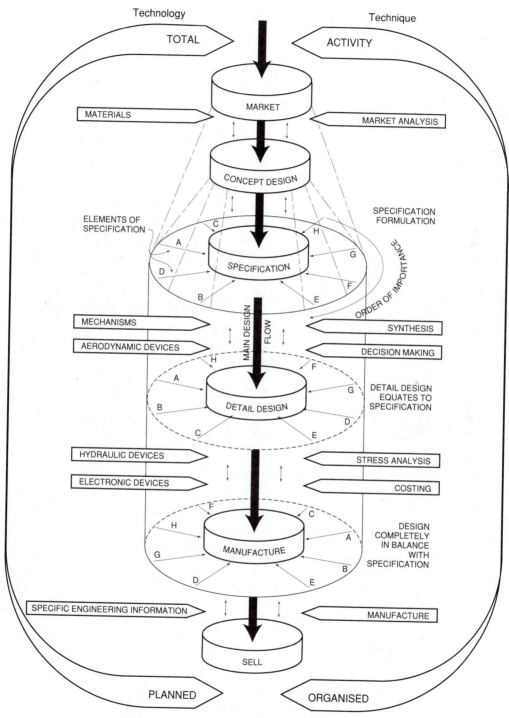

FIGURE 9.4. Design Core Bounded by Product Design
Specification: Static or Fixed Concepts

namic, that many differing designs (conceptually) are yet to emerge, and that one should practice as if this were the case. Thus, Figure 9.5 then applies. Care must be taken in writing specifications (when this happens) that do not make the assumption of a fixed or static concept. This is a particularly important point as computers now start to become highly relevant, since efficient operation and application of computers in an integrated manner depends entirely upon the conceptual choice having been made (Pugh, 1984). Is the marine industry therefore in danger of suppressing new ship concepts because of increasing computer utilisation—acknowledged as being necessary for efficient design and production once a conceptual choice has been made? I am completely at one with Erichsen (1982) when he states, 'Furthermore, the use of computer programs—that are exact in their calculations—may create a false feeling of correctness by the user, making him forget uncertainties in the basic assumptions of the design.'

Thus, if the assumption is made that all ship concepts are dynamic, at least to some degree, then the systematic generation of alternatives should be carried out against the controlling background of the PDS. Having generated alternatives (about which much has been written), the question then arises of logical reduction to the choice of the best one and, probably more important, knowing why the rejects are not as good as the chosen one. This is perhaps a suitable point to differentiate between evaluation and optimisation, which again causes confusion and concern to designers.

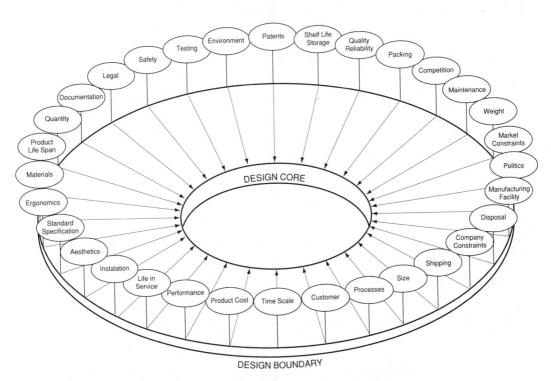

FIGURE 9.5. Elements of Product Design Specification

Evaluation Before Optimisation

Analytically orientated engineers always wish to optimise, whilst others are usually happy to use intuition, experience, and guesswork. None of these methods are deemed acceptable when answering the question of conceptual choice.

System integration becomes more attainable, smoother running, and more comfortable for most people, so the tendency in the engineering industry generally is to work with the assumption of a static, frozen concept. There is some evidence of this in the marine industry. Day (1982), discussing developments in marine activities in the twenty-first century based upon the evolutionary pattern of the current century, says, 'I have therefore selected four areas which I believe will continue to receive considerable attention and thus are indicative of likely future change. The genuine breakthrough, the unexpected change, or the "revolution," as contrasted with continuing development or evolution, are precluded as, by definition, they are hardly foreseeable.'

Evaluation is the activity to be carried out in selecting the best choice of design concept with which to proceed—assuming, of course, that alternatives to meet the PDS have been generated. Experience with some fifty projects over the past fourteen years, carried out in a variety of industries, shows that in areas of high uncertainty, combined with parameters that cannot be quantified, essentially nonnumerate evaluation methods work well in that they cause the real relationships to unfold and the logically correct relationships to be established.

To summarise, proper evaluation to select the best concept lays a very firm foundation for subsequent optimisation (numerate) procedures.

The difference between evaluation and optimisation was discussed by Pugh (1983c). It is a great temptation for all engineers, whatever their orientation, to get into numbers, usually the quicker the better. Yet numbers, as such, are not of much use in nonquantifiable situations (Pugh and Smith, 1976b). Historians, artists, geographers, and the like make good decisions quite happily without numbers, but engineers in general and designers in particular need to acquire skills and techniques to achieve a proper balance between the numerate and nonnumerate. To this end, extensive use is now made of a matrix method of analysis which enables the proper mix to be established and which allows many concepts to be considered in a systematic manner, measured against the PDS. This is the procedure of controlled convergence (Pugh, 1981). Essentially, it is a procedure which proceeds from the macro to the micro—from broad outlines with little detail to more closely defined designs with much more detail—in a controlled manner without stifling creativity. It is simultaneously a convergent-divergent procedure. A marine engineer became aware of this sort of problem in discussing optimisation techniques in the marine field (Kreitner, 1966): 'The algorithmic subsets are capable of being programmed for computer execution; the creative though, of course is not.'

The essential differences between evaluation and optimisation are slowly being recognised, particularly in Scandinavia and Italy, although the Italians tend to follow a fuzzy set line.

Any technique of evaluation should present not only the opportunity but also, through its inbuilt method of operation, the absolute priority to cause new valid concepts to be generated. Usually, following the exhaustive group working, conceptual saturation is achieved in that no more valid concepts emerge. This is the time to stop and evaluate. Interestingly, however, it is not unknown for a new concept to emerge later, thus demonstrating that it is difficult to get all the cats in the bag at the same time.

It might be thought that matrix evaluation allied to the complexity of the marine industry is a nonstarter. It is suggested that this is not so, since whatever the artefact or system, it must be broken down to portions of manageable size and operated upon accordingly. Recombination into the whole then comes later.

The Computer as the Integrating Tool

It was suggested earlier that for the effective and efficient use of the computer, a systematic structure of operation is essential in the first place, in order that integration of the various stages of the design activity becomes not just a possibility but a probability. Many companies, both in shipbuilding and industry generally, are using computers efficiently and effectively to speed up the process of design and production. Do they, in fact, harness them with such effectiveness in variable, dynamic concept situations and even earlier in trying to resolve the market need situation? From the evidence available, it is suspected that the answer is no.

In a recent study of the majority of the world's CAD systems (the expression *CAD* being used in the loosest definition possible), it was shown that the evolution of such systems has been backwards from the production end, which has produced efficient drawing and analysis programs but has had a minimal effect on the front end of the design activity.

Examination of the comments of system users again draws a parallel with those in the marine industry. In fact, irrespective of the nature of the product, everyone is experiencing the same dilemma—limited usage at the front end of design, except in a staccato manner, *unless* the conceptual choice has been made. This step at least enables data bases to become established. However, the degree of iteration required of the design activity, even in a fixed-concept situation, is leading to some disquiet amongst users striving to achieve integrative working. So it may be done faster, but is it better and as thorough?

Designers in many industries, aided and abetted by computer scientists, fondly believe that they cater for conceptual design using integrated CAD systems (Lind, et al. 1983): 'Ship design has traditionally been described as an iterative search for one consistent solution. The process was graphically illustrated as a converging spiral [or diverging, depending upon whose model one chooses] where some calculations were carried out repeatedly with increasing accuracy as the work proceeded until the design was consistent,' and what is more enlightening, 'From the designer's point of view, ship design may also be illustrated as a stepwise process comprising conceptual design, design and detailed design.' Whatever happened to user needs? Would the programmes be relevant to a vessel having twin screws at the bow, none at the stern, with rigid aerodynamic programmable sails fitted? One suspects not. This would be a different concept to that assumed by the programme builders.

Again, marine industry traumas are no different in nature to those of other industries. This is not surprising, since the study revealed that none of the turnkey CAD systems have themselves been designed with the user designer in mind (Pugh, 1984). Figure 9.6 illustrates this in relation to the design core.

The problem of conceptual suppression by computers was recognised by Kreitner (1966) and quite separately by Pugh and Smith (1976a). It is still a problem in all industries and product areas. However, providing that an industry is aware of this, it can do something about it, such as not commit all its design work to computer systems. In fact, the leading effective U.K. user of CAD systems is now installing another twenty workstations (making fifty in all) and another twenty drawing boards for 'creative work.' It is not surprising to find that computer scientists are now doing a rapid volte face. In 1984, they are turning their attention to studying the design process; they consider it is high time to consider the user or need situation and the design process in particular. It is suggested that computer system evolution has slowed down and is in danger of stagnating at the barrier between conceptual design and detail design shown in Figure 9.6. The very recent call for papers for the conference titled Design Theory for CAD indicates an urgency to understand the design activity to a greater depth. This is to be welcomed as such research should lead to better equipment that is more designer friendly.

CONCLUSIONS

It is to be hoped that a close parallel has been established between design in the marine field and that in other fields and that this has been adequately demonstrated. Correlations have been made from user need (the start) to detail design, and it can be seen that the problems and practices are common. Perhaps participation by designers of all kinds, irrespective of industries or products, should meet more often to discuss these areas of obvious common ground. All design, to be effective, is multidisciplinary. Marine design is no different: 'From the barges of the Egyptians to Columbas's Spanish galleons to the modern day navel frigates, basic ship design problems have remained about the same' ('Shipbuilder's Graphics,' 1983). One wonders!

APPENDIX: DESCRIPTION OF THE DESIGN ACTIVITY MODEL

If one considers the design activity as having a central core as in Figure 9.3—the essential prime phases of which are market, specification, conceptual design, detail design, manufacture, and selling—then one has an iterative process which proceeds generally from market to selling. It must, of necessity, proceed this way otherwise nothing would ever be produced. The outcome—the selling of a product or system—itself influences the market

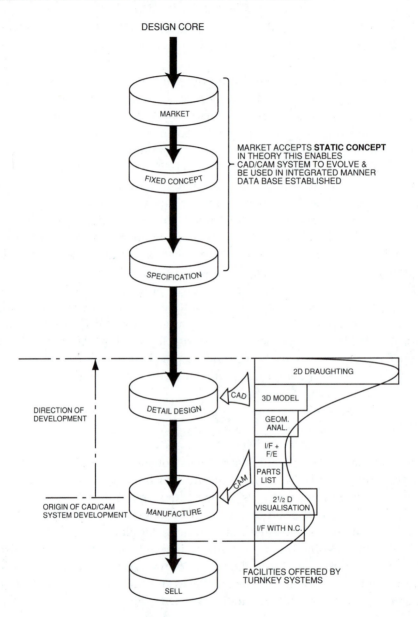

FIGURE 9.6. Design Core

and therefore sets up another cycle of design activity which interacts with the market and causes it to change. This flow of activity is, or should be, carried out within the mantle of an evolving specification, which embraces many factors as well as functions and which itself changes and evolves as the design proceeds—due to the interaction with the core activity. The specification is therefore the design boundary, the elements of which are shown in Figure 9.4.

In order to be able to bring the core to life and make it work, one needs not only an understanding of the iterative nature of the core activity but also to have knowledge and awareness of the techniques available to aid its working. In other words, the tools available not only to set it in motion but also to improve its function and performance. These are called techniques related to the design core. Typical of them might be costing, ergonomics, creative methods, and the like. Again, these are inputs to the design core and interact with it. It may be interesting here to note that, as yet, we have made no mention of technology. At this stage of understanding, technology is not only unnecessary but may cause confusion and mitigate against a clear understanding of design. Still considering techniques related to the design core, there is a finite situation in view. However, considering the whole of technology as being potentially available as input to the core leads to an almost infinite situation. Thus, according to the product or process area in which a design is being undertaken, it will determine the mix of technologies needed for success. For any successful product they will always be multidisciplinary. The whole of this activity should be carried out under a management umbrella of planning and organisation. Place this total package within a corporate business structure, and one has a picture of how design fits, or should fit, into the modern business.

It is always interesting to retrofit previous designs to the model and is very revealing: it should make a person more aware of what is or is not happening in a particular organisation. It is against a progressive build-up of this model from design core to organisation and planning that we not only teach our postgraduate students in engineering design, but we structure our course content to match the model and practice this way professionally.

REFERENCES

Atkinson, J.R. (1972). 'An Integrated Approach to Design and Production.' *Phil. Trans. Royal Society London* A273, pp. 99–118.

Day, J.G. (1982). 'Selected Likely Development in Marine Activities During the Twenty-first Century.' Proceedings of the International Centenary Conference. Strathclyde: University of Strathclyde.

Eames, M.C., and Drummond, T.G. (1976). 'Concept Exploration: An Approach to Small Warship Design.' *Proceedings of the Royal Institute of Naval Architects* pp. 29–54.

Erichsen, S. (1982). 'Design of Transport by Sea.' *Proceedings of the First International Marine Systems Design Conference* (pt. 1, pp. 11–22). London: MSDC.

Gallin, C. (1976). 'The Realities of Present Day Ship Design.' *Proceedings of Europort '76, International Maritime Conference.* Amsterdam: IMC.

Gallin, C. (1977). 'Inventiveness in Ship Design.' *Trans North East Coast Institute of Engineers and Shipbuilders* 94(1), pp. 17–32.

Heirung, E. (1982). 'Opening Address.' *Proceedings of the First International Marine Systems Design Conference* (pt. 2, pp. i–ii). London: MSDC.

Kreitner, G.W. (1966). 'Discussion of Paper: "Optimisation Methods Applied to Ship Design."' Mandel & Leopold, Trans SNAME, pp. 506–508.

Langenberg, H. (1982). 'Methodology and System Approach in Early Design in a Shipyard.' *Proceedings of the First International Marine Systems Design Conference* (pp. 31–40). London: MSDC.

Lind, B.O., Bakke, E., and Reinertsen, W. (1983). 'Shipshape: A New Ship Design Program for Norway.' *The Motor Ship* (December), pp. 40–42.

MacCallum, K.J. (1982). 'Creative Ship Design by Computer.' *Computer Applications in the Automation of Shipyard Operation and Ship Design IV*. Amsterdam: North Holland.

Mandel, P., and Chryssostomidis, C. (1972). 'A Design Methodology for Ships and Other Complex Systems.' *Phil. Trans, Royal Society London* A273, pp. 85–98.

Meek, M. (1972). 'The Designer's Response from the Owner's Side.' *Phil Trans Royal Society London* A273, pp. 45–58.

Nordenstrøm, N. (1976). 'Economic Appraisal and Market Potential of Ships.' Paper presented at Europort '76, International Maritime Conference, Amsterdam.

Pugh, S. (1974). 'Engineering Design: Towards a Common Understanding.' *Proceedings of the Conference on Information Systems for Designers* (pp. D4–D6). Southampton: CISD.

Pugh, S. (1977). 'The Engineering Designer: His Tasks and Information Needs.' *Proceedings of the Conference on Information Systems for Designers* (pp. 63–66). Southampton: CISD.

Pugh, S. (1981). 'Concept Selection: A Method That Works.' *Proceedings of ICED '81* (pp. 497–506). Rome: ICED.

Pugh, S. (1982a). 'Debate: Design—The Integrative, Enveloping Culture, Not a Third Culture.' *Design Studies* 3(2) (April), pp. 93–96.

Pugh, S. (1982b). *Engineering Design: Time for Action, Changing Design* (pp. 85–97). London: Wiley.

Pugh, S. (1982c). 'A New Design: The Ability to Compete.' *Proceedings of the Design Policy Conference* (pt. 4, pp. 14–16). London: DPC.

Pugh, S. (1983a). 'The Application of CAD in Relation to Dynamic/Static Product Concepts.' *Proceedings of ICED '83* (vol. 2, pp. 564–571). Copenhagen: ICED.

Pugh, S. (1983b). 'Research and Development: The Missing Link—Design.' *Proceedings of ICED '83* (vol. 2, pp. 500–507). Copenhagen: ICED.

Pugh, S. (1983c). 'State of the Art on Optimisation in GB.' *Proceedings of ICED '83* (vol. 1, pp. 389–393). Copenhagen: ICED.

Pugh, S. (1984). 'CAD/CAM: Hindrance or Help to Design?' *Proceedings of the Conference on Conception et Fabrication Assistes par Ordinateur*. Brussels: Université Libre.

Pugh, S., and Smith, D.G. (1976a). 'CAD in the Context of Engineering Design: The Designer's Viewpoint.' *Proceedings of CAD '76* (pp. 193–198). London: CAD.

Pugh, S., and Smith, D.G. (1976b). 'The Dangers of Design Methodology.' Paper presented at the First European Design Research Conference on Changing Design, Portsmouth.

Rawson, K.J. (1976). 'The Art of Ship Designing.' Paper presented at Europort '76, International Maritime Conference, Amsterdam.

'Shipbuilder's Graphics Goes to the Seeing of Its Ships.' (1983). *Computer World (USA)* 17(30) (July) p. SR/41–2.

10

Design Activity Models: Worldwide Emergence and Convergence

ABSTRACT

This chapter was prompted by the proceedings of the International Design Participation Conference. In particular, there seemed to be evidence of a convergence towards
a design activity model which would be an acceptable basis for most, if not all, design
professions. This topic is pursued here by cross-referring from some of the conference
papers to deductions made from the marine and general engineering fields.

The programme for the International Design Participation Conference whetted my
appetite, since it appeared that many of the problems in architectural design were similar in nature to those experienced in other design areas. A subsequent reading of the
proceedings confirmed this earlier view in that there were parallels to be drawn with
these other fields where design activity is fundamental to success and progress.

From *Design Studies* 7(3) (July 1983), 167–173.

PARALLELS CAN LEAD TO PROGRESS: A DISCUSSION

The introductory paper by Beheshti (1985) is particularly revealing. In speaking of the design coalition team he says, 'It is simply defined as all those who are involved in, or affected by, the process of designing the built environment.' Now, to me, this means not only buildings and the like, since all the artefacts designed and built by humans are, in effect, the built environment. Yet how are we to communicate across boundaries without common ground, without a common understanding?

Habraken (1985), discussing participation, says, 'Used as a label for a common attitude I can, for instance, applaud the idea of a participation conference like the one we are engaged in now.' So are we to take it that the search is on for a common attitude? Further clues: 'With the introduction of the computer, we find an increased interest among researchers to find out what designing is about. If we do not know what we are doing, how can we make a computer help us, or take over some of our tasks? At MIT we find an increasing interest amongst architectural students to connect their design studies to other disciplines.'

One finds that design students, independently of discipline, quickly realise that design, per se, is interdisciplinary. Also, the realisation of these interconnections is somewhat stifled in the academic environment because of the way it is almost universally structured around specialisms, which in many instances have now been diffused with departments carrying out what might be called 'partial' design. This has led to further conflict with academia and added to the confusion of both students and staff alike. This aspect is, however, a separate but nonetheless important issue of this paper.

I also fully agree with Habraken when, in talking of coming across particular pointers relevant to the area under discussion (serendipity?), he says, 'Signals like this have appeared for years to those who would listen.' I consider my foray into this topic at this point in time, to be just another one of those signals. Interestingly, nowhere in Habraken's paper does a diagram or graphic model appear to aid communication; yet in architecture, as in engineering, the primary form of communication is by drawing, augmented and supplemented by the written word. Building anything to a word description can be done, but it is a laborious and inefficient process!

Swinkels (1985) is very much aware of the need for a design model: 'At last, I discuss my own ideas about a possible way of working that I shall refer to as "the integrated design model."' He then goes on to describe such a model: 'the conference should have reached consensus about the design process proper, e.g., as laid down in this model.' He then provides a drawing of the integrated design model, as shown in Figure 10.1. The same might be said of the marine industry, where verbal models have been latterly punc-

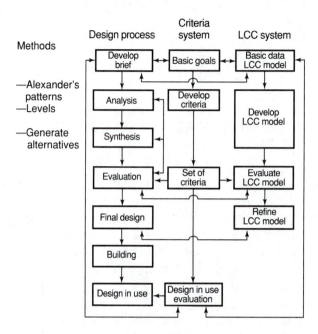

FIGURE 10.1. Design Model After Swinkels (1985)

tuated by graphic models depicting the design process (Pugh, 1985). It is my unequivocal opinion that the use of graphic models of the design activity greatly accelerates the understanding of that activity, although, as soon as one produces such a model, the critic puts forward an alternative which aids communication and understanding and is not just different for the sake of it. As one who has been designing such models for many years, I am very conscious of this fact. So one recognises the structure of Swinkels' model. One feels familiar with it and likens it to other models found outside architecture.

Bleker (1985), in discussing the project management concept, provides yet another model—Figure 10.2, which, again, one can recognise and which achieves its objective of aiding communication. The conclusions drawn by Yoell (1985) are very interesting, revealing, and worthy of full repetition. He says, 'In 1974 some of the best association football was played by the Dutch national team, producing and executing the concept of "total football." The team was made up of skilled specialists in the normal team positions, yet they were able to interchange with one another as they had an awareness of the disciplines attached to one another's roles. This was a beautiful design concept arising out of coalition and, to me, this is the model analogy for the successful coalition of design in education and practice. If students and staff of the various disciplines can understand one another in the same way, better design must result.'

This is very true and something which becomes a reality with a suitable design model. We have pursued a path of 'total design' at the Engineering Design Centre at Loughborough for the past eighteen years, but to achieve real progress, we have to practice, and, in order to achieve efficiency of both teaching and practice, we have had to design models which manifest an efficient design activity.

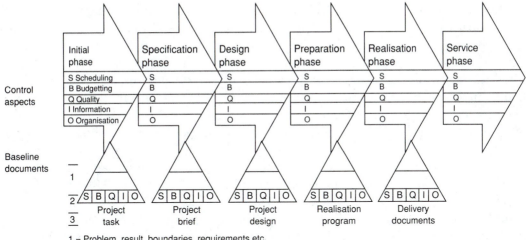

1 = Problem, result, boundaries, requirements etc.
2 = Activities
3 = Standards, monitoring, control systems

FIGURE 10.2.   Design Model After Bleker (1985)

My own criteria of any model of the design activity are as follows:

- All must be able to relate to it,
- All must be able to understand it,
- All must be able to practise more effectively and efficiently as a result of using it,
- It must be comprehensive, and
- It should preferably have universal application, owing allegiance to neither traditional discipline, industry, nor product.

Perhaps the process rules outlined by Frieling (1985) and the verbal descriptions of Hinrichs (1985) and Pirnie (1985) would benefit from, and become much more useful and pertinent, were they to adopt a sound activity model. Pirnie appears aware of this problem: 'There is a need to make the general public aware of the ways in which architects can help them.'

Ertürk and Ertürk (1985) in their excellent paper, inadvertently say it all when they talk of 'visual models which are traditionally common tools for reaching decisions and communication.' Yet nowhere do they use their no doubt considerable visual skills to attempt to communicate their view of the design activity. Van Loon (1985), in describing the design process at Delft between 1982 and 1983, has really put into words what is, in my view, more adequately described in parallel with a visual model. He talks of 'boundary conditions and different points of view.' Yet, as will be discussed later, boundary conditions as such have formed the basis for design activity models since the early 1970s. Pugh (1974), in giving a description of the design activity as being analogous to a circus act, is very much in agreement with the views of Maver (1985): 'It (design) is less like solving a logical puzzle and more like riding a bicycle, blindfold, while juggling!'

Again, the almost persistent lack of sound design activity model manifests itself in yet another description of the design process given by Botma (1985), which in 1985 must be considered naive. He says, 'The design process can be divided into three stages: analysis, synthesis and appraisal.' It may be satisfactory to accept this view in architectural design, but in product design all I would say is 'Try adopting this view and competing on the open market.' Such a view, however, supports current computer systems design and the application of computers to the design activity. In fact, one begets the other from logical necessity, but it can be extremely dangerous and misleading (1984).

By now, readers will no doubt be aware of my view that lack of a structured approach to the design activity is as commonplace in architectural design circles as elsewhere, although interestingly, in the light of my previous statement, Hibino (1985), in listing planning and design approach (PDA) strategies and processes step by step, comes very close to understanding the problem. He names eight steps with which I entirely agree, as they closely match my own major design core stages. He highlights the 'facilitator,' by which I take him to mean the designer. Boekholt (1985) suggests that 'designers are different, so design processes are different. Still, observing all these individual different processes, it will be clear that all these processes have something in common. This we will call the "structure" of the design process. Knowledge of this structure gives us a standard for the judgement and management of all the different design processes.' There then follows a simple model of the design process in words:

- Phase 1: formulation of the design problem,
- Phase 2: generation of (intermediate) solutions, and
- Phase 3: evaluation of (intermediate) solutions,

together with a correspondingly simple model of the design process (Figure 10.3). Again we have flirting with the visual model. Beheshti (1985) proposes a model which separates out building construction and systems development without saying why the activity has been split into two. In fact, his paper is concerned with a consideration of different models of design process, all of which are recognisable to some degree. All are striving to aid communication and, in their own way, succeed to some extent.

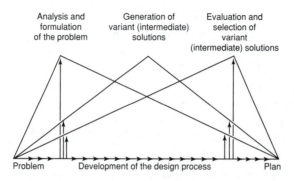

FIGURE 10.3. Design Model After Boekholt
(1985)

I make no apologies for having laboured the conference papers for so long, representing as they do a very fruitful and wide spectrum of architectural design opinion whilst simultaneously crossing national boundaries. Collectively, they enable me to put forward a cogent rationale for a design activity model which may be of assistance to the architectural design profession. If it is not, then discard it. Beheshti nicely makes and reinforces a major point of my argument in that his design model researches do not appear to have taken him outside the architectural sphere of influence. Should he have ventured into other disciplines, he may have found equally fertile pastures. The common features of design activity models have been found to be as follows:

- The models show an increasing tendency to use the same words, although with somewhat different meanings and arrangements;
- Verbal descriptive models are used which can be understood only with difficulty by those with experience;
- The graphics used to depict models generally tend to confuse the nonfamiliar user;
- Very few models are self-descriptive or comprehensive; variants and gaps are widespread;
- Very few give the user either understanding or confidence;
- Many assume that design is a linear, somewhat iterative process; this lends support to the systematic, integrative properties of the computer;
- Many researchers' models vary widely in logic, rationale, and graphics;
- They vary from the simple and understandable to the complex and incomprehensible;
- They tend to be mechanically oriented and do not fit or appeal to a catholic audience or products—that is, they do not have universal application, from teatowels and towers, to toasters and televisions; and
- They all strive to explain the design activity and in so doing, will aid the efficient application of computers.

Nevertheless, in spite of these criticisms, they are all tending towards a single model in verbal terms—that is, the same semantics. It is in the graphic depiction of such models that the pendulum keeps swinging—or at least it did until recently, when some damping has become evident in engineering.

DESIGN ACTIVITY MODEL: THE RATIONALE FOR INTEGRATION

Models, as with everything, have to be designed. Hence, a design activity model has to be designed. In doing so, the opportunity has been taken, at least in theory, to remove the artificial barriers between the sections of any business producing artefacts. Thus everyone in a business must, and does, contribute towards design: they make an input and, that input should be visible, rational, and not contrived. The design activity model shown in Figure 10.4 has been designed. Interestingly and excitingly, however, one tries

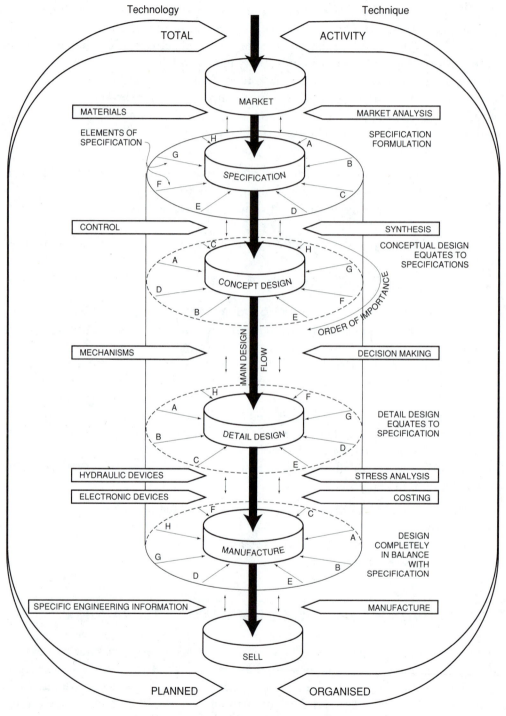

FIGURE 10.4. Pugh Design Model Design Core
Bounded by Product Design Specification

to destroy it, or conversely, whatever the situation to which it is related, it always seems to fit. Its structure is obvious, and, no doubt by pure chance, it relates to the building industry not only in content, which many would recognise, but also in structure.

Rawson (1976) in discussing the art of ship designing, says, 'This leads to the third feature of design, important enough to distinguish from the objective although it is often contained within it. It must be bounded. A painting must be contained within a physical space, an historical survey within certain dates or ethnic influences, an engineering design within cost or behaviour constraints. Such boundaries introduce into the design process the ideas of discipline. Wide as the imagination may wander, it must be disciplined to remain within a boundary that has been judged most appealing to society or those who are sponsoring the design activity.' He then goes on to talk of iteration and so on. Perhaps, he was unknowingly describing the above design activity model. Erichsen (1982) also refers repeatedly to design boundaries, as does Van Loon (1985).

The outcome of this worldwide activity, irrespective of origin, seems destined to converge on variants of three-dimensional design activity models, and they seem to be becoming commonplace. Our experience in using such a model for the past twelve years is that it greatly enhances understanding of just what total comprehensive design entails and that the achievement of this understanding leads to better teaching and practice and, more important, to better designs. Ehrlenspiel (1984) also seems to take this view: 'the three-dimensional model of the design process we are describing here has the following advantages over the usual two-dimensional plans . . . and: three-dimensional models are most easily remembered by engineers who work with three-dimensional concepts.'

But architects also work entirely with three-dimensional concepts! So it appears that perhaps stability in design activity models is imminent. This, in my opinion, will mean that a universally accepted view of just what design activity entails in the modern world—one, in particular, that embraces factors of human significance, including the technical—will lead to better designs, better understanding of design, and better communication between design and disciplines. I will now attempt to look further into the design of the design activity model.

DESIGN ACTIVITY: THE MODEL ISSUE

In evolving the model of the design activity shown in Figure 10.4, it has become increasingly apparent that since this work commenced, our rate of understanding of design has accelerated almost exponentially (Pugh, 1985). In fact, as the model became simpler and more visible, understanding increased at an alarming rate. In my opinion, the criticality of models is no longer in doubt. In fact, it is suggested that many of the mismatches between humans and machines, and in particular, CAD systems, arise because of the lack of an appropriate design model. Cooke (1984), in trying to resolve and reconcile electronic system design, resorts to a model in an attempt to unscramble the situation. This is shown in Figure 10.5.

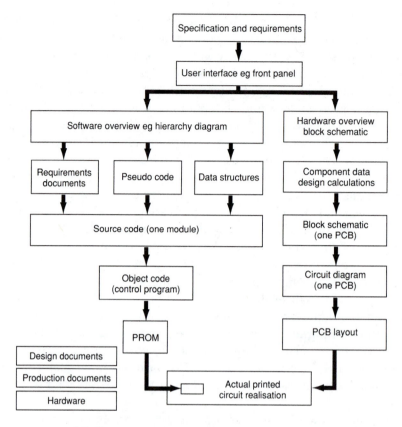

FIGURE 10.5. Design Model After Cooke (1984)

It is my experience that, without a structured approach to design, there is no way that the user and need situation will ever be satisfied. Such a structure, whilst imparting a discipline of systematic working, should nevertheless allow for variations to that systematic working, whilst retaining discipline and also imparting comprehensiveness. In other words, the system flexibility within a framework must not impose the rigidity and linearity suggested by many models of the activity and, in particular, those now emerging from the computer industry. Model evolution is also inextricably linked to design practice, and the additional complications of product complexity and status must be taken on board in such evolution, so as not to lose sight of what little understanding we have at the moment (Pugh, 1983, 1984). Additionally, and most important, it is recognised that model evolution has centred around the product, whilst dealing only peripherally with the total business activity. The wider issues, however, are now brought into the equation through the integration of the product design activity model, into the overall business structure, the combination of which I shall refer to as the business design activity model.

THE BUSINESS DESIGN ACTIVITY MODEL

The business design activity model has as its central theme the product design activity model, which is logical, since without a product (and, in the broadest definition, the product might be a building or a service), we have no business and therefore nothing to manage. Products are thus central to business, and models should emphasise this fact: it should be obvious. The product design boundary remains, and the outer perimeter now represents the business design boundary. It is in the outer ring that we can incorporate all other parts and specialisms of a business, to depict their relationship and interaction with the product design activity model.

The general case model is shown in Figure 10.6. If it is accepted that the product design specification forms the control for the product design, then the outer boundary

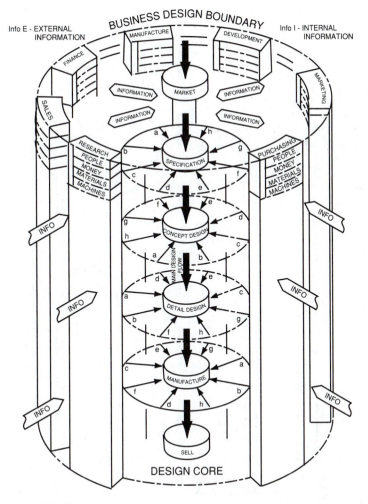

FIGURE 10.6. Business Design Activity Model: Pugh

may be said to represent the business design specification—the one encompassing the other. The outer ring constituents, made up for a typical business, are shown containing the activities and resources of specialist people and money, which, in terms of level and timescale, need to be provided to cause the business to function efficiently. It is recognised that architecture/building/contractor functions rarely come together under one roof, with the possible exception of the large contracting companies. There is, however, a distinct similarity with the real requirements of the different industries, which may be seen through the model.

In order to further encourage and foster understanding, the reader is encouraged to consider Figure 10.7, which has been left almost entirely blank in word descriptive terms. Might not Figure 10.8 apply to the architectural/building/civil engineering-type industry?

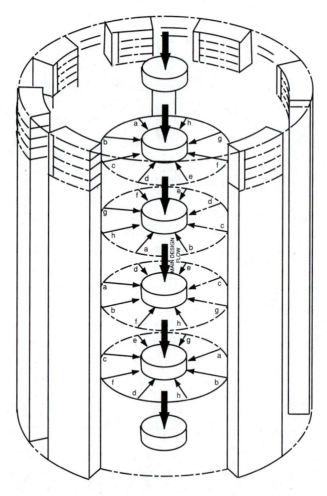

FIGURE 10.7.   Business Design Activity Model: Blank

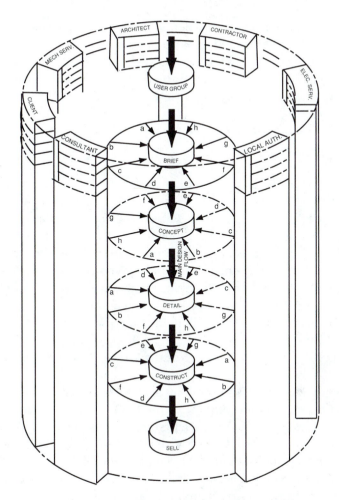

FIGURE 10.8.   Business Design Activity Model:
Architecture of Shipbuilding

Perhaps if the elements of the product design specification which form the control in engineering product design (brief, in architectural terms) were consistently adhered to within a structured framework of design (not necessarily systematic, per se), then many of the customs and practices which came to light in the conference to discuss architectural competition (Jenkin, 1984) would have been avoided. The Right Honourable Patrick Jenkin, M.P., in his opening address, said, 'If the design briefs are inadequate and unsuitable, the entries will not fulfil the client's expectations: a lot of time and effort—the client's and the architect's—will have been wasted for no purpose. The more the RIBA can provide the framework for satisfactory relationships between client and architect, the greater will be the support for competitions.' Peter Murray, discussing the National Gallery extension in London said, 'The brief for the gallery itself was clearly much too loose in relationship to the demands that were subsequently made by the National Gallery representatives on the panel of advisors.'

I could go on, but instead, I will leave you with these thoughts: the model shown, in practice, allows creative work within a framework of disciplines which is essential to sound competitive design. It helps us. Maybe it can help you.

REFERENCES

Beheshti, M.R. (1985a). 'The Development of an Architectural Database.' *DPC 85* (vol. 2, pp. 299–319).

Beheshti, M.R. (1985b). 'Introduction.' *Proceedings of the International Design Participation Conference (DPC 85)* (vol. 1, pp. vi–xiv).

Bleker, B. (1985). 'The Prospects of Design Participation, Now and in the Future, When Applying the Project Management Concept.' *DPC 85* (vol. 1, pp. 32–45).

Boekholt, J.T. (1985). 'The Architect as a Skilled Participant.' *DPC 85* (vol. 2, pp. 249–259).

Botma, E.F. (1985). 'CAD in Support of User Participation.' *DPC 85* (vol. 2, pp. 124–142).

Cooke, P. (1984). 'Electronics Design Demands a New Approach.' *Engineering* 224(4) (April), pp. 315–317.

Erhlenspiel, K. (1984). 'Denkmodel des Konstruktions Prozesses.' *Schweizer Maschinemarkt* 51.

Erichsen, S. (1982). 'Design of Transport by Seas.' *Proceedings of the First International Marine Systems Design Conference* (pt. 1, pp. 11–22). London: IMSDC.

Ertürk, Z., and Ertürk, S. (1985). 'Participation in Architectural Design Activity: An Experimental Study in Processes and Visual Models.' *DPC 85* (vol. 1, pp. 394–403).

Frieling, D.H. (1985). 'Designs Originate in People.' *DPC 85* (vol. 1, pp. 154–168).

Habraken, N.J. (1985). 'Who Is Participating?' *DPC 85* (vol. 1, pp. 1–10).

Hibino, S. (1985). 'A Participatory Planning and Design Meeting (PPDM) Through the Planning and Design Approach (PDA).' *DPC 85* (vol. 2, pp. 196–206).

Hinrichs, C.L. (1985). 'Designing the Design Coalition Team: Owner/User Participation in Architecture. *DPC 85* (vol. 1, pp. 236–251).

Jenkin, P. et al. (1984). 'Architectural Competitions.' *Proceedings of the Journal of the Royal Society of Arts,* London (March).

Maver, T.W. (1985). 'CAD: A Mechanism for Participation.' *DPC 85* (vol. 2, pp. 89–105).

Pirnie, B. (1985). 'The Skarne Project: Methods and Techniques Used in Participative Design in a Low Let and Demand Local Authority Housing Estate. *DPC 85* (vol. 1, pp. 337–354).

Pugh, S. (1974). 'Engineering Design: Towards a Common Understanding.' *Proceedings of the Information Systems for Designers* (pp. D4–D6). Southampton: ISD.

Pugh, S. (1977). 'The Engineering Designer: His Task and Information Needs.' *Proceedings of the Information Systems for Designers* (pp. 63–66). Southampton: ISD.

Pugh, S. (1983). 'The Application of CAD in Relation to Dynamic/Static Product Concepts.' *Proceedings of ICED 82* (vol. 2, pp. 564–571). Copenhagen: ICED.

Pugh, S. (1984a). 'CAD/CAM: Hindrance or a Help to Design?' *Proceedings of CFAO (Conception et Fabrication Assistées par Ordinature).* Brussels: Université libre de Bruxelles.

Pugh, S. (1984b). 'Further Development of the Hypothesis of Static/Dynamic Concepts in Product Design.' *Proceedings of ISDS Conference* (pp. 216–221). Tokyo: ISDS.

Pugh, S. (1985a). 'CAD/CAM: Its Effects on Design Understanding and Progress.' *Proceedings of the Conference on CAD/CAM, Robotics and Automation* (pp. 285–289). Tucson, AZ: CCRA.

Pugh, S. (1985b). 'Systematic Design Procedures and Their Application in the Marine Field: An Outsider's View.' *Proceedings of the Second IMSDG Conference* (pp. 1–10). Copenhagen: IMSDG.

Rawson, K.V. (1976). 'The Art of Ship Designing.' *Proceedings of Europort 76, International Maritime Conference.* Amsterdam: IMC.

Swinkels, T. (1985). 'New Demands: The Design Coalition Team.' *DPC 85* (vol. 1, pp. 11–31).

Van Loon, P.P. (1985). 'A Computer-Aided Urban Design Decision-Making Process for Participation: A Case Study.' *DPC 85* (vol. 2, pp. 65–88).

Yoell, B. (1985). 'The Design Coalition in Education: The University of Bath School of Architecture and Building Engineering Joint Course.' *DPC 85* (vol. 1, pp. 123–137).

CHAPTER

Integration by Design
Is Achievable

ABSTRACT

This chapter discusses the confusion which arises in design education, particularly between design teaching and the teaching of drawing. This necessitates distinguishing between design (engineering) and drawing—the manifestation of the design. The topic of total design is introduced in which total design is made up of many partial designs—the basis of much engineering teaching (of necessity). Only by these means is integration truly achievable.

From *Engineering Design Education Supplement* (Autumn 1986), 30–31.

For several decades there has been an increasingly vociferous clamour for design in industry to be improved—for design, especially British design, to be enhanced and for the design content of engineering and other courses to be increased. Yet to attempt these shifts in attitude and practice without getting down to the question of just what design is has always seemed to me to be putting the cart before the horse. In engineering, for example, it would be accepted without a second thought that we cannot improve the standard of, say, bicycle performance without knowing a lot about bicycles.

Where then does this leave design in teaching and practice? My understanding of the nature and extent of design has evolved through a combination of experience in both teaching and practice. This understanding has enabled students to practice efficiently in the design environment. It has also indicated many pointers towards effective methods for use in industry: the 'bicycle' has thus become tangible. Consequently, understanding and development of the design process are taking place at an increasing rate.

The nucleus of my work with Douglas Smith has been adopted by Sharing Experience in Engineering Design (SEED). The adoption of our work is now on a national scale, quite simply because design practitioners relate to it.

This chapter is therefore not one of reminiscence but an attempt to accelerate the necessary procedures even faster and further.

Let me state what to many will be truisms:

- There is very little wrong with the content of traditional engineering courses. It is the way in which they are taught that is the problem.
- Ad hoc design and projects are not substitutes for traditional subjects.
- The major problem is to harness the traditional subject material to produce an effective design whole.
- Avoidance of mechanism and means for making design visible to students has been a major inhibitor to progress on achieving integration. Exhortation has little effect without understanding.

A comparison of the educational environment for design with the industrial environment for design reveals some quite startling facts. These must be understood if the current confusion about design thinking is to be avoided.

Education and training in a design environment is, for most people, the precursor to practice in industry. It is also true that, for the majority of students, exposure to actual design practice puts them in a better position to recognise and influence good design later on in their careers.

Let us examine the differences between design in education and design in industry.

First and most important, design in industry should be a total process. Industry is, or should be, concerned with all aspects of design, leading to products which are complete in a total design sense.

Analysis of any product will reveal that it contains contributions from a mix of traditional design areas. There are inputs from both technological and nontechnological areas. The former might include some mechanisms, stress analysis, and electronics; the latter may include ergonomics and aesthetics. The typical total product may therefore be said to have required many inputs in order to produce the final whole. So total design is the metier of industry, and industry should thus be geared towards this objective.

TOTAL DESIGN THINKING

In higher education the situation is somewhat different. The departmental structure of universities, polytechnics, and colleges is such that, in a total design sense, it may be said to be anti-integrative. If one takes the departmental or subject structures of mechanical, electrical, and civil engineering as examples, it may be safely said that they all teach and carry out partial design to varying degrees. The design of a machine element such as a cam or shaft is very much partial design—essential to the success of the whole but nevertheless partial. Thus, systems of partial design teaching have evolved with little cross-talk or diffusion across departmental or subject boundaries. So ask any engineers in a university if they are concerned with design and they will say yes if they teach drawing and no if they specialise in the analysis of thermodynamic processes. In my view the former answer is right, and the latter is wrong, since everyone contributes, or should contribute, to design in a partial sense. It is these many contributions that go to make up the whole.

If one breaks down the total design product model in Figure 11.1 into its constituent parts, one achieves an approximation to the partial design characteristics of higher education as depicted in Figure 11.2. As stated above, there is nothing wrong with the essential elements of the partial specialisms. Indeed, they are the very fabric of the engineering profession. It is the context or lack of context in which they are taught that is the main problem.

Students have to be given the opportunity of understanding total design at the earliest opportunity. Teachers should recognise that students cannot begin to practice effectively over the total spectrum until well into the final year or postgraduate studies.

A structured approach to total design should be given to students. This will enable them to relate their traditionally taught material to total design and minimise confusion on the meaning of design. The traditionally taught material—exercises and projects—should be presented as partial design and should be placed in a total design context. It must be disciplined and structured; it should not be loosely piled together into a topic called design merely to satisfy the almost unanimous appeals from engineering institutions. The establishment of design as the integrating theme of all engineering courses has my wholehearted support, but it is rather easier to say than to put into practice.

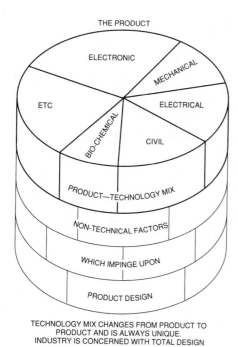

FIGURE 11.1. Total Design Product Model

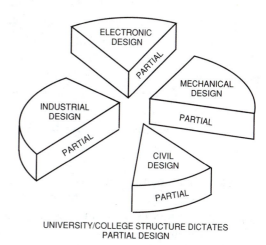

UNIVERSITY/COLLEGE STRUCTURE DICTATES
PARTIAL DESIGN

FIGURE 11.2. Partial Design Characteristics
of Higher Education

The structure and framework for total design should be of a form which automatically leads not only to integration but to a natural feeling of integration within the participant, irrespective of discipline. Each contributor should be able to see how partial contributions fit into the total whole.

Total design should be taught and practised in a progressive manner with enhanced information, knowledge, and techniques leading progressively to increased rigour in a total design sense. This should lead to the breaking down of the twin barriers of inefficiency and misunderstanding. This is why the SEED initiative is so important; it provides structure and visibility to students in a total design context.

EFFICIENCY

There have been moves to allow diffusion to occur across departmental boundaries in some educational institutions. At Strathclyde University, for instance, we now have a Faculty of Engineering which should permit much more interaction across subject boundaries. It is my view that a structured approach to total design can greatly accelerate the diffusion of ideas and emphasise the commonalities between different subject areas. It thus becomes possible to increase the efficiency of the total design activity in teaching.

In order to make any noticeable impart on the national situation, it will be necessary for all institutions to adopt approaches of this kind.

An additional contribution to improved efficiency will come from the design of curriculum material that relates recognisably to the model of total design. Teaching material generated on an ad hoc basis has little tangible effect on efficiency. The teaching material of the future will improve teaching efficiency, create enthusiasm for design and engineering, and simultaneously will contribute to the demolition of the boundaries between disciplines.

Thus we come to the interface between curriculum development in total design and the traditional engineering curriculum. It is clear that the interface itself needs to be designed in order that it relates to total design through the summation of the partial designs, and the modus operandi is available to achieve this. It is currently being developed, but as a large topic area it probably warrants a paper in its own right.

Design Techniques
and Methods

COMMENTS ON PART III

After reading Parts I and II, the reader can grasp the big picture—the scope of total design and its role in integrating arts and sciences, university, and industry and the many diverse engineering disciplines. Now in Part III, the reader is introduced to the specific generic methods within the broad scope of total design that enable actual practice.

This part contains the paper for which Stuart Pugh is most famous in Chapter 14, Concept Selection: A Method That Works. First presented in Rome in 1981, it had already been forged in many trials under Stuart's leadership. A video tape made at MIT in 1990 shows Stuart facilitating a team from the Digital Equipment Corporation in applying the process that had already long been known as the Pugh concept selection process. It is a marvel to behold. Stuart developed this process during the 1970s and beyond by trial and error, guided by the principles that design should be creative and teams should be empowered—and capable. He eliminated mind-numbing numerology so that the process would lead to creative work, and the results proved the great value of this improvement. (The reader should beware of academics who wish to reintroduce the mind-numbing numerology—apparently guided by the dubious principle that more numbers are better.) I have called Eureka concepts one of the ten cash drains of product development, and they are a feature of partial design. Using the Pugh concept selection process produces strong concepts in which the team has confidence—a major advantage of total design over partial design.

Stuart was a vigorous proponent of design methodology, but he was also aware that much pseudomethodology was published—primarily by design academics—that had not been tried in battle. When Stuart tried many of them on design projects, they did not work. Methodology must be grounded in applications before it can be considered to be ready to be pushed onto practitioners.

This part also describes methods that Stuart developed to prepare for concept selection, such as parametric analysis, which is presented in Chapter 13, A New Design: The Ability to Compete, first published in 1982. In 1984 I learned about quality function deployment (QFD), and a few months later I met Stuart Pugh. (Both events took place in Tokyo, giving me fond memories of the city.) In our discussions during the next few years we compared QFD with the methodology that Stuart had developed. In 1990 we integrated them in a paper presented in Chapter 16, Enhanced Quality Function Deployment, which we first presented at the international conference in Honolulu in 1991. Thus Chapter 16 integrates Chapters 13 and 14 along with QFD. As the culmination of our discussions during the previous six years, Stuart and I worked together with great intensity for one week to achieve this integration. It looks very simple now, but it was not at all obvious to us before we worked it out.

Chapter 15, State of the Art on Optimisation in Great Britain, is closely related to concept selection. It makes the basic point that optimization starts with concept selection. By the time the design is constrained to the point where quantitative optimization methods can be applied, much design freedom has already been given up. The basic message is to not quantitatively optimize a poor concept but to carry the basic optimization thinking back into concept selection to help select the best concept.

As Chapter 16 integrates Chapters 13 and 14, it can be read first. Then Chapters 13 and 14 can be read to gain a firmer grasp of methods that are only briefly introduced in Chapter 16. Chapter 15 is an extension of the role and context of Chapter 14. Chapter 12, The Dangers of Design Methodology, is a warning about pseudomethodology and offers a process for sorting out the chaff from the wheat.

In this part look for concept selection and its role relative to optimization, EQFD—the complete concept development process, parametric analysis and quantitative benchmarking, and the dangers of ungrounded methodologies and how to recognize them. These competencies make the design activity model operational. This part enables the reader to do the front end of product development well—far better than in partial design.

—Don Clausing

INTRODUCTION TO PART III

Part III commences with a cautionary note entitled Chapter 12, 'Dangers of Design Methodology'—a paper given in 1976 to a most receptive audience who were convinced that most, if not all, design methods worked. I, with my colleagues, had been trying to use most of the methods then available (independent of product or technology) without any degree of success in producing better designs than traditional methods. This statement with a few exceptions is valid today.

What follows in Chapter 13, A New Design: The Ability to Compete, Chapter 14, Concept Selection: A Method That Works, and Chapter 16, Enhanced Quality Function Deployment, are the descriptions and exemplars of methods designed by me and now in regular use worldwide. These have been proven to work by me and many others and now form the basis of enhanced quality function deployment, which augments standard quality function deployment (QFD) with contextual, parametric analysis, structured specification, concept generation, and selection, all at recognisable multiple levels (total system, subsystem, or piece parts).

The necessity for completeness of design activity cannot be overemphasised, and the power of concept selection and generation as both analytic and generative tools is discussed.

A return to the more conventional is made in Chapter 15, State of the Art on Optimisation in Great Britain, the point being made that optimisation must not be confused with evaluation.

The methods discussed are all in current usage, but in this part reference has been made only to those methods attributed to the author. Part V throws more light on methods and attempts to reconcile the above with Japanese methods.

CHAPTER

12

The Dangers of
Design Methodology

ABSTRACT

This chapter is concerned with the application of nonspecific-technology-oriented design methods, within a structure of quantitative and qualitative boundaries. Examples are given where the various methods discussed are highlighted, the pros and cons of the methods are highlighted, and conclusions are drawn as to their effectiveness.

With Douglas G. Smith, paper presented at the First European Design Research Conference on Changing Design, Portsmouth Polytechnic, April 1976.

INTRODUCTION

The main objective of this chapter is an attempt to place in perspective design methods in the sphere of engineering design. It is, we feel, significant that whilst there are a great many design methods available, there are doubts as to the effectiveness of such methods in given design situations. This is indeed a strange state of affairs: mathematical methods are developed and proved by reference to examples using the methods; analytical methods in chemistry are developed and again are proven by usage and examples to demonstrate the effectiveness of the methods. Why then in engineering design is there a dearth of information on the effectiveness of design methods? Is it because designers do not publish their works, or is it because the methods themselves do not work? The papers and books which review the methods are legion: they tell you how to use them but never give practical examples of usage.

We have had the opportunity over a number of years to try out almost all the currently known methods in real-life design situations, both in postgraduate project work within the Engineering Design Centre of Loughborough University of Technology (now, 1992, extinct) and also in professional consultancy work. To say that we have had mixed results would be an understatement of the facts. However, having tried the methods, we consider ourselves to be qualified to comment from this basis of fact and feel that such comment is called for, if only as a guide and pointer to designers about to embark upon their careers and also to practising designers.

We review the methods concerned in our postgraduate teaching programme and comment upon our own experience of them. It is interesting to note that in broad terms qualitative methods seem to be of limited use and effect whilst quantitative methods may act as a positive restriction upon the emergence of the best design solutions. These differences we illustrate by reference to examples taken from a number of real-life design activities, some successful and others less so. We conclude by drawing upon the examples and our experience in order to provide a set of ground rules for the application of creative design methods. In an attempt to avoid confusion and misunderstanding, we set out, in simple terms, our understanding of interactive engineering design.

INTERACTIVE ENGINEERING DESIGN

To attempt to define an extremely complex topic in a complex manner is, in our opinion, likely to lead to misinterpretation and ultimately lack of understanding. Just as with a typical design situation one must start from simple premises in order to gain understanding and build a base from which to expand one's thinking, which sometimes does lead to complex solutions. To do otherwise is to court disaster: to proceed from the simple to the complex with effect is difficult; to attempt to go from the complex to the simple is not only difficult, but in certain situations it becomes impossible.

We would define successful engineering design as being the outcome of an activity (not a process) which achieves the best balance between all the factors inherent in any design—factors such as market, consumer, weight, aesthetic appeal, cost, production, facility, tooling, standards and specifications, and patents. It is, in other words, a totality of interactions rather than specific interactions as, say, with interactive computer graphics. If the designer fails to achieve the best balance or compromise between all of the factors, then his or her design may be said to have failed (Pugh, 1974).

In order to have the best possible chance of achieving a successful, balanced design, at the onset of the design—the conceptual stage—the designer must produce a multitude of ideas and possible solutions. Some of these will ultimately form the basis for his chosen design. We thus return to the methods which give rise to these ideas and assess their value and effectiveness.

REVIEW OF THE COURSE

One of the main activities within the EDC is the postgraduate course in engineering design leading, upon successful completion, to the award of a master of technology degree. The course lasts a full year, commencing with the normal academic year in October but continuing until the following September. Inevitably the structure and content of the course reflect our view of engineering design and the way in which it should be taught. One predominant factor in regard to the latter aspect is that design cannot be taught in a series of lectures alone, it must be developed in the individual by practice. However, the manner in which it is practised is also of vital importance. Students take a very different attitude towards a design finishing up as a paper exercise compared with a design which culminates in a piece of hardware. (This is not to say that the paper exercise has not got its place.) Furthermore, to finish a design as a piece of hardware is one thing, to end with hardware which is to form part of a factory production system or which must sell upon the open market against competition is quite another. It is the practice of design with 'real-life' projects which forms the main feature of the course. For purposes of carrying out these projects, which commence in January and carry on until the course ends, the students are divided into syndicates of four or five members, each syndicate being under the tutorial guidance of a member of the EDC staff. In order to give adequate supervision it is essential that staff members have considerable industrial design experience in industry and also maintain their professional expertise which they do by acting as consultants to industrial companies. What medical student would have confidence in a surgeon who did not practice what he was teaching?

The projects (Pugh and Smith, 1974) are usually industrially based and are selected such that they fit the time scales and resources available as closely as possible, and also allow the maximum scope for the students in two ways. Projects are preferred which firstly, start with a market need and finish with hardware and secondly, allow the student scope to practice in as many facets of design as possible from appearance to mathematical analysis. All the design work is carried out by the students, manufacture and testing generally being undertaken by the sponsoring company under the supervision of the students.

The nature of the projects, basically real-life, coupled with the environment of the EDC provides an ideal opportunity for testing out design methods and selected examples are given later in the paper.

We may have given the impression that we denigrate lectures in the teaching of design—this is not the case. Albeit, the majority are conducted in an informal way with considerable interchange between lecturer and students. During the first term, lectures form the main part of the course but thereafter are reduced progressively as the projects proceed. In one series of lectures design methods are introduced and discussed and assignments given so that they can be practised and to some extent evaluated by the students.

SCOPE OF METHODS

The current scope of design methods is well documented, and there is little point in offering a further summary here. The main methods, with one or two exceptions, are introduced to students attending our course.

Design methods may extend from pure unconstrained thinking to structured mathematical analysis. The former we call the qualitative boundary and the latter the quantitative boundary. The methods are shown categorised in Figure 12.1, those which are nonnumerate being at the qualitative boundary and those which include numeracy being at the quantitative boundary. It will be noted that the majority of the methods are of the qualitative type. Certain others include AIDA, decision trees, and value analysis, extending across the spectrum. Lateral thinking is also being developed this way.

For purposes of classifying these methods we have basically used an independent source in a recently published review of design methods (Turner, 1975).

It is necessary to make an observation at this point with regard to the place of evaluation. We consider this to be in the central area between the quantitative and qualitative boundaries, whereas it will be noted that with the exception of value analysis, methods embracing quantification extend to the quantitative boundary. We return to this point later in our discussion on decision matrices after reviewing our experience in using most of the other methods.

SUMMARY OF THE EFFECTIVENESS OF METHODS

To discuss fully the effectiveness of each of the above methods could well form the main theme of a separate conference. In this instance we comment only briefly on our

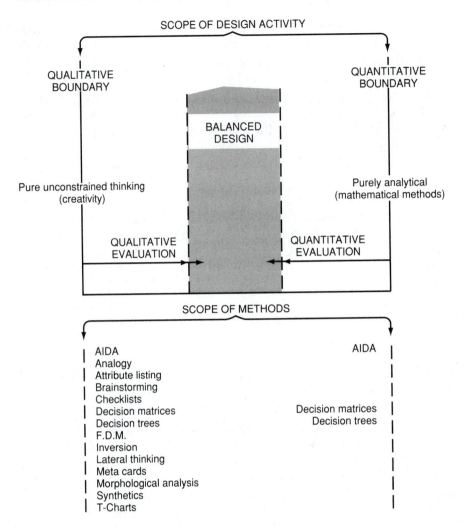

FIGURE 12.1. Scope of Design Methods

experience, which embraces most of the methods. We then discuss at length specific examples in the use of decision matrices since we believe that if any area is full of pitfalls for the unsuspecting, it is this.

Qualitative Methods

The methods we find most useful are the various types of analogy, attribute listing, inversion, and T-charts. This, however, is not surprising since they are the stock-in-trade of many good design engineers for whom creative and evaluative methods are as unknown language. The main advantage of highlighting them as aids to creative thinking is that because one is made more aware of the underlying principles, one is more likely to use them in practice.

We cannot find an example where brainstorming, checklists, lateral thinking, or synectics have led directly or indirectly to a used solution. It is interesting to note that if one uses these techniques both in paper exercises and real-life projects, one is likely to draw different conclusions. We often find that some students claim that these methods enable them to generate more ideas than would otherwise be the case in the paper exercise situation. However, when it comes to the real-life situation, it is a different story. We are of the opinion that these methods may help in comparatively simple problems—such as, 'In how many ways can I use a paper-clip?'—but cease to be of real value in complex design problems subjected to real-life constraints. One positive value we have found in regard to brainstorming is that it can be a mechanism for removing formality in a group and getting people to say what they really think, but there are, we believe, better ways of achieving this.

In no case where we have tried to use morphology as a creative tool can we trace that it has led directly or indirectly to a used solution. On the contrary, we have found it to be a positive hindrance as students who have used the method stop thinking, since they are given the impression that all possible solutions should be contained in the matrix when correctly formulated. However, we have found the technique to have some usefulness as an aid to classification and analysis of existing designs. An extremely valuable principle in the approach is the requirement to isolate the basic functions required, but there is no need to employ morphology to do this.

We have no experience of FDM or META cards and therefore cannot comment upon their effectiveness.

Qualitative and Quantitative Methods

AIDA is really a type of morphology at the qualitative end and a type of decision matrix at the quantitative end. We have not yet found an effective use for the technique, but, again, it embraces some important principles. For example, a potential option in a decision area which is highly interactive with other options in other decision areas is usually not a desirable choice. One keeps one's options as confident as possible. As a general rule this is a fact of good design practice which does not require AIDA.

We have found decision trees impossible to use in a design situation. In fact, the only design decision trees of which we are aware have been drawn after the design's completion. Marples (n.d.) also confirms this view. This is not to say that such exercises are of no value but that they are of little help to a designer setting out to design a new product. Decision trees may be of value for business decisions where the choice of decisions and range of possible options can be more readily defined beforehand. In design the choices and their outcomes are not predictable at the outset and usually reveal themselves only as the project proceeds.

With regard to value analysis we consider that if design and redesign are properly carried out with due regard to all the relevant factors, including costs, the necessity for value analysis diminishes (Rhodes, 1975).

Quantitative Analysis

As mentioned earlier, the only purely quantitative method is the decision matrix, which is discussed at some length in the next section.

Summary

Of all the methods discussed, those which we have found most useful are simple in nature and frequently used, albeit unknowingly, by practising designers—namely, analogy inversion, attribute listing, and T-charts. However, even then the methods are of use only in the hands of those competent to use them. We do not make people mathematicians by giving them formulas in which to put numbers. Why then should we think we can make designers by giving people some methods, even useful ones?

However, the methods often embrace sound principles, and it is the principles which need to be highlighted and taught.

Further, to make our position absolutely clear, we are not against method in design! But each design is different, and whilst there may be a helpful method to forward that design, the converse may also be true. It has been our experience that the most effective methods are those evolved to suit the particular design task in hand.

Considering the list of creative methods, it is significant that the most productive and effective method is not even mentioned. It is a group of people thinking and discussing the design in a relaxed and informal manner with even way out ideas discussed seriously and positively. From such discussions, methods of approach, appropriate to the task in hand evolve. One of the examples given in the next two sections underlines the effectiveness of this method; we have many others.

DESIGN OF A BRUSH-MAKING MACHINE

This first example is concerned with the design of a machine to manufacture small brushes of the cosmetic/mascara applicator type. Carried out as a student project some years ago, different designs of brush had been considered, not in isolation, but in an interactive manner with the possible design of a brush-making machine. The design chosen was a twisted wire type with filaments arranged in a helical pattern obtained naturally from the twisting of the wire.

Having taken the decision on the brush design, the next steps in the design activity were to consider alternative machine concepts based upon different principles, and from these select a concept as the basis for a full design programme which would lead ultimately to the building, testing, and development of a machine.

In all, four different concepts emerged as follows:

Machine Design	Operation	Wire form
M/c 1	continuous, linear	two continuous
M/c 2	continuous, rotary	staple
M/c 3	noncontinuous, two station, rotary	staple
M/c 4	noncontinuous, four station, rotary	staple

Without going into the detail of the different concepts, which we feel will cloud the issue, the fundamental problem was how to select a machine concept for further work. Each design had advantages and disadvantages technically; projected manufacturing costs were different, etc.

The students decided to utilise a decision matrix as an aid to their decision taking, although in practice it is possible that the matrix took the decision out of their hands; this matrix is shown in Table 12.1.

This type of matrix, where the main objectives are defined and listed, being then considered on a numerical basis against alternative designs by means of weighting the objectives (ie order of importance in overall design spectrum) and rating each design's chances of achieving these objectives.

The totals of (weight factor x rating) are interesting in that M/c 4 emerges with the highest value at 112, and that the three highest totals are for machines employing the use of a staple in the production of the brush. We shall return to this point later on.

If a detail comparison is made between M/c's 1 and 4 (the highest and lowest scores) only two criteria have different individual scores, ie end of brush safety and good filament density. Therefore, on the assumption that the judgement made for the other criteria are sound and reasonable, it is interesting to speculate and consider these two criteria in detail. Taking an extreme view, if it were possible and feasible to solve the problem of end of brush safety and in so doing raise the zero to a score of 25 and

TABLE 12.1.   A Decision Matrix for a Brush-Making Machine

Objective	Weight Factor	Rating				Weight Factor x Rating			
		M/c 1	M/c 2	M/c 3	M/c 4	M/c 1	M/c 2	M/c 3	M/c 4
Make 20,000 brushes per 8 hr shift	4	5	5	0	5	20	20	0	20
End of brush safety	5	0	5	5	5	0	25	25	25
Ease of manufacture of machine	3	4	0	3	4	12	0	9	12
Reliability of operations	5	4	2	4	4	20	10	20	20
O/all size of machine	2	3	4	3	3	6	8	6	6
Cost	3	3	2	3	3	9	6	9	9
Good filament density	4	3	5	5	5	12	20	20	20
Total						79	89	89	112

M/c 1: Continuous, linear
M/c 2: Continuous, rotary
M/c 3: Noncontinuous, two station, rotary
M/c 4: Noncontinuous, four station, rotary

also raise the score on filament density from 12 to 25, then M/c 1 achieves the same total score as M/c 4, ie 112.

Examination of the conceptual design schemes of the four machines considered indicated very clearly that the linear machine is apparently the simplest in concept; in that it is continuous and utilises a device known in the brush industry as a 'knot picker.' Devices of this type have already been developed by the industry. Against this background the score of 12 for filament density does therefore seem to have been an arbitrary and possibly incorrect decision.

Protection of the end of the brush after cutting was considered in its own right, it is however significant that combined methods of cutting and fusing the ends were not to our knowledge considered.

The conclusions to be drawn from the above, are that a fuller investigation in the two areas mentioned could possibly have led to M/c 1 having an equal score to M/c 4— would the same decision (ie to proceed with M/c 4 because it had the highest score) then have been made? Would a machine designed to M/c 1 concept have proved better in practice? Perhaps we shall never know the answer to this question as M/c 4 has now been designed and manufactured. A statement from the students' project report is worthy of note and may throw some light on the initial design selection: 'However, the final concept of machine (M/c 4) proved to be more complicated than had been expected.'

There are in our opinion several lessons to be learned from the foregoing:

1. The use of decision matrices of the type described above is fraught with untold dangers since they depend so heavily on the selection of the criteria which in turn hinges upon judgement. To use them in 'anger' even with post-graduate students is therefore problematical.
2. The use of such a method instils a confidence into the user which in fact may be ill-founded, this we think is due to the transposition of judgement decisions into numbers. Engineers generally are very much at home with numbers, and rightly so as they are a vital and necessary part of any design activity.
3. The use of numbers in engineering—analysis, costing etc. are what we would define as factual numbers (eg Young's Modulus for steel 207,000 N/mm^2 cost of steel £300/tonne, pressure in a hydraulic system $20N/mm^2$). All these are factual numbers obtained and utilised in and by calculation.

   The numbers used in decision matrices are entirely different; they are what we would define as 'judgement numbers' and may have no relationship to fact.
4. We have observed that engineers using this technique tend to treat the total scores as being absolute—just as if they were stressing a beam or calculating a bearing load. Having done the 'calculation,' the method has given the answer, they then move on and tend to stop thinking.
5. The use of decision matrices even by experienced designers with a record of good judgement gives results which again are extremely problematical and a later example will highlight this point.

METHOD OF APPLYING HEAT

For this example, we consider a decision matrix as applied to the subsystem for a machine of novel concept—a machine for processing metal. The subsystem is concerned with the means of applying heat to a rotating ingot of metal. A number of possible methods were identified and each thoroughly investigated:

- Resistance heating (by passing current into the metal via brushes),
- Induction heating (by means of induction coils),
- Tungsten inert-gas torch,
- Fuel burner (direct) (heating the metal directly with a flame),
- Fuel burner (indirect) (by heating a conductor which in turn transmits heat),
- Heat to the metal,
- Friction (analogous to friction welding), and
- Plasma torch.

A decision matrix shown in Table 12.2 gives the alternative methods of applying heat on the horizontal axis and the choice criteria on the vertical axis.

On the basis of the highest score, the plasma torch leads with a score of 212; in fact, this was used in the final machine. As numerical gradings are subjective and therefore tolerance sensitive, on the basis of the numbers alone we should reject only the methods with the significantly lower scores—resistance heating at 129 and fuel burners (indirect) at 144. How could we possibly discriminate between the remainder, as we could not accurately assess the tolerance, and the spread of scores is very narrow—from 179 to 212? We remain therefore in a situation of indecision.

In a paper exercise, this situation may be allowable; in practice, it is most definitely not. How then did we decide? Briefly and simply, by returning to balanced subjective judgement and relooking at the relationships of the criteria. Size constraints did not allow brushes of adequate area to carry the required current; therefore, resistance heating was eliminated. Size constraints also ruled out induction heating. Direct fuel burners were not acceptable due to excessive contamination of the processed metal, and indirect fuel burners were not feasible on technical grounds. Friction was unacceptable due to its uncontrollability.

What we have left is only the tungsten inert gas torch and the plasma torch. Table 12.1 shows that they differed on a number of criteria. Why do they differ? Are the differences important? The numbers do not help us; we have to make a judgement decision, which was to proceed with the tungsten inert gas torch on the basis of heat distribution. The tungsten torch has a narrow-track input characteristic, which was what we required.

There is an interesting comment in one student's report concerning the decision matrix: 'An evaluation chart was used to decide which was the most suitable method. The most difficult parts in the use of this technique proved to be the selection of appropriate criteria and the allocation of weighting values to each. However the method greatly assisted the choice and gave confidence to the decision to use an electric arc method of heating.'

TABLE 12.2   A Decision Matrix: Methods of Applying Heat

Description	W	Resistance Heating	V	WV	Induction Heating	V	WV	Tungsten Inert Gas TIG	V	WV	Fuel Burners						Friction	V	WV	Plasma	V	WV
											Direct	V	WV	Indirect	V	WV						
Size	5	Current density problem	3	15	Flux density problem	2	10	No problem	5	25	No problem	5	25	No problem	5	25	No problem	5	25	No problem	5	25
Adaptability for various materials	7	Low power low Temperature	1	7	Least suited to material	3	21	Very good	5	35	Very good	5	35	Limited by material and temperature	2	14	Reasonably good	4	28	Very good	5	35
Control	9	Good but limited due to speed and temperature at contact point	2	18	Very good	5	45	Good within a current range	4	36	Good within a range of gas flow	4	36	Time lag	2	18	No direct control, external cooling required	1	9	Good within a current range	4	36
Contamination	10	Brush wear shielding required to prevent oxidation	3	30	No problem	5	50	Possible tungsten inclusion	4.5	45	Oxidation	1	10	No problem with copper on copper, poor otherwise	2	20	No problem	5	50	Negligible	5	50
Capital cost	2	Moderate	3	6	High	2	4	High	2	4	Low	4	8	Low	4	8	Low	5	10	Very high	1	2
Running cost	2	Moderate	3	6	Moderate	3	6	Moderate to high	2	4	High	1	2	High	1	2	High	1	2	Moderate	3	6
Physical limitations	4	Large but short conductors insulation required	2	8	Large H.F. coils, inflexible shape, problems in machine design	2	8	Possible need for local insulation to stop arc-blow	4	16	Access required for piping	4	16	Access required for piping	4	16	Die cooling necessary	3	12	See T.I.G.	4	15
Safety	5	Moderate	3	15	Good	4	20	Moderate	3	15	Good	4	20	Good	4	20	Very good	5	25	Moderate	3	15
Experience	3	Very little	2	6	Little	3	9	Little	3	9	Well known	5	15	Very little	1	3	Very little	2	6	Little	3	9
Heat distribution	6	Long brush spreads contact	3	18	Long coils	3	18	Narrow track	3	18	Narrow track	3	18	Spread contact	3	18	Heat in wrong place	2	12	Spot heat unless ocillated	3	18
Totals				129			191			207			185			144			179			212

The lessons to be learned from this decision matrix are somewhat different from the previous case:

- We believe that in this instance the methods of heating were thoroughly investigated and also that the correct criteria were selected, notwithstanding this the decision did not give the 'right' answer.
- The numerical grading system causes factual numbers and other facts to be turned into numbers of judgement. For example, we knew the actual capital and running costs of the tungsten inert gas and plasma torches. This factual information is much more useful than subjective numbers. Again, we knew that we could not get the induction heating coil in the volume available; turning such facts into subjective numbers only adds to confusion.
- We are not suggesting that alternatives should not be judged against criteria; they must be. But the skill lies, first, in identifying these criteria, often an extremely difficult task, and second, in properly evaluating the alternative against these criteria, which is equally demanding. Further, critical criteria are peculiar to a design; there is no blanket checklist. For example, contamination of the processed metal outside certain limits was not acceptable, a criteria which had to be satisfied. The direct fuel burner alternative received a high score of 185!
- Decisions are usually based partly on factual numbers, partly on objective facts, and partly on opinion and hence lie between the qualitative and quantitative boundaries shown in Figure 12.1. To try to push decision artificially to the qualitative boundary is to court disaster. It is essential to recognise facts and opinions for what they are.

ENERGY-SAVING DEVICE

For our final example we refer to two figures developed as decision-making aids in the design of an energy-saving device. At the commencement of the project we decided to investigate trends in basic energy resources and also trends in the types of system used for domestic heating purposes.

In order to be able to better understand each situation and to distil the essential information, we developed Figures 12.2 and 12.3, which are peculiar and unique to the situation in hand.

From these figures we were able to arrive at an inescapable conclusion. Armed with all the facts and the answer, we tried to produce a quantitative decision matrix of the type previously described. In effect, we were retrofitting. Suffice to say that we could not even obtain a useful answer from this exercise, let alone the answer already obtained.

This is an example of a form of decision matrix developed specifically for the particular project which we found to be extremely valuable.

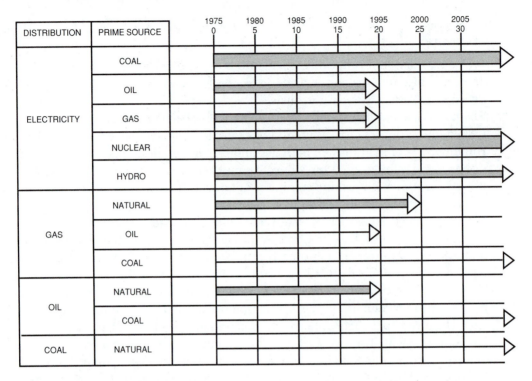

FIGURE 12.2. Prediction of Future Available Energy Sources

ENERGY SOURCE	WATER SYSTEM PRESENT (0 YEARS 15)	WATER SYSTEM FUTURE (15 - 30)	AIR SYSTEM PRESENT (0 - 15)	AIR SYSTEM FUTURE (15 - 30)	DIRECT SYSTEM PRESENT (0 - 15)	DIRECT SYSTEM FUTURE (15 - 30)
GAS	INCREASING TO INCREASING	INCREASING TO LEVELLING	LEVEL TO FALLING	FALLING TO FALLING	INCREASING TO INCREASING	INCREASING TO LEVELLING
ELECTRICITY	LEVEL TO INCREASING	INCREASING TO INCREASING	INCREASING TO INCREASING	INCREASING TO INCREASING	LEVEL TO LEVEL	LEVEL TO INCREASING
OIL	LEVEL TO LEVELLING	LEVEL TO LEVELLING	FALLING TO FALLING	FALLING TO CEASING	LEVEL TO LEVELLING	LEVEL TO FALLING
COAL	FALLING TO LEVELLING	LEVEL TO LEVELLING	FALLING TO FALLING	FALLING TO CEASING	LEVEL TO FALLING	FALLING TO LEVELLING
SOLAR	LEVEL TO LEVELLING	LEVEL TO INCREASING				

KEY: PRESENT CONSUMPTION TREND AT START OF PERIOD TO ESTIMATED CONSUMPTION TREND AT END OF PERIOD

FIGURE 12.3. Energy Consumption Trends for Various Distribution Systems

CONCLUSION

In discussing the various design methods and specific examples of usage, we have inevitably drawn certain conclusions as to the methods themselves but also conclusions specifically related to the examples given.

For the sake of clarity we summarise our main conclusions as follows:

- Design methods may be classified as qualitative, quantitative, or a combination of both. Generally speaking, creative methods are qualitative and analytical methods are quantitative.
- The qualitative methods which we find most useful are essentially simple in nature—namely, analogy, inversion, attribute listing, and T-charts and decision matrices of the nonnumerical type. Even so, they tend to be most useful in the hands of those competent to use them.
- Creative methods of the quantitative type impart a sense of certainty and confidence which often does not exit and can confuse facts with judgement.
- Creative methods which tend towards quantification we have found in many instances to be an inhibiter to further thought.
- Many of the methods embrace sound principles, and it is these which designers should be made aware of and taught to use.
- Methods in design can be useful but in our experience of usage often require to be developed for the problem in hand in order to obtain the best results (view of the situation in 1976).

We trust that from the above that we have given our experience of methodology and have done so in an unbiased manner. If, in so doing, we have given an insight into the use of methods which will prove useful both to engineers about to embark upon a design career and also, hopefully, to practising designers, then our major objective will have been achieved.

REFERENCES

Marples, D.L. (n.d.). 'The Decisions of Engineering Design.' Discussion paper published by the Institution of Engineering Designers.

Pugh, S. (1974). 'Engineering Design: Towards a Common Understanding.' *Proceedings of the Second International Symposium on Information Systems for Designers* (pp. D4–D6). Southampton: University of Southampton.

Pugh, S., and Smith, D.G. (1974). 'Dumper Truck Design Highlights Industrial/Academic Co-operation.' *Design Engineering* (August), 27–29.

Rhodes, C. (1975). 'Where Design Is Independent and Self-contained.' *Engineering* 215(1) (January), 38–42.

Turner, B.T. (1975). 'Creative Approaches to Engineering Design.' *Chartered Mechanical Engineer* 22(10) (November), 85–89.

13

A New Design: The Ability to Compete

ABSTRACT

This chapter discusses the necessity for a thorough review and analysis of all directly and indirectly competing products as a prelude to embarking on the design of new products in that field.

In seeking to design, manufacture, and market such products, it is, in my opinion, essential to gain an insight into likely competing products well before a commitment is made to a particular design or production route. The chapter examines and discusses 'the market research plus' required in order that designers of the new product can evaluate their designs against those of the competition using criteria of their own choosing but being influenced by such competition—that is, parametric analysis.

The techniques of such parametric analysis are discussed in detail here, and examples are given of the results of such analyses, albeit in the main based on exercises designed to give practice and confidence in the technique and carried out without the intention of producing new competitive designs in the particular product fields. Comment is made on the outcome of these exercises, and reference is made to a sample taken from a competitive design.

From Design Council, *Proceedings of the Design Policy Conference, 1982, Vol. 4, Evaluation* (London: Design Council, 1984).

DESIGN AND MARKET INTERFACE

The design and market interface was and always will be important but never more so than today, with increasing world competition. To quote from a paper I gave in 1974 (Pugh, 1974), 'With the rapidly increasing complexity of markets and hence market situations, it is becoming more and more difficult to design the right product for the right market. Concorde may yet turn out to be the epitome of this statement. It is because of this evolving situation that companies are placing much more emphasis today on the marketing side of their operations. What they do and how they do it are vitally important to the design area.'

At that time, I had become interested in the analysis of historical cost data related to specific production situations, since the analysis of such data always seemed to yield a mine of information useful to the designer. In parallel with research activity in this field and practice in the establishment and use of cost patterning, ongoing cooperation with and the direction of market researchers concerned with laying down foundations for new products led rapidly to the conclusion that there was a void of information between the final output of the market researcher, usually a comprehensive report, and the basic essential needs of the designer concerned with the new design. One must assume that the intention would not always be just a new design, but one given every chance of being better than the competition in all aspects—performance, cost reliability, and so on. In fact, extension of the analysis of historical cost data into the wider analysis of parameters other than cost, on a 'let's see what happens' basis, yielded startling results. Companies, by and large, in carrying out sophisticated market research activities invariably fall short in the quality and the quantity of information given to the designer at the onset of a new design venture and very rarely, if ever, analyse the competition in terms useful to the designer—and yet it is the designer who has to 'get it right.'

The fundamental question to my mind is, 'How can the designer possibly get it right if the "design foundation" (for want of a better phrase) is incomplete and incorrect, with massive gaps and unknowns?' The answer is plain: he or she cannot, and no amount of systems analysis, stress analysis, or application of the latest production technology will recoup the situation. It is a major concern to me that whilst the latest panacea for all our ills—robots allied to flexible manufacturing systems—is becoming all the rage, very little thought or support is being given to the activity which precedes this manufacture—which is the front end of the whole system—the design and market interface. Lorenz (1982), in an article on production, discusses Harvard and MIT and talks of the great divide in the United States: 'With only a few exceptions, Harvard's professors are concentrating only on the "How?" of manufacture (production strategy and efficiency, management and the rest), at the expense of the "What?" (product design and development). If, allied to these modern procedures, we err at the front end, the rubbish we produce economically, at high speed and minimum cost, will not solve anyone's problem since any product, regardless of its technology, has in my opinion, zero value if no one wants it. This is not to decry the use of modern technology but rather to put in the repetitive plea for balance; one must give consideration to the horse that is expected to pull the technological cart.'

Companies do carry out market research, and in so doing they also carry out analysis of the competition, usually on the basis of comparative functions such as finance, product market share appraisal, resources, and a whole multitude of additional comparisons. Porter (1980), in his book *Competitive Strategy*, discusses how, why, when and what one does to analyse and assess one's competitors. Interestingly, very little mention is made of the product as being central to any business activity, and design is mentioned only perfunctorily. His 'wheel of competitive strategy' (Figure 13.1) is made conspicuous by its missing element 'design.' The United States, as with the majority of the United Kingdom, cannot or will not unscramble that ball of cotton wool called R&D.

It is quite simple: in order to design effectively, you need to carry out or have carried out research of all ilks—technical, market, and so on—and the artefact is then designed. Having reached this stage, it can be developed. The fact that it is missing from the 'wheel' reinforces the Lorenz view.

I am concerned with the teaching and practice of new product design, and, as has already been stated, there is a large gap in the quality and quantity of information given to the designer. Failing alternative involvements, designers must set out to establish and

FIGURE 13.1. The Wheel of Competitive Strategy,
Porter (1980)

stabilise the front-end design foundation if others will not do it for them or with them. Working with industry on student projects and in a professional capacity over the past twelve years, we have yet to experience the situation where such a gap did not exist. It varies from small fissures to enormous chasms but always is present.

In the Engineering Design Centre, where we practise what we preach, many analyses of competitive products over widely differing fields have been carried out in terms useful to the designer (it also transpires that such analyses are also useful to other functions in a business). The outcome of such analyses have *always* revealed the frailty of market research information (or alternatively, the project brief) compiled by the company. From construction equipment to medical equipment, from processes plant to robots, the story is always the same—incomplete design foundation.

What can be done about it? It is recommended that a major contributor to any design foundation is parametric analysis of the competition from all angles—a synthesis of parameters.

PARAMETRIC ANALYSIS: A SYNTHESIS OF PARAMETERS

Parametric analysis (PA) means just what it says—analysis of competing products from a consideration, at least initially, of published catalogue data. Such analysis is primarily concerned with the seeking out of relationships between parameters for the particular product area under consideration, and this by definition means cross-plotting such parameters to find the existence (or lack of existence) of relationships between them. In a realistic situation, cross-plots of vast amounts of data are made—usually manually since it is impossible to program the computer without knowing whether relationships exist between parameters and what those relationships are. To be of full value, one does in practice produce literally hundreds of parameter cross-plots, although in most instances probably half of them exhibit no relationship whatsoever, and the other half shows strong relationships and almost invariably they were unknown at that point in time.

A word of warning at this point. In order to carry out a comprehensive analysis, it is an absolutely essential prerequisite that the participants be happy, willing, and able to carry out the multitude of cross-plots necessary without, in the *first instance*, being able to deduce and understand the reason for so doing. In other words, the particular cross-plot may—prior to carrying it out—appear to be entirely illogical.

This illogicality problem causes a big problem for many engineers: if they cannot see the logic of a situation, then they cannot or will not start work and will dismiss the exercise as trivial. This is not to say that parameters having traditional, logical relationships should not be plotted. Usually the greatest amount of useful information stems from the apparently illogical plots. In such cases, having 'seen' the relationships established, the logic has been retrofitted—but without the visual parametric painting, the discovery of the relationship would be unlikely. (Many companies are now doing PA as part of their activity.)

What are the rules and guidelines for parametric analysis?

1. Assemble as much information as can be mustered on one's own and competitive products.
2. Manually cross-plot the data in graphical form and look for patterns and relationships between parameters—the parametric painting.
3. Ensure that the data being used relates to comparable products at the same generic level. Motor cars cannot meaningfully be compared with sewing machines except within a narrow band of parameters; they are not the same.
4. By all means start with logical relationships; these are usually quickly covered even by people without experience in the product area. Then plot, plot, plot, and do not worry about the logic.
5. Do not try to be clever: use the raw data in its basic state and then having gained greater understanding from the initial plots, identify any useful combined parameters that may emerge.
6. Question why plots are exhibiting strong but unexpected relationships.
7. Place the promising plots side by side on a wall or blackboard (preferably very large), and compare one with the other; they must be viewed collectively. This step is essential since it invariably helps to establish the missing logic and gives rise to explanations which, not surprisingly, are then considered obvious.
8. Look for exceptions to strong patterns. Usually this happens because (a) the plot is wrong, (b) the particular model is exceptional (why?), or (c) the model is not of the same generic base.
9. Draw conclusions from the plots; do not leave them as being of only academic interest.
10. Correlate the findings and conclusions with traditional market research data.
11. Use the information gained to influence and abet the product design specification—the fundamental design foundation.

Some interesting examples of cross-plots taken from a multitude of plots relating to cars are given in Figures 13.2 through 13.8, and with the exception of Figure 13.8 I will refrain from comment. If I were a car designer, I would be extremely interested not only in the relationship itself but particularly as to why models A and B are exceptions to what appears to be the norm.

Additionally, a selection of plots for food processors and kitchen scales also shows interesting relationships (see Figures 13.9 through 13.13). I would comment as follows on several of these. Regarding Figures 13.9 and 13.10, why is model A out of line, or alternatively is it out of line? Should the pattern be different? In Figure 13.12, what are the reasons for models A and B being apparently exceptional?

One wonders whether people in these industries and particularly designers know of these apparent relationships?

Figure 13.14 shows one such plot established in 1972 before the design of the Giraffe site placement vehicle was started, at the end of the market research phase (Pugh, 1977). The predicted unladen weight of the new vehicle was between 12,000 and

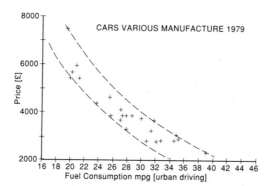

FIGURE 13.2. Price Versus Fuel Consumption: Cars, Various Manufacturers, 1979

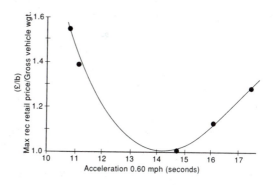

FIGURE 13.3. Retail Price/GVW Versus Acceleration 0–60 mph: Cars, One Manufacturer, 1979

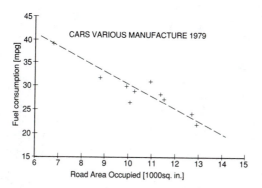

FIGURE 13.4. Fuel Consumption Versus Road Area Occupied: Cars, Various Manufacturers, 1979

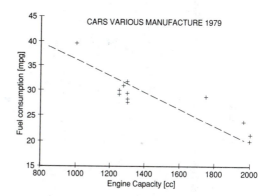

FIGURE 13.5. Fuel Consumption Versus Engine Capacity: Cars, Various Manufacturers, 1979

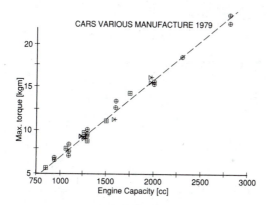

FIGURE 13.6. Maximum Torque Versus Engine Capacity: Cars, Various Manufacturers, 1979

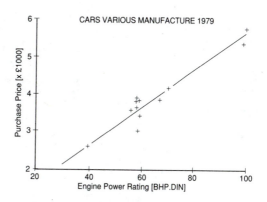

FIGURE 13.7. Purchase Price Versus Engine Power Rating: Cars, Various Manufacturers, 1979

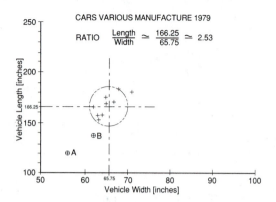

FIGURE 13.8. Vehicle Length Versus Vehicle Width: Cars, Various Manufacturers, 1979

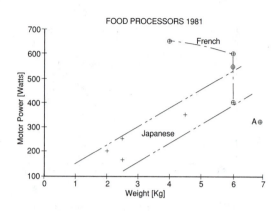

FIGURE 13.9. Motor Power Versus Weight: Food Processors, 1981

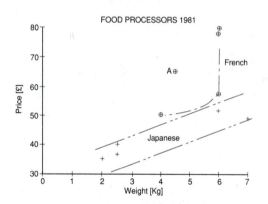

FIGURE 13.10. Price Versus Weight: Food Processors, 1981

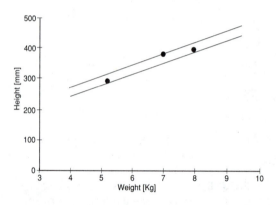

FIGURE 13.11. Height Versus Weight: Food Processors, 1981

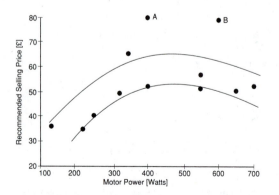

FIGURE 13.12. Selling Price Versus Motor Power: Food Processors, 1981

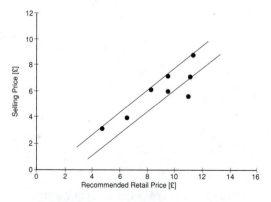

FIGURE 13.13. Selling Price Versus Recommended Retail Price: Kitchen Scales, 1981

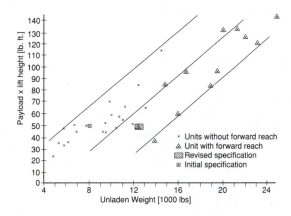

FIGURE 13.14.   Payload x Lift Height Versus Unladen
Weight: Rough Terrain Fork Lift Trucks, 1972

13,000 lbs. In 1973, when the vehicle was designed and built, it weighed 12,800 lbs. The logic of the chosen parameters was retrofitted and now seems commonplace—that for a given payload, the higher one raises the load and the further one reaches forward with it, the heavier the handling device is likely to become.

The examples given are, in the main, taken from exercises carried out to give practice in parametric analysis, and, whilst some comment has been given, readers can speculate and draw their own conclusions. They are examples of interesting relationships. Also in order to avoid controversy, all reference to models and manufacturers has been removed.

On recent projects, in one case some thirty-three plots proved to be useful, showing relationships which in the main were unknown to the manufacturer and, what is more significant, totally unknown to the consultant market research organisation that had just completed their report. In another case, some fifty-six plots were of significant value, at least to the designer.

CONCLUSIONS

If you are in a competitive field, you cannot afford not to analyse and fully understand such competitors in terms of other than just knowing of and about them. Such analysis, if soundly structured and carried through, yields immense amounts of information and forces approaches to and viewpoints of products which otherwise would be impossible to trigger by other means (Pugh, 1981).

REFERENCES

Lorenz, C. (1982). 'Why America Is Rushing to Learn About Production.' In *Financial Times*, 2 April.

Porter, M.E. (1980). *Competitive Strategy: Techniques for Analyzing Industries and Competitors*. New York: Free Press, Collier MacMillan.

Pugh, S. (1974). 'Manufacturing Cost Information: The Needs of the Engineering Designer.' In *Proceedings of the Second International Symposium on Information Systems for Designers*. Southampton: ISD.

Pugh, S. (1977). 'Manufacturing Cost Data for the Designer.' In *Proceedings of the First National Design Conference* (pp. 17-1–17-16). London: NSD.

Pugh, S. (1981). 'Design Is the Biggest Exposure.' *Proceedings of the Conference on the Improvement of Product Safety*. London: International Press Centre.

CHAPTER

14

Concept Selection:
A Method That Works

ABSTRACT

One of the most difficult, sensitive, and critical problems in design, both in teaching and in practice, is the selection of the best concept with which to proceed to detail design and ultimately manufacture. This chapter deals with the question of conceptual choice and discusses in detail a method which has been evolved over a period of years and which has been tested and used effectively by postgraduate students in pursuance of competitive designs.

Primarily concerned with the avoidance of conceptual vulnerability brought about by lack of thoroughness in conceptual approach, concept selection, by definition, means the emergence and selection of the best and strongest concepts. The effectiveness of the method, which is strictly nonnumeric, is illustrated by an example of its use. The chapter concludes with a discussion of the method against the background of others purporting to have the same objective. (Note: You can carry out conceptual design at any level.)

From *Proceedings of the International Conference on Engineering Design* (Rome: ICED, 1981), 497–506.

The postgraduate course in engineering design has been in operation at Loughborough University for some twelve years and is primarily concerned with providing students with a sound professional approach to and competence in design suitable for today's needs. Based firmly on design in practice, a major component of the course is a group design project provided by industry, the outcome of which is an attempt to produce a competitive product design (Pugh and Smith, 1978). During the evolution of the course particular attention has been paid to the testing and evaluation of existing design methods in the real-life situation, and many of the so-called methods have been found wanting when applied in these situations (Pugh and Smith, 1976). A repetitive, particular difficulty always occurs at the point in the design activity when solutions to the particular problem in hand have been generated and the question arises as to how to select the best concepts with which to proceed. One thing is certain: it is extremely easy to select the wrong concept and difficult to select the best one. If the wrong one is chosen, the design may be said to suffer from conceptual weakness, and the design may be said to be conceptually vulnerable. This chapter is concerned with a method developed to minimise such vulnerability.

EXISTENCE OF CONCEPTUAL VULNERABILITY

Conceptual weakness in any design usually manifests itself in two ways:

- The final chosen concept is weak due to lack of thoroughness in conceptual approach. Thereafter, no amount of attention to detail requirements, technical requirements, and the like will recoup the situation.
- The final chosen concept is strong and the best possible within the constraints, but, due to lack of thoroughness in conceptual approach and selection, alternatives suggested, say, by others, cannot be refuted by sound technical argument and debate. In other words, the concept is the best available, it is strong, but the reasons for its strength are not known or fully understood.

So here we have, by definition, two cases of conceptual weakness—the former being truly weak, the latter being strong but lacking thoroughness in approach and apparently weak. The above hypothesis has been arrived at in creating and evaluating designs for products covering a wide range of industry, and, as a result of continual evolution and refinement, an approach has been formulated in many design situations.

Bear in mind that in the absolute sense it is impossible to evolve and evaluate all possible solutions to a particular problem. In order to minimise the possibility of the wrong choice of concept, it becomes essential to carry out concept formulation and evaluation in a progressive and disciplined manner. This disciplined approach necessitates a number of rules and a procedure, which, if followed, leads to significant improvements not only in concept formulation but also in selection.

PHASE 1: PROCEDURE FOR MINIMISING CONCEPTUAL VULNERABILITY

In the teaching phase of our course students have this discipline and procedure impressed upon them and obtain practice in usage, based upon relatively (apparently) simple examples. The basic rules and procedure are as follows:

1. Establish a number of embryonic solutions to the problem in hand, and produce these solutions in sketch form to the same level of detail in each case.
2. Establish a concept comparison and evaluation matrix which compares the generated concepts, one with the other, against the criteria for evaluation. A skeleton of the matrix is shown in Table 14.1.
3. Ensure that the matrix has all the visuals (sketches) of all the concepts incorporated into it in order that the participants can witness the patterns of emergence.
4. Ensure that the comparison of the different concepts is valid: all are to the same basis and at the same generic level.
5. Choose criteria against which the concepts will be evaluated. Usually these are based upon the detailed requirements of the product specification and are established before solution generation commences (Pugh, 1978). It is essential that the criteria chosen are unambiguous and understood by all participants in the evaluation.

TABLE 14.1. Evaluation Matrix

Criteria	Concept										
	1	2	3	4	5	6	7	8	9	10	11
A	+	−	+	−	+	−	D	−	+	+	+
B	+	S	+	S	−	−		+	−	+	−
C	−	+	−	−	S	S	A	+	S	−	−
D	−	+	+	−	S	+		S	−	−	S
E	+	−	+	−	S	+	T	S	+	+	+
F	−	−	S	+	+	−		+	−	+	S
Σ +	3	2	4	1	2	2	U	3	2	4	2
Σ −	3	3	1	4	1	3		1	3	2	2
Σ S	0	1	1	1	3	1	M		1	0	2

6. Choose a datum with which all the other concepts will be compared. If a design or designs already exist for the product area under consideration, these must be included in the matrix and always form a useful first datum choice.

7. In considering each concept or criteria against the chosen datum, use the following legend:
   - A plus sign (+) means *better than, less than, less prone to, easier than,* and so on, relative to the datum.
   - A minus sign (−) means *worse than, more expensive than, more difficult to develop than, more complex than, more prone to, harder than,* and so on, relative to the datum.

   Where any doubt exists as to whether a concept is better or worse than the datum, use the following:
   - An "S" means *same* as the datum.

8. After selecting a datum, make an initial comparison of the other concepts using step 7. This establishes a score pattern in terms of the number of plusses, minuses, and S's achieved relative to the datum.

9. Assess the individual concept scores. If certain concepts exhibit exceptional strength, rerun the matrix with the strengths removed. If, as a result of running the matrix several times, the initial high scorers persist—such as (1), (1), (1), (1)—they are likely to be the best concepts with which to proceed.

10. If a strong pattern of concepts does not emerge in step 9—for example, if all appear to have uniformity of strength (which is very unusual)—change the datum and reassess the pattern.

11. If, for example, one particular concept persists, change the datum and repeat. If the result remains the same, let the emergent strong concept assume the role of datum, rerun the matrix, and again assess the result.

It is preferable that the course of action detailed in steps 1 through 11 be carried out on a large blackboard or similar display unit as it is considered essential that all participants to the evaluation take part in the usually hectic discussions on each point and that each sees the complete picture for the whole of the period. (Note: Today (1992) we make extensive use of electronic whiteboards.)

Having completed what might be called the initial evaluation comprising a number of runs, the participants will have acquired

- Greater insight into the requirements of the specification,
- Greater understanding of the problem,
- Greater understanding of the potential solutions,
- An understanding of the interaction between the proposed solutions, which can give rise to additional solutions,
- A knowledge of the reasons why one concept is stronger or weaker than another.

If additional solutions arise, the comparison and evaluation procedure should be repeated. Note: In conducting the comparison and evaluation just described, the tutor or project leader will have to control the questioning of concepts and maintain a tight discipline on the participants. Reaching this point in an evaluation should be construed as being the end of the first phase.

PHASE 2

The decision is taken to proceed to develop the strongest concepts emerging from the initial evaluation in phase 1. This entails further work on these concepts to engineer them to a higher level and in more detail than was carried out in phase 1. Again, care should be taken to ensure that each one remains comparable with the others. The additional work, involving finer detail, results in even greater understanding of the problem and its projected solutions, and such understanding also leads to a refinement and expansion of the criteria for evaluation. The matrix is reformed to incorporate the enhanced concepts and also the revised or expanded criteria, and the mechanism of the first phase is repeated. The outcome of this reevaluation will either confirm the pattern established previously or give rise to a reordered set of concepts. In each case, the reasons for the emergent pattern and the relative strengths and weaknesses should be questioned deeply.

It is interesting to note that students invariably develop a critical awareness of the whole procedure. Particularly if a strong concept emerges and stays, as it were, in the lead, they now begin to have doubts over the whole matter and usually suggest that many of the concepts generated initially are far better than the emergent leader. This becomes a particular problem with strong-willed individuals whose initial concepts have not emerged in the final selection; they then commence a defence based upon emotion and may prefer to ignore the facts of the situation.

It is recommended that at this stage all concepts are placed in a reformed matrix with the revised criteria, and a comparison is carried out and assessed. Almost without exception, the results of phase 1 and 2 will be confirmed, which inspires confidence in the procedure. It should always be borne in mind that with the method just described, the choice of concept remains with the participants. The matrix does not make the decisions: it is simply a procedure for controlled convergence onto the best possible concept and is not composed for absolutes in the mathematical sense; the decisions remain with the user.

SUBSEQUENT PHASES

As design work proceeds on the chosen concepts, it may become necessary to repeat the continually refined procedure several times in order to confirm the approach adopted. Depending upon the complexity of the project, it is not untypical in our experience to carry out five or six evaluations and comparisons before a single concept emerges, which is then carried through to final design, detailing, and manufacture. A single typical matrix run may take anything up to a whole day to complete, and in a recent project thirty-two different concepts were evaluated in arriving at the solution. What is particularly heartening about the whole procedure is that, in dealing with a variety of companies, no approaches to the solution of the problem have been able to seriously fault or cause to be altered the chosen concept which leads to the final design. Bearing in mind the expertise and experience present in those companies, if a student group can defend their designs in a sound and logical manner when confronted by such expertise and not be faulted conceptually, then the procedure underwriting this situation may be deemed not only to have some merit but also to have minimised conceptual vulnerability.

AN EXAMPLE OF USAGE

In the first term of our course as a first exercise in the procedure, students are given the task of generating embryonic solutions (concepts) for an audible means of approach for a motor car (car horn). They develop their own initial product specification and using this as a basis, produce a variety of concepts with the ultimate capability of effectively satisfying this specification. Figure 14.1 shows fourteen comparable concepts produced by a student group. Eleven others generated at the same time have not been included since they all, in principle, required a major additional component to achieve the same function and therefore were evaluated separately.

Within the limited space allowed for this chapter it is not possible to portray the matrix including drawings of the concept in the same figure and remain meaningful. Table 14.2 shows the evaluation chart for the fourteen comparable concepts. A word of explanation; the chosen datum, concept 1, is the traditional motor horn which has been developed over many years and is fitted to millions of cars today. Whilst it is debatable whether the relative order of merit for the different concepts is correct, the almost overwhelming strength of concept 5 is not only extremely interesting; it is perhaps a foretaste of future car horn design and development. The reader is recommended to ask the question, 'Does a vehicle audible means of approach, as outlined in concept 5, exist at the present time?' Certainly not to my knowledge.

Subsequently, further exercises are carried out on more difficult and complex products in order to gain familiarity with the confidence in the procedure and to prepare the students for the main group project where it will be used with intent in a real situation.

DISCUSSION AND CONCLUSIONS

The establishment of the procedure described has been found necessary and desirable for a number of reasons, including the following

- The inadequacy of existing methods to produce effective and best solutions,
- The constraints to creativity imposed either knowingly or subliminally by these methods,

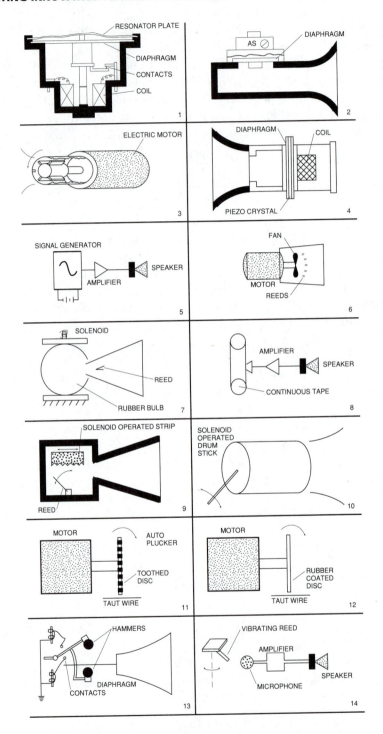

FIGURE 14.1.   Comparable Car Horn Concepts
Produced by One Student Group

TABLE 14.2. Evaluation Chart for a Motor Horn

Criteria		Concept												
	1	2	3	4	5	6	7	8	9	10	11	12	13	14
Ease of achieving 105–125 DbA		S	–		+	–	+	+	–	–	–	–	S	+
Ease of achieving 2,000–5,000 Hz		S	S	N	+	S	S	+	S	–	–	–	S	+
Resistance to corrosion, erosion, and water		–	–	O	S	–	–	S	–	+	–	–	–	S
Resistance to vibration, shock, and acceleration	D	S	–	T	S	–	S	–	–	S	–	–	–	–
Resistance to temperature	A	S	–		S	–	–	–	S	S	–	–	S	S
Response time	T	S	–		+	–	–	–	–	S	–	–	–	–
Complexity: number of stages	U	–	+	E	S	+	+	–	–	–	+	+	–	–
Power consumption	M	–	–	V	+	–	–	+	–	–	–	–	S	+
Ease of maintenance		S	+	A	+	+	+	–	–	S	+	+	S	–
Weight		–	–	L	+	–	–	–	S	–	–	–	–	+
Size		–	–	U	S	–	–	–	–	–	–	–	–	–
Number of parts		S	S	A	+	S	S	–	–	+	–	–	S	–
Life in service		S	–	T	+	–	S	–	–	–	–	–	–	–
Manufacturing cost		–	S	E	–	+	+	–	–	S	–	–	–	–
Ease of installation		S	S	D	S	S	+	–	S	–	–	–	S	–
Shelf life		S	S		S	S	–	–	S	S	S	S	S	S
Σ +		0	2		8	3	5	3	0	2	2	2	0	4
Σ –		6	9		1	9	7	12	11	8	13	13	8	9
Σ S		10	5		7	4	4	1	5	5	1	1	8	3

- A desire to formalise and establish a procedure that not only works effectively in practice but is also teachable, and
- The necessity to avoid the false confidence which many methods give to the user (Pugh and Smith, 1976).

In using this procedure many times in practice to establish new or improved products, I have found that constraints to creative thinking are minimised. The fundamental difficulty, particularly in a teaching situation, of lack of convergence on the best solution is avoided whilst the participants remain open-minded in approach. In other words, having converged on to a solution in a controlled manner, they are still capable of divergent thinking about the same problem.

A particular difficulty which arises in practice comes with a group which contains people with long experience of design in industry. They exhibit an impatience 'to get on with it' and may consider that the procedure holds them back from arriving at a solution. However, exposure to the full procedure usually convinces them of the error of their ways, particularly when asked if the solution, as it emerged, was ever likely to have arisen with their random, intuitive approach. The procedure positively stimulates creative, unconstrained thinking due to its lack of rigorous structure. For instance, it takes a lot of thought to establish in the first place, and yet in my experience avoids the rigidity and false confidence instilled by numeric rating and weighting matrices.

Whilst this chapter has been concerned with a procedure for minimising conceptual vulnerability, it should be pointed out that as an approach it can be used at any level in design from the most complex concept to the simplest detail component. In conclusion, it offers many opportunities to improve design teaching and practice and will remain under continual scrutiny and development.

REFERENCES

Pugh, S. (1978). 'Quality Assurance and Design: The Problem of Cost Versus Quality.' *Journal of the Institute of Quality Assurance* 4(1), 3–6.

Pugh, S., and Smith, D.G. (1976). 'The Dangers of Design Methodology.' Paper presented at the First European Design Research Conference, Portsmouth Polytechnic, (UK).

Pugh, S., and Smith, D.G. (1978). 'Design Teaching Ten Years On.' *Engineering Design Education Supplement* 2, 20–22.

CHAPTER

State of the Art on Optimisation in Great Britain

ABSTRACT

The question of optimisation in Great Britain is considered against the background of the design activity. A hypothesis is put forward which relates the enhanced usage of mathematical optimisation techniques to the fundamental elements present in any artefact. The transition during design from evaluation, essentially qualitative, to optimisation, essentially quantitative, is used as a framework for this hypothesis.

From *Proceedings of the International Conference on Engineering Design* (Copenhagen: ICED, 1983), 389–394.

The title of this chapter, without detailed investigation, may lead to a narrow view of optimisation related to design which would prove only of mild interest to the mathematician, and be of possibly greater interest to a designer with an intractable problem. Yet it may completely mask one of the major issues in the whole design activity—the selection of the best design concept, which can then be operated upon with a myriad of mathematical techniques in order to attempt to produce the optimum design.

In the engineering sense, the word *optimisation* has in the main long been taken to mean the mathematical optimisation of sets of parameters pertaining to artefacts or systems.

Whilst in the United Kingdom, as with other countries, the latter has been the predominant order of the day, since it represents tangling with a set of visible tangibles, decisions can be made on the basis of mathematical technique, which gives the user a comforting feeling of confidence, safety, and rightness.

In reality, in any new design situation, one is initially in an area of almost complete intangibles, which by definition, since the situation is intangible and not fully understood, cannot be meaningfully or usefully addressed (yet) by the application of mathematical optimisation techniques.

The word *optimum*, meaning 'best or most favourable,' implies no allegiance to mathematics nor should it, since in practice, proceeding from the intangible to the tangible means moving from the qualitative to the quantitative (Pugh and Smith, 1976). Therefore, in seeking to achieve the best and most favourable design, one is always proceeding from the evaluation of alternative approaches to meet the specification, to the point where one or more has been selected and, as has already been stated, where valid and erudite mathematical methods can be applied to refine and enhance the chosen design.

It is vitally important, therefore, to make the distinction between evaluation and optimisation. My experience over a wide project field indicates that in evaluating alternative designs to meet a given set of circumstances (defined in a specification) (Pugh, 1981), mathematics may be of little assistance (as yet) and can, in fact, hinder and set the whole design activity off course, particularly if one is determined to apply mathematical technique.

Therefore, at this stage in the development of techniques varying from evaluation (nonmathematical) to full optimisation, in the generally accepted sense of the word mathematical, it is essential that this differentiation is made. There is some evidence that in Great Britain the transitional nature of methods to achieve the optimum is at last being recognised.

According to Rogers (1979), 'The design of the complex, heavily constrained optics required for the majority of electro-optical systems nowadays demands a reasonable sized computer; a large versatile optimisation programme—preferably interactive in operation—and, most of all, an experienced designer capable of indulging in the occasional bit of lateral thinking.'

Steele (1980), in discussing the optimum energy usage in water distribution systems, recognises the constraints of mathematical optimisation: 'Optimisation must therefore, besides the main objective of minimum energy costs, take into account certain constraints imposed by the system and satisfy some secondary conditions which can be in opposition to the optimisation.' One wonders whether in any system, if a condition *must* be satisfied, whether in fact it should be considered as secondary?

Westlund (1977) considers the design of pneumatic systems for optimum performance utilising the Sanville formula: 'The reported formulae cannot be derived from established physical laws. Instead they are, like many other formulae in the flow technics, the result of fitting mathematical expressions to observed facts.'

So we have a recognition that in a system selection sense—or, say, in the selection of the best design concept for a machine with which to proceed and hence the greater possibility of mathematical optimisation—a large difficulty has to be overcome before one can rationally apply mathematical technique. Ritchie (1975) confirms the difficulty.

Caldwell and Hewitt (1976), in considering the cost-effective design of ships structures, calls for 'methods of synthesising the design of real structures to meet whatever criteria (weight, cost, reliability, etc.) may be appropriate. This in turn calls for, (a) examination and definition of alternative criteria for optimisation, or "objective functions" and (b) development of efficient design optimisation procedures capable of handling many variables of form, materials and geometry.'

It is my firm opinion that the only methods currently available to achieve success in the optimisation of large systems or equivalently in a single machine with a large number of elements which may be considered to be a large system at the choice or selection of system or machine stage (the conceptual stage) rely heavily on the qualitative approach (Pugh, 1981), leading to the application of quantitative techniques as certainty and rationale loom larger.

The establishment of laws and relationships upon which to base new designs cannot wait upon the manifestation of such laws by logical procedures. Selection, choice, and the evaluation of alternatives can be enhanced by the application of parametric analysis (Pugh, 1982), which can bring to light whole new vistas of relationships, leading to better choices and therefore in the global sense, more optimum choices, which can then be refined by enhanced mathematical techniques. As Westlund (1977) said, 'The optimisation that cannot be derived from physical laws' (as yet).

As one descends from the global question of selection, choice, and evaluation to the elements contained within that selection, the picture becomes more stable and more thickly populated with research activity and hence more papers. Since conceptually the choices have been made, the concepts are fixed and methodologies proliferate over the whole spectrum of engineering. Russell and Barnard (1979), Morley and Gulvanessian (1977), McKenzie (1980), Castellano and Adie (1980), and Bleay and Fells (1979) are all concerned with the optimisation of designs where conceptual choices have been made for the system or machine, in what I call the middle ground between evaluation and optimisation. Further delving into the discrete single elements of these machines or systems greatly expands the opportunity for the necessary further optimisation of these elements, and the field widens greatly. Ellis (1977), Cole (1976), Holland (1976), Spence (1977), Rowbottom (1981), and Dowson and Ashton (1976) all concentrate on a variety of elemental optimisation.

So the optimisation scene in Great Britain, which is currently probably not very different from that in other countries, ranges from a dearth of activity at what I call the front end of design (the conceptual choice end), to strong and varied activity on fixed concepts and elements once such choices have been made. Thus the more fundamental the elements, the more effective mathematical optimisation. We are concentrating our activity at the front end of design since I consider that others are probably better fitted to carry out the work at the elemental end of the spectrum. For the new, exciting innovative designs of the future the success and thoroughness of the front end work becomes a necessity. Only by considering the questions in this manner can I possibly conceive of how to separate out the vexed question of the optimisation (elementally) of the wrong concept (system or complete machine) as opposed to what we are all striving to achieve in the optimisation of the best concept choice.

Evaluation at the intangible end of the spectrum, therefore, is in a sense more important than strict mathematical optimisation at the tangible end, since the latter may be rendered useless by any earlier selection mistakes. In my experience, there is no such thing as the optimum design in the mathematical sense, unless one chooses to artificially constrain the optimisation criteria to suit the mathematics: the successful design is always the sum of the best compromises.

Who should carry out research into evaluation and optimisation techniques? I do not think there is any one answer to this question. It is doubtful that the mathematician can contribute much at the front end of design until by some manner or means we can explain better just what it is we are seeking to do, hence the proliferation at the elemental end referred to earlier. On the other hand, designers are in the main interested in designing artefacts and are not too concerned about how they do it as long as what they do is successful. So maybe the answer is the designer or researcher with special interests in this problem. Are there such people about? Not, I suspect, in Great Britain since design and research are completely different activities.

REFERENCES

Bleay, J.A., and Fells, I. (1979). 'Optimisation of the Design of Combined Heat and Power Schemes: A Linear Programming Approach.' *Journal of the Institute of Energy* (September), 125–126.

Caldwell, J.B., and Hewitt, A.D. (1976). 'Towards Cost-Effective Design of Ships Structures.' *Proceedings of the Conference on Structural Design and Fabrication in Shipbuilding* (pp. 203–214).

Castellano, E.J., and Adie, J.F. (1980). 'Optimisation of the Design of a High Capita, Cost Parcel Handling Machine by Computer Simulation.' *Proceedings of the Institution of Mechanical Engineers* 194, 145–156.

Cole, B.N. (1976). 'Some Overlooked Aspects of Winkler's Theory of Curved Beams.' *IJMEE* 4(3), 243–246.

Dowson, D., and Ashton, J.N. (1976). 'Optimum Computerised Design of Hydrodynamic Journal Bearings.' *International Journal of Mechanical Sciences* 18, 215–222.

Ellis, J. (1977). 'An Introductory Problem in Mechanical Design Optimisation.' *IJMEE* 5(2), 121–126.

Holland, M. (1976). 'Pressurised Member with Elliptical Median Line: Effect of Radial Thickness Function.' *Journal of Mechanical Engineering Science* 18(5), 245–253.

McKenzie, A.B. (1980). 'The Design of Axial Compressor Blading Based on Tests of a Low Speed Compressor.' *Proceedings of the Institution of Mechanical Engineers*, 194, 103–111.

Morley, C.T., and Gulvanessian, H. (1977). 'Optimum Reinforcement of Concrete Slab Elements.' *Proceedings of the Institute of Civil Engineers* (pt. 2, 63, pp. 441–454).

Pugh, S. (1981). 'Concept Selection: A Method That Works.' *Proceedings of the International Conference on Engineering Design* (pp. 497–506). Rome: ICED.

Pugh, S. (1982). 'A New Design: The Ability to Compete.' Paper presented at the Design Policy Conference, Royal College of Art.

Pugh, S., and Smith, D.G. (1976). 'The Dangers of Design Methodology.' Paper presented at the First European Design Research Conference on Changing Design, Portsmouth Polytechnic.

Ritchie, G.S. (1975). 'The Prediction of Journal Loci in Dynamically Loaded Internal Combustion Engine Bearings.' *Wear* 35, 291–297.

Rogers, P.J. (1979). 'Aspects of Contemporary Optical Design.' In *Optics and Laser Technology* (pp. 203–211). London: IPC Business Press.

Rowbottom, M.D. (1981). 'The Optimisation of Mechanical Dampers to Control Self-Excited Galloping Oscillations.' *Journal of Sound and Vibration* 75(4), 559–576.

Russell, C.M.B., and Barnard, L. (1979). 'Method Optimises Steam Turbine Air Cooled Condenser Design.' *Combustion* (May), 37–39.

Spence, J. (1977). 'Type 1 Energy Analysis for the Creep of Smooth Circular Pipe Bends Under Combined In-plane Bending and Internal Pressure.' *Journal of Strain Analysis* 12(1), 1–7.

Steele, K.A. (1980). 'Optimisation of Energy in Water Distributions Systems by Effective Measurement and Control.' *Brown Boveri Review* 6, 350–355.

Westlund, B. (1977). 'The Sanville Formula for Pneumatic Components.' *Design Engineering* (June), 65–68.

16

Enhanced Quality Function Deployment

ABSTRACT

Quality function deployment (QFD) was first developed in Japan about twenty years ago. QFD emphasises cooperation, convergent consensual decision making, and systematic linkage of engineering activities. The need for this has increased markedly during the past few decades, as a strong trend has developed towards cloistered groups of specialists, focused inward on their specialty, only inadequately linked towards producing a desirable product. QFD has now been practised in the United States for about five years, with Ford, General Motors, and Procter & Gamble among the most active companies.

Although QFD as transferred from Japan to the United States during the middle 1980s is very powerful, it has always had some obvious opportunities for enhancement. As one of us (Clausing) has helped to introduce QFD into the United States, it has usually been linked with the Pugh concept selection process. Now the two of us have made this linkage more explicit and added other linkages and enhancements. It is this that we are calling enhanced quality function deployment (EQFD).

With Don Clausing, in *Proceedings of the Design Productivity International Conference* (Honolulu: DPIC, 1991), 15–25.

QUALITY FUNCTION DEPLOYMENT

Basic quality function deployment (QFD) is summarised in Figure 16.1. (For a more detailed description the reader is referred to Hauser and Clausing, 1988). QFD is a systematic process by which a multifunctional team deploys from the voice of the customer to operations on the factory floor. It apparently grew out of value analysis/value engineering (VA/VE) (Miles, 1972), which was first developed about forty years ago, although its basic precepts have always been present to some extent in good engineering design. The basic deployment steps are as follows:

Deployment	Matrix
Voice of customer—System specifications	House of quality
System specifications—Detailed design decisions	Design
Detailed design decisions—Process decisions	Production process engineering
Process decisions—Operations decisions	Production operations planning

The word *deployment* signifies a combination of translation from one language to another and team decision making. QFD is used primarily for decisions where individual experience is insufficient but where the collective experience base of a multifunctional team is capable of guiding sound decisions.

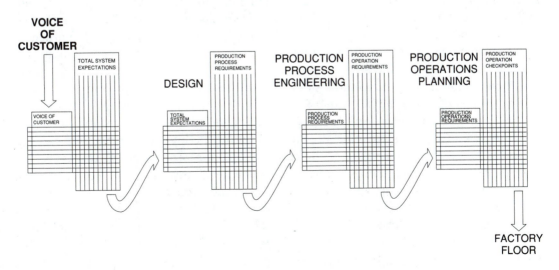

FIGURE 16.1. Basic Quality Function Deployment

There is an obvious opportunity for enhancement between matrix 1 and matrix 2. The output from the house of quality (matrix 1), which is the input to matrix 2, represents the system-level specifications (corporate expectations) for the total product system. The output from matrix 2 is detailed design decisions and is feasible only for products that are simple and are held conceptually static during the current design cycle. (The concept could become dynamic for the next variant of the product.) For products that are conceptually dynamic, a concept selection process is needed. For products that are complex, we must take into account the multiple product levels, such as system, subsystem, and component. Thus, one tightly integrated set of opportunities for enhancements consists of (1) concept selection, (2) static/dynamic (S/D) conceptual status, and (3) multiple levels of a system.

Design experience has shown that there is a great benefit to a sound, early analysis of the context of the new design with respect to the business and technological environment. QFD places great emphasis on the voice of the customer. Usually the only structure for the voice-of-the-customer input to the house of quality is imposed afterwards by forming logical affinity groups. Experience has shown that specifications are more quickly and certainly defined with the help of some a priori structure. Thus, the two remaining enhancements to be described are (4) analysis of context and (5) structure of requirements.

INTRODUCTION OF ENHANCEMENTS

The Pugh concept selection process (Pugh, 1981) has been developed and applied in design classes and in industrial practice during the past fifteen years. It has been highly successful and to some extent has always been integrated with the introduction of QFD in the United States. The Pugh concept selection process and QFD have the same characteristics: (1) multifunctional team, (2) systematic, convergent decision process, and (3) a matrix with criteria as rows and decision elements as columns. That this consistency exists between two systematic methods, one developed in Japan and the other in Great Britain, suggests quite strongly that these characteristics have a fundamental role in the improved design process.

The evaluation of the conceptual static/dynamic (S/D) status of a product (Pugh and Smith, 1976; Pugh, 1977, 1983, 1984; Hollins and Pugh, 1990) is closely related to the use of the Pugh concept selection process. If the conceptual status is static, then the concept is known from existing products. Dynamic conceptual breakout can be achieved only by the application of the concept selection process or left to chance. In the case of the Marathon dump truck (1972) (Pugh and Smith, 1974), it was not found possible to break out at the TSA level; at this level the overall concept is static. Break out was possible at the subsystem level—the four-wheel drive transmission.

In the case of the Giraffe (1973) (Pugh, 1979)—a rough terrain fork lift truck with telescopic boom—breakout with a new concept occurred at the TSA level as well as the SS level, setting a new static concept for many others to follow. Buggywhips are the classical example of an excessive static outlook.

At the opposite extreme, random unstructured attempts to be disfunctionally dynamic usually lead to chaos and confusion, which greatly extend the development schedule, with accompanying failures in cost and quality. Concept selection should be used to check out S/D status but always in a sensible controlled manner; this is critical to success.

Complex systems, such as cars, copiers, and computers, must be considered at several levels, such as total system, subsystem, and component. The actual number of levels will depend upon the complexity of the system and the preferences and traditions within each corporation. Here we will use three levels as sufficient to exemplify the principle. It is rather directly obvious that there is a need to deploy from one system level to the next. Also, the issues of S/D status and concept selection arise for each level. Therefore, the status and multiple levels are tightly interrelated and will be discussed together in the next major section of this chapter.

Analysis of the technological and business context for the new product is critical to success. This analysis should occur at the beginning of the product development process, often before a specific development programme has been formally defined. A worldwide search for relevant information should place the new product in a realistic technological perspective. Particularly important are parametric analysis and reverse concept selection, which define the relative characteristics of existing products. It is often very beneficial to also analyse analogous products. It is critical to develop a clear specification for a new product, which will provide strong guidance to subsequent activities. To facilitate rapid and certain development of the specifications, it has been found helpful to have a generic structure for the specifications.

The S/D status has a profound influence on the best style for utilising both the analysis of context and the structure of requirements. For static concepts the analysis of context and the development of specifications should be largely completed before anything else is done. However, for highly dynamic products this is not possible. Initially, some analysis of context and specification preparation (house of quality) are done based on the existing notions about the concept for the new product. Then some concept development is done. This leads to further refinement of the contextual analysis and specification definition and so on in an iterative process until the contextual analysis, specifications and concept are fully developed and compatible.

CONCEPT SELECTION, LEVELS, AND STATIC/DYNAMIC STATUS

The basic process of enhanced QFD (EQFD), when the system has multiple levels, is displayed in Figure 16.2. Here there is the opportunity to perform concept selection at the level of total system architecture (TSA), subsystem (SS), and component or piece-part (PP).

Now, instead of deploying directly from the system specifications that appear in the columns of the house of quality down to detailed piece-part decisions, there is instead deployment from the house of quality to decisions at the system level, then decisions at

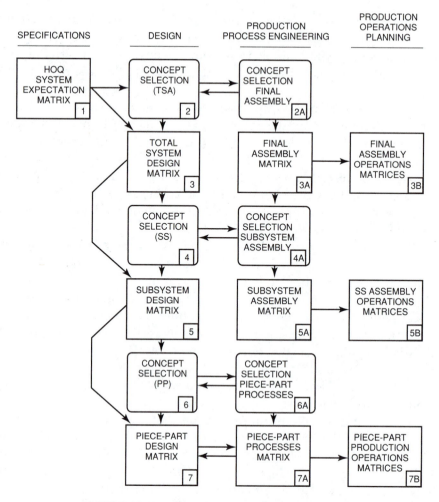

FIGURE 16.2. The Basic Process of Enhanced QFD

the subsystem, and finally, decisions at the piece-part level. The evaluation of S/D status is appropriate for all three levels. For example, the TSA might be static, but one or more subsystems could be dynamic. (As a design cycle proceeds, all concepts must become fixed for the duration of that specific product. Most dynamic work must be done early, in advanced development—more about that later.) This will now be described by using a Xerox paper feeder as an example. This is a prescriptive case study, using an actual design and knowledge of its development to describe how it would be done today by using EQFD.

The heart of the house of quality for this example is shown in Figure 16.3. The entire house of quality would actually be done as described in Hauser and Clausing (1988). Here we show only the main features of the house of quality that are sufficient to exemplify the present subject. The voice of the customer includes the ten characteris-

FIGURE 16.3. Core of House of Quality (1)* Example: Xerox Copier

TOTAL SYSTEM EXPECTATIONS

VOICE OF CUSTOMER	MISFEED RATE (A)	MULTIFEED RATE (B)	JAM RATE (C)	COPY RATE (D)	JAM CLEARANCE TIME (E)	PAPER DAMAGE RATE (F)	UMC (G)
1 ALWAYS GET A COPY	O						
2 NO BLANK SHEETS		O					
3 NO JAMS TO CLEAR		O	O				
4 MEDIUM SPEED				O			
5 COPIES ON CHEAP PAPER	O	O	O				
6 COPIES ON HEAVY PAPER	O		O				
7 COPIES ON LIGHT PAPER		O	O				
8 EASY TO CLEAR JAMS					O		
9 NO PAPER DAMAGE						O	
10 LOW COST							O
	$<50/10^6$	$<50/10^6$	$<100/10^6$	70 ± 2 CPM	<20 SEC	$<100/10^6$	$<\$6000$

tics that are shown as examples. The first example, always get a copy, means that if we programme the copier for five copies, we want to get only the five copies, without any blank sheets interspersed among them. In the columns, these requirements are deployed into corporate expectations for the total system. The relationship matrix shows the relationship between the corporate expectations and the voice of the customer. For example, misfeed rate is the corporate expectation that corresponds to the customer's desire to always get a copy. At the bottom of the house of quality, the corporate expectations are quantified, for example the misfeed rate should be less than $50/10^6$.

In this example, this is the TSA for the Xerox 1075 copier, which was introduced into the market in 1981 (Figure 16.4). The TSA for a complex product includes the location of the subsystems (often referred to as packaging), the initial selection of the basic concept (technology) for each subsystem, and a small amount of system engineering to give credibility to the feasibility of the TSA.

Based on the selected TSA, a total system design matrix is prepared in which decisions are made at the level of the total system, and the total system expectations from the house of quality are deployed into subsystem expectations (Figure 16.5). Three types of deployment are illustrated in Figure 16.5. An example of the first type is the deployment of misfeed rate. As misfeeds in the copier can occur only in the feeder, the total numerical requirement for misfeed rate is simply in its entirety deployed to the expectations for the feeder. Thus the total system expectation that the misfeed rate will be less than $50/10^6$ becomes the numerical expectation for the feeder itself. An example of the second type of deployment is jam rate. Jams can occur throughout the paper path of the copier, and therefore the required jam rate for the feeder can be only a fraction of the total ex-

*Number refers to step in Figure 16.2.

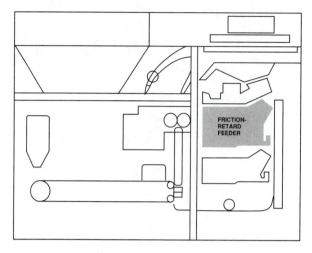

FIGURE 16.4.   Concept Selection: Total System Architecture (2) Example—Xerographic Copier

pectation for the copier. In Figure 16.5, by way of illustration, the jam rate for the feeder is shown to be 30 percent of that for the entire copier. The third type of deployment is illustrated by paper speed and delivery time. These are system engineering decisions that are made by doing system engineering at the level of the complete system. In this case, decisions stem primarily from the analysis of the timing diagram for the copier. The de-

	TOTAL SYSTEM EXPECTATIONS		MISFEED RATE (1)	MULTIFEED RATE (2)	JAM RATE (3)	COPY RATE (4)	JAM CLEARANCE TIME (5)	PAPER DAMAGE RATE (6)	UMC (7)	PAPER SPEED (8)	DELIVERY TIME (9)
A	MISFEED RATE	$<50/10^6$	O								
B	MULTIFEED RATE	$<50/10^6$		O							
C	JAM RATE	$<100/10^6$			O					O	O
D	COPY RATE	70_{-3}CPM				O				O	
E	JAM CLEARANCE TIME	<20 SEC					O				
F	PAPER DAMAGE RATE	$<100/10^6$						O		O	
G	UMC	$<\$6000$							O		
			$<50/10^6$	$<50/10^6$	$<30/10^6$	70_{-3}CPM	<20 SEC	$<40/10^6$	$<\$250$	11.7 ± 0.3 IPS	141 ± 10mSEC

SYSTEM DESIGN DECISIONS, SUBSYSTEM EXPECTATIONS

FIGURE 16.5.   Total System Design Matrix (3)

cision is that the paper speed must be 11.7 inches per second in order for the copier to meet all of its expectations. Within the total cycle time to make one copy, the system analysis led to the decision that the delivery time of the sheet from the feeder to the transport that carries it to the photoreceptor should be 141 milliseconds to enable the sheet of paper to match up with the image on the photoreceptor.

The expectations for the subsystem are then used to select the concept at the level of the subsystem. In this example, the basic concept of the feeder was already selected to be that of a friction-retard feeder as part of the concept selection for the TSA. In the example of the friction-retard feeder that was developed initially for the Xerox 1075 copier, three features are conceptually dynamic. Concept selection is done for these dynamic features, using the Pugh concept selection process, with the criteria (row headings) taken from the columns of the total system design matrix, in which have been recorded the subsystem expectations. One example is the rotating retard roll, which is shown in Figure 16.6. Previous friction-retard feeders had retard elements that were almost stationary. They moved only very slowly, so as to distribute wear over the entire surface. In this feeder, the selected concept has a friction brake inside of the retard roll. When there is only one sheet of paper in the nip between the feed belt and the retard roll, the friction is sufficient to overcome the brake torque, and the roll rotates. This minimises wear on the roll. If two sheets of paper enter the nip, the friction between the two sheets is inadequate to overcome the brake torque. The retard roll becomes stationary and retards the second sheet of paper so that it does not become feed into the copier. This is an example of concept selection at the subsystem level.

Based on the selected concepts for the subsystem, next the subsystem design matrix is prepared, as shown in Figure 16.7.

Here design decisions are made at the level of the subsystem, deployed from subsystem expectations to piece-part expectations. The rows that are the input to the subsystem design matrix are the columns from the total system design matrix. The columns in the subsystem design matrix are piece-part expectations. An example is the retard fric-

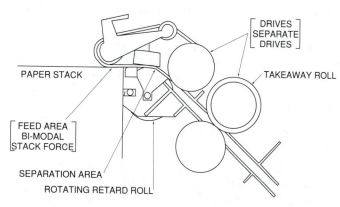

FIGURE 16.6.   Concept Selection: Subsystem (4)

SUBSYSTEM EXPECTATIONS		RETARD FRICTION COEFFICIENT (A)	RETARD BRAKE TORQUE (B)	RETARD RADIUS (C)	NORMAL STACK FORCE (D)	ENHANCED STACK FORCE (E)	TRIGGER TIME (F)	TAR ACTION TIME (G)	TAR SURFACE SPEED (H)	JAM CLEARANCE STRATEGY (I)	UMC BREAKDOWN (J)
1	MISFEED RATE	<50/10^6	O		O		O					
2	MULTIFEED RATE	<50/10^6	O	O	O		O	O		O		
3	JAM RATE	<30/10^6					O					
4	COPY RATE	70+?CPM							O			
5	JAM CLEARANCE TIME	<20 SEC									O	
6	PAPER DAMAGE RATE	<40/10^6	O									
7	UMC	<$250										O
8	PAPER SPEED	11.7±0.3 IPS								O		
9	DELIVERY TIME	141±10mSEC							O			
			1.50±0.25	40±41N-OZ	0.880±0.005IN	0.3 LB	0.7 LB	100mSEC	120mSEC	11.7±0.3 IPS	REF. Y	REF. Z

FIGURE 16.7.   Subsystem Design Matrix (5)

tion coefficient. The friction coefficient, between the retard roll and the paper being fed, must have the correct value to avoid excessive misfeeds and multifeeds. In the example the retard friction coefficient is shown to be 1.50. This particular decision exemplifies a critical aspect of design decision making. Even the collective experience of the multifunctional team that is utilised in QFD is not always sufficient to make the most critical design decisions. Such an example is the retard friction coefficient. Because of its critical impact upon the feeding function, the value must be very carefully determined. In the case of such highly critical design parameters, it is essential to optimise the values to achieve the maximum robustness of the functions (Taguchi and Clausing, 1990; Phadke, 1989).

The piece-part expectations that are in the columns of the subsystem design matrix are next used to perform concept selections for the piece-part. The piece-part expectations are used as criteria (row headings) in the Pugh concept selection process. An example of a selected piece-part concept is the microcellular polyurethane material for the retard roll (Figure 16.8). The decision is made in this example that microcellular polyurethane material will best meet all of the piece-part expectations that have been defined in the subsystem design matrix.

Based on the selected concepts for the piece-part, the detailed design decisions are made for the piece-part in the piece-part design matrix (Figure 16.9). The decisions that are shown in the columns of the design matrix are the type of detailed design decisions that are usually shown in a single design matrix in basic QFD. In this example, concept

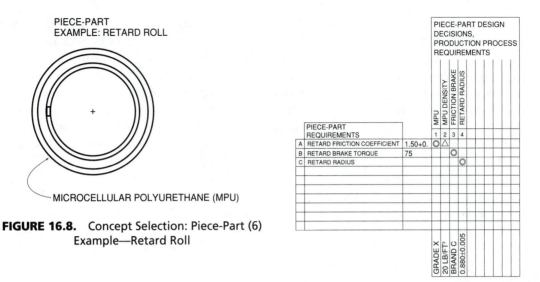

PIECE-PART
EXAMPLE: RETARD ROLL

MICROCELLULAR POLYURETHANE (MPU)

FIGURE 16.8. Concept Selection: Piece-Part (6)
Example—Retard Roll

PIECE-PART DESIGN
DECISIONS,
PRODUCTION PROCESS
REQUIREMENTS

PIECE-PART REQUIREMENTS			MPU	MPU DENSITY	FRICTION BRAKE	RETARD RADIUS					
			1	2	3	4					
A	RETARD FRICTION COEFFICIENT	1.50+0.	O	△							
B	RETARD BRAKE TORQUE	75			O						
C	RETARD RADIUS					O					
			GRADE X	20 LB/FT³	BRAND C	0.880±0.005					

FIGURE 16.9. Piece-Part Design Matrix (7)

selection has been used at the total system architecture level, the subsystem level, the piece-part level, and the requirements have been deployed through the total system design matrix and the subsystem design matrix, all in order to enable us to reach the point of making detailed design decision.

The example that has been described demonstrates the deployment through the different levels of a complex product, with the use of concept selection at each level. Now the static/dynamic (S/D) status will be described in more detail. Here we switch the case study mode from prescriptive to descriptive. The example is the Xerox 1075 copier, which includes the paper feeder that has been used in the above example (Figure 16.10).

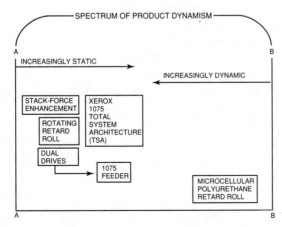

FIGURE 16.10.   Static/Dynamic Status Examples from
the Xerox 1075 Copier

The Xerox 1075 copier was developed during the late 1970s and early 1980s and was put into production in 1981. Its TSA is shown in Figure 16.4. This system architecture was a significant departure from previous copiers and was the outgrowth of a major advanced development activity that was known in Xerox as the third-generation activity. It is reasonable to think of this as having been at the time a rather dynamic system architecture. It has since been incorporated, with relatively small changes, into other Xerox copiers and printers, such as the 1090, 4050, and 5090. It is still today given serious consideration for new copiers and printers. Therefore, we can describe this TSA as having been dynamic in the late 1970s and early 1980s and having remained relatively static since then.

In the TSA decisions for the 1075, it was decided to use a friction-retard feeder. At this level of decision, the feeder could not be thought of as dynamic. Friction-retard feeders had been long used, probably having first originated in the late nineteenth century. At the time of the development of the 1075, the most recent and comparable friction-retard feeder was in the Xerox 9400 copier/duplicator. The 9400 had been put into production in 1976.

From 1973 until 1980, there was much advanced development work done on the friction-retard feeder. At the time that the product engineering work was started on the 1075, three functional areas had emerged as dynamic as a result of successful activities in advanced development. These three functional areas were the feed area, the separation area, and the drives area, as shown in Figure 16.6. The feed area is the area where the feed belt rests upon the paper stack; the separation area is the area between the feed belt and the retard roll, where the second and any ensuing sheets attempt to come along with the first sheet are retarded and prevented from going into the copier prematurely. The drive system drives the feed belt and the takeaway rolls which are used to pull paper away from the separation region and feed it into the downstream part of the copy handling module. The three dynamic functional areas are described in more detail below.

The first dynamic concept was for the separation functional area. It consisted of the retard roll mounted on a slip brake, which has already been described. This conceptual enhancement intrinsically increases the robustness of the system. In this case, the robustness is increased with respect to the noise of time and use.

Detailed studies in advanced development had led to the conclusion that the feed force is a very critical parameter and is often either too large or too small. When the feed force is too large, it tends to cause multifeeds; when the feed force is too small, it tends to cause misfeeds. It might seem simple to find an optimum feed force which will avoid both. However, this is tremendously complicated by the noises presented by paper and environmental conditions. With some kinds of paper and environmental conditions, the paper feeds very easily, and there is no danger of misfeeds, but there is a significant danger of multifeeds. For such paper one would prefer to have a small stack force. However, other papers and conditions tend to have the opposite effect. They require a strong force to avoid misfeeds but have little tendency to multifeed. For the first type of paper and environmental condition, one would like to have the stack force set at a small value—for example, 0.3 lb. With the second condition, one might like to have the stack force set a significantly higher value, such as 0.7 lb. To have the customers like the machine, we don't want to ask the customer to determine the conditions at the time they are making the copy and set the normal force accordingly. These evaluations led to the second dynamic concept, U.S. patent No. 4,561,644. The essence of this invention is that there is a sensor at the separation zone. The stack force is normally at a low value, approximately 0.3 lb. If the sensor in the separation zone does not see the lead edge of the sheet by the critical time in the cycle of the copier, a solenoid is activated which increases stack force to approximately 0.7 lb. This then avoids the incipient misfeed. Thus, this invention keeps the stack force low, minimising misfeeds, except in those critical situations when larger stack force has been proven by the actual operation to be required. The system, in essence, makes its decision on a sheet-by-sheet basis. This conceptual enhancement made the basic subsystem concept inherently much more robust.

The third dynamic concept was in the area of the drives. The conventional system had the takeaway roll and the feedbelt driven together. The problem with this is that when the trail edge of the first sheet pulled out from the area where the feedback contacts the stack, the belt was exposed to the second sheet and moved it forward. This, of course, was tending to aggravate any tendency for the second sheet to be fed into the downstream portion of the copy-handling module during the same cycle with sheet one—in other words, a multifeed. Evaluation of this led to the conclusion that there ought to be separate drives for the feedbelt and the takeaway roll. Then, during the operation of the takeaway roll, the feedbelt would not be operated, thus creating a higher potential gap between the trail edge of the first sheet and the lead-edge of the second sheet. Note that this too provided conceptual enhancement of the robustness of the paperfeeder.

Going from the subsystem level down to the component or piece-part level, an example is the material for the surface of the retard roll. For such a critical element, concept selection is certainly warranted. However, in this case the material that was selected was microcellular polyurethane, which was verified to be relatively static.

The S/D that have been described are summarised in Figure 16.10.

There is a critical question for the paper feeder. Its basic concept, friction-retard feeder, was static, but because it had three highly dynamic areas, the engineers felt that they were being very dynamic. The feeder was certainly very dynamic with respect to previous friction-retard feeders, but what about alternative concepts, such as vacuum feeders? The only safe approach is to take the viewpoint of the customers, who care little for the engineers' technical brilliance in creating the three dynamic concepts. In the final analysis it is dynamic cost-performance improvements that are needed. Otherwise, innovation may lead to differences, not improvements. In this example, the paper feeder that has been used as an example has continued to be used in the other Xerox copiers and printers, so it was clearly quite dynamic in 1980. The friction-retard feeder has been held static since then and has vied for applications with a slightly newer concept, the top-vacuum-corrugation feeder (TVCF).

In concluding this section on *concept selection, levels, and S/D status,* we note that production process engineering is done at the same time as the design activities, as indicated in Figure 16.2. Then the production process engineering decisions are deployed into production operations characteristics.

In summary, this section has described the systematic deployment of specifications through the levels of a complex system, the evaluation of static/dynamic (S/D) status for each level, and concept selection at each level.

ANALYSIS OF CONTEXT

In the beginning, as the new product development programme is being formulated, it is essential to analyse contextual information from the world economy. Two techniques that have been found especially useful are *parametric analysis* and *matrix analysis (reverse concept selection)*. Both can be thought of as forms of competitive benchmarking (CBM) and supplement and assist the competitive benchmarking that is already imbedded in basic QFD.

Parametric analysis (Pugh, 1982) consists of plotting two characteristics against one another. An example for aerial access platforms (cherry pickers) is shown in Figure 16.11. This was prepared in the early 1980s and shows the parameter *payload multiplied by height* plotted against *gross vehicle weight.* The data are plotted for very many (as close to 100 percent as possible) products. The data tend to fall into two categories. Furthermore, the better performance (higher line) tends to be associated with one concept, the articulated boom. The lower performance tends to be associated with an alternative concept, the telescopic boom. However, some articulated booms are on the lower line, which immediately reveals that they have failed to achieve the better performance of most articulated-boom products.

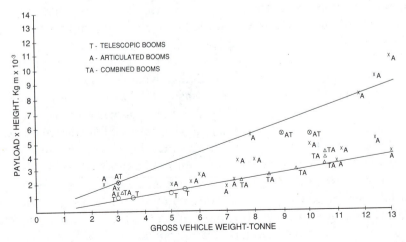

FIGURE 16.11.   Aerial Access Platforms

This example helps to illustrate the uses of parametric analysis: to evaluate S/D status, help guide concept selection, and guide major quantitative design decisions.

If every product lies on one straight line, it indicates that the product is being treated as conceptually static. Of great interest are products that lie above the conventional line, which indicates some form of change from convention. This could be simply due to an increase in efficiency (perhaps from using a new material) or conceptual breakout. The significance is that parametric analysis makes you investigate further. In doing concept selection, a concept should be sought that will be superior in performance, as revealed by the parametric analysis. Once a concept has been selected, the trend line guides the definition of the critical parameters. For example, if we select the articulated concept and require payload multiplied by height of 8 kg-m x 10^{-3}, then we should plan on a vehicle weight of 12 tons.

In performing parametric analysis, it is important to make many plots. Raw data should be plotted. Also, rational parameters are formed on the basis of dimensional analysis and simple analyses based on the engineering sciences. It is very useful to mount all of the plots on a wall (or walls), so that they can be viewed in relationship with each other. Unsuspected patterns often emerge. It is also valuable to analyse analogous products. An example of an analogous pair is the VCR and the audio tape recorder/player.

Matrix analysis is essentially the application of the Pugh concept selection process to existing products. The evaluation is done for existing products to select the competitive benchmark product(s) that are then used in the subsequent CBM activities in the house of quality (Hauser and Clausing, 1988) and in design CBM. This reverse concept selection activity also provides the concept team with insights into the strengths of existing designs, which is very helpful in the subsequent concept development and selection for the new product. An example for microscope manipulators for neurosurgery is shown in Table 16.1.

A slight variant on the direct use of reverse concept selection is shown in Table 16.2. Here there is not any evaluation but, rather, a simple checking of the presence or absence of possible features in the leading products. This feature map helps to identify the features that are required and those that might provide competitive advantage. Both parametric analysis and matrix analysis must be done early in a new development programme. For static concepts this contextual analysis should be completed before the concept development and selection for the new product is begun. For dynamic concepts, the contextual analysis and new concept development and selection will be iterative. They will guide each other and converge to compatibility as they are mutually developed.

STRUCTURE OF REQUIREMENTS

The rows and columns of the house of quality (Hauser and Clausing, 1988) contain the requirements for the new product system. The rows are in the voice of the customer, and the columns are in the language of corporate expectations. In basic QFD these are structured after the fact by the organisation into logical affinity groups.

TABLE 16.1. Matrix Analysis: Microscope Manipulators for Neurosurgery

No.	Criteria	Zeiss-Cont.	Zeiss-Univ.	Weck-Opmi 5	Wild-MSC	Wild-MSB	Wild MSF	Moller-Univ.	Amsco-Amscope	E.D.C. Micron 160	Cooper Vision	Zeiss (J) Slit Lamp	Zeiss Univ.S3	Wild MSC	Wild MSB	Wild MSF	Weck 10108	Weck (Ceiling)	Amsco	Keeler	Zeiss E.M.C.	Zeiss (J) 110	Zeiss (J) 150	Zeiss (J) 212	Zeiss (O) UNIV. S2	Zeiss CONT E.N.T.	Zeiss STD 1	Weck Mark	Zeiss Opmi 9	Zeiss Opmi 99	Zeiss T	Zeiss S	Olympus O.C.S.	Zeiss 80	Zeiss 6CFC/M
	Field	\<--- Neurosurgery ---\>									\<--- Ophthalmology ---\>										\<--- E.N.T. ---\>							\<--- Diagnostic ---\>							Colpo
1	Number of major mechanical parts (less the better)	–	D	S	–	–	–	–	–	S	S	+	S	–	–	–	–	–	–	S	–	–	–	S	S	–	S	+	+	+	+	S	S	S	S
2	Motorized movements (less the better)	–	A	–	S	S	–	–	–	+	S	S	S	S	S	–	–	–	–	+	–	+	+	+	S	+	+	+	+	+	+	+	+	+	+
3	Energized lockings (less the better)	–	T	S	S	S	S	–	–	–	S	–	S	S	S	S	S	S	–	S	–	+	+	+	S	+	S	+	+	+	+	+	+	+	+
4	Ease of manufacture (more the better)	–	U	–	+	+	–	–	–	–	S	+	S	+	+	–	–	–	–	+	–	+	+	+	S	+	+	+	+	+	+	+	+	+	+
5	Number of attachment carried by the manipulator (less the better)	S	M	+	+	+	+	S	S	S	+	+	S	+	+	+	S	S	S	+	S	+	+	+	S	+	+	+	+	+	+	+	+	+	+
6	Plug in sockets within the manipulator (less the better)	S	–	+	+	+	+	S	S	S	+	+	S	+	+	+	S	S	S	+	–	+	+	+	S	+	+	+	+	+	+	+	+	+	+
7	Number of control knobs (less the better)	+	D	S	–	–	–	+	+	+	S	+	S	–	–	–	S	S	+	S	+	+	+	–	S	+	–	+	+	+	+	+	+	+	–
8	Lightness of construction (more the better)	–	A	S	+	+	+	–	–	–	S	+	S	+	+	+	S	–	–	+	–	+	+	+	S	+	+	+	+	+	+	+	+	+	+
9	Number of nonmanual controls (less the better)	+	T	+	+	+	+	–	–	+	S	S	S	+	+	+	S	S	–	+	–	+	+	+	S	+	+	+	+	+	+	+	+	+	+
	Number of pluses	2	U	3	5	5	4	1	1	3	2	6	0	5	5	4	0	0	1	6	1	8	8	7	0	8	6	9	9	9	9	8	8	8	7
	Number of sames	2	M	4	2	2	1	2	2	3	7	2	9	2	2	1	6	5	0	3	1	0	0	1	9	0	2	0	0	0	0	1	1	1	1
	Number of minuses	5	–	2	2	2	4	6	6	3	0	1	0	2	2	4	3	4	8	0	7	1	1	1	0	1	1	0	0	0	0	0	0	0	1
	Placing order										6										5	3	3	4		3	6	1	1	1	1	2	2	2	4

It has been found useful to have a generic structure that is used to identify and compile the needs. The generic structure helps to improve completeness and speed in preparing the needs. A generic structure that has been found to be very useful in practice is shown in Figure 16.12.

The product design specification (PDS) shown in Figure 16.11 contains two types of elements: (1) most of the elements are the type of characteristics that are used in the traditional house of quality; (2) some elements of the PDS have not traditionally been included in the house of quality but have usually been included in the business strategy, product strategy, and product programme plan. The PDS should be used to structure both the rows and columns of the house of quality, at least initially. The product development team can decide to include all elements or include some in other related documents. The important necessity is to address all of them in an integrated way.

TABLE 16.2. Matrix Analysis: Microscope Manipulators for Ophthalmology

No.	Feature and function	Cooper Vision	Zeiss (J) Slit Ia.	Zeiss Universal	Wild MS-C	Wild MS-B	Wild MS-F	Weck 10108	Weck (Ceiling)	Amsco, Amscope	Keeler K-380 FW	Zeiss E.M. Ceiling	Graphic Representation of Percentage	%
1	Coaxial illumination				√	√	√	√	√	√			xxxxxxxxxxx	54
2	Moto. vert. course movement						√	√	√	√	√	√	xxxxxxxxxxx	54
3	Moto. vert. fine movement		√						√				xxxx	18
4	X–Y attachment	√	√	√				√	√	√		√	xxxxxxxxxxxxx	64
5	Motorised zoom (microscope)	√	√	√	√	√	√	√	√	√	√	√	xxxxxxxxxxxxxxxxxxxx	100
6	Motorised focus (arm)	·						√	√	√			xxxxx	27
7	Auto. step magnification		√	√	√	√	√			√	√	√	xxxxxxxxxxxxxx	73
8	Hand switch controls			√	√	√	√	√	√	√	√	√	xxxxxxxxxxxxxxx	82
9	Mouth switch													0
10	Foot switch	√	√	√	√	√	√	√	√	√	√	√	xxxxxxxxxxxxxxxxxxxx	100
11	Counterbalanced by weight									√			xx	9
12	Counterbalanded by spring	√		√	√	√	√				√		xxxxxxxxxxx	54
13	Energised locking		√							√			xxxx	18
14	Friction locking	√		√	√	√	√	√	√		√	√	xxxxxxxxxxxxxxx	82
15	Stepped locking				√	√	√						xxxxx	27
16	Rotation free around stand	√	√	√	√	√	√	√	√	√	√	√	xxxxxxxxxxxxxxxxxxxx	100
17	Vert. course mov. of assy.		√			√	√			√	√		xxxxxxxxx	45
18	Course mov. in horiz. plane	√	√	√	√	√	√	√	√	√	√	√	xxxxxxxxxxxxxxxxxxxx	100
19	Vertical mov. of the arm	√		√	√	√	√				√		xxxxxxxxxxx	55
20	Tilt of micros. attachm.	√	√	√	√	√	√	√	√	√	√	√	xxxxxxxxxxxxxxxxxxxx	100
21	Rotation of micros.attachm.	√	√					√	√	√		√	xxxxxxxxxxx	55
22	Yaw of micros. attachm.				√	√	√	√					xxxxxxx	35
23	Connection facil. for attachm.	√	√	√	√	√	√	√	√	√		√	xxxxxxxxxxxxxxxxxx	91
24	Stand levelling facilities			√									xx	9
25	Stand braking facilities	√	√	√	√				√		√		xxxxxxxxxxx	55
26	Fibre optic illumination	√		√				√	√	√	√	√	xxxxxxxxxxxx	64
27	Down limiting stop	√	√	√						√		√	xxxxxxxxx	45
28	Manual step magnification		√	√	√	√	√			√	√	√	xxxxxxxxxxxxxx	73
29	Slit illumination		√	√				√	√	√	√	√	xxxxxxxxxxxx	64
30	Manual zoom													0
31	Horiz. mov. on plane of arms		√					√		√		√	xxxxxxx	36
32	Floor mounted	√	√	√	√	√		√			√		xxxxxxxxxxxx	63

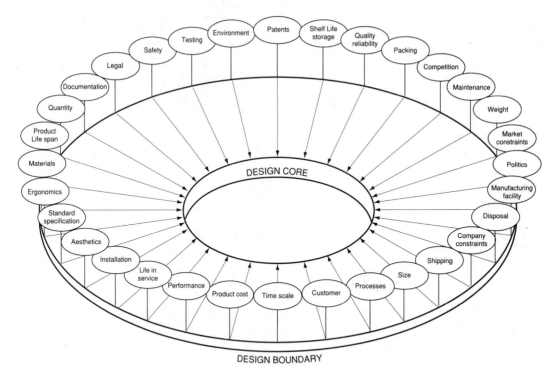

FIGURE 16.12.   Elements of Product Design Specification

It seems to be possible to further analyse and structure the specifications. We can distinguish between requirements that are imposed by the role of the product in carrying out a process (camera in making a picture, for example) and intrinsic characteristics, such as mass. More research and development is needed before this further analysis and structuring becomes a clear design method. However, designers are encouraged to attempt this on their own initiative, to the extent that seems useful.

As with contextual analysis, the specifications should be completed at the beginning of the development process for static products. For dynamic products some iteration will usually be necessary to achieve convergence of the concept and the specifications.

ADVANCED DEVELOPMENT VERSUS PRODUCT ENGINEERING

Product development work on the dynamic aspects of a new product is done primarily in advanced development, and static work is done primarily in product engineering. It is the purpose of having an advanced development organisation to develop the dynamic new concepts that cannot be readily done on the schedule and in the environment of product engineering for a specific product that is aimed at a specific market segment at a fixed date of introduction. Advanced development, the work on the dynamic concepts, is carried out until the new concepts are sufficiently mature so that it is judged safe to include them in a specific new product development programme for a new product. (The new concepts will then be held static for the duration of the specific product variant.) At this time technology transfers occurs from advanced development to product engineering. This technology transfer is usually best done by transfer of people. First one or more development engineers transfer from product engineering to advanced development to form a team with the advanced development people already working on the new concept. Then this team is transferred into product engineering and remains there until the new technology is completely transferred.

There are two modes of EQFD: one mode is for advanced development, and the other mode is for product engineering. In the advanced development mode, a few key improvements are selected for concentrated activity. EQFD is then used to help develop these few critical improvements. In the product engineering application, EQFD is used essentially as a product management tool.

In product engineering a specific product is designed and developed to be introduced to a previously chosen market segment, with a fixed date of introduction. The date of introduction is selected to coincide with a large, open market window for the market segment. To avoid missing the open market window, and to achieve the best possible improvement in cost and performance, the product development activity must be carried out in a very disciplined style with emphasis upon holding a very short and tight schedule. This usually precludes the complete development of major dynamic improvements within product engineering. The dynamic improvements typically take longer than is prudent or necessary for the development of a specific product. Also they have to be done in a somewhat different style, with more uncertainty and ambiguity and therefore more iterative trial and error. It is this distinction that has led to the formation of advanced development groups. These groups operate on a longer schedule and in a mode of iterative improvements to achieve sufficient maturity for the new concept so that it can now be treated as being essential static and incorporated in the short time schedule of the product engineering group.

When this distinction is successfully made, the primary activity of the product engineering group is to very quickly develop the new product, its accompanying production capability, and field support capability, with robust quality included. It is the primary focus in such product engineering activity to maintain a very short schedule with a minimum number of mistakes. It is not the purpose of such product engineering activity to invent dynamic new concepts, which would inevitably increase time and mistakes. The incorporation of some dynamic improvements that have been developed to maturity in advanced development will require changes in the surrounding static parts of the product. Also, additional engineering will be required at the next level of detail. For example, if a subsystem was developed dynamically in advanced development, then many piece-parts will need to be finally defined during product engineering.

Product engineering must be able to design and develop such accommodating changes and new piece-parts in a very short time with zero mistakes. There has been considerable confusion about this in the United States and Western Europe. Most product engineering has suffered from a desire to appear dynamic, with resulting significant shortfalls in the capability of being very good at the tremendous challenge of developing static concepts in a very short time with hardly any mistakes.

One of the greatest disfunctions with respect to static/dynamic status is the failure to leave a static concept alone. When a new product is being developed, certain functional areas should be selected to be dynamic and worked on in a mode to achieve dynamic improvement or to prove its static status. Everything else in the product should be kept as static as possible. There is an unfortunate tendency on the part of the design engineers to redesign the areas of the product that should remain static simply because they think that the role of the design engineer is in all situations to produce a new design. Clarification of the distinction in the role of the design engineer and in the design process for dynamic functional areas and static functional areas would go far towards making major improvements in cost and time.

It is the objective of the work in advanced development to produce generic improvements. The improvements are generic in the sense that, after being developed to maturity in advanced development, they can then subsequently be applied to more than one product programme. In the past we have typically thought of these improvements in advanced development as being entirely new inventions, and we often called this activity *technology generation*. However, it is possible to make very important generic improvements that do not include new inventions. It is important to have an open mind about the nature of generic improvements that are most desirable at any given point in time. Enhanced quality function deployment, applied during advanced development, can be very helpful in identifying the generic improvements that will be most beneficial.

In developing generic improvements that are primarily new inventions, the process usually starts with strategic technology planning, which incorporates major needs of the customer with the emerging physical concepts that determine where the next major improvements in the technology probably lie. This is followed by the invention of some new concepts which seem capable of meeting those improvements. These are enhanced by laboratory experimentation and analyses, which are best thought of as continuations of the invention process. The Pugh concept selection process, contextual analysis, and matrix analysis can be used most effectively at this time to help lead us to the dynamic concept that will be most beneficial at improving product function and/or cost/performance ratio.

An example of such new concepts that were developed in an advanced development activity are the three functional areas which made the friction-retard feeder a dynamic subsystem for the Xerox 1075 copier. All three of these improvements were aimed at improving the feeding and separating performance of this friction-retard feeder relative to earlier friction-retard feeders, such as those that had been incorporated in the Xerox 9200 and 9400 copier-duplicators.

A generic improvement that in some sense may seem a little less dynamic than the functional areas of the Xerox 1075 feeder is the improvement in rust prevention that was made in Toyota Autobody in the late 1970s. This was also done as a generic development activity. Rust prevention was one of the four major areas that had been selected for concentrated improvement. This selection was made with the use of quality function deployment, especially QFD1. This activity led to huge improvements in the rust resistance of Toyota vehicles.

Both the dynamic improvement of the three functional areas in the Xerox 1075 feeder, and the perhaps somewhat more static but major and comprehensive improvement in Toyota rust prevention, are good examples of different styles of activity in creating generic improvements during advanced development. Both were very generic; the Toyota improvement was initially developed for vans but has subsequently been applied to all vehicle bodies produced by Toyota Autobody. The Xerox paper-feeder was initially in the 1075 but has since been in the 1090, 4050, and 5090.

The three inventions that were in the Xerox 1075 feeder illustrate another point about the maturation of dynamic concepts to enable their selection for a specific product. After their invention, the feeder was optimised for robustness (Taguchi and Clausing, 1990) in advanced development. Verification of robustness was a key step in evaluating the maturity of the technology and concluding that it was eligible to be selected for inclusion in a specific product.

The three inventions in the Xerox 1075 were done in advanced development, the result of considerable study and evaluation of the basic functioning of a friction-retard feeder. This was applied research. This demonstrates the need for more time than is appropriate for the development of a product in product engineering.

The Pugh concept selection process is used in both advanced development and in product engineering. In advanced development all concepts are eligible. In product engineering the only concepts that should normally be considered are (1) existing concepts in the marketplace and (2) concepts that have been matured in advanced development.

To summarise, checkout for dynamism is best done in advanced development. The work that is done on the product areas that are judged to be static is done primarily in product engineering. Both types of work are extremely challenging to the practitioners, with the challenges being quite different in the two activities.

SUMMARY

The enhancements that have been described are contextual analysis, structured requirements, static/dynamic (S/D) status evaluation, Pugh concept selection, and deployment through the levels of the product. Contextual analysis, S/D status evaluation, and concept selection can be done for production processes as well as for the product. In addition, the relative roles of advanced development and product engineering, and the interaction of these with the enhancements, have been described.

The benefits of basic QFD are very strong: shorter development time, smoother entry into production, features that appeal to customers, lower manufacturing costs, and better quality. The improvements are commonly large—30 to 200 percent. Enhanced QFD (EQFD) greatly increases the probability that the right product will be developed. It also makes QFD explicitly applicable to complex products with multiple levels in the system.

EQFD also further reduces development time. This is especially true if the development time is averaged over all of the development programmes. EQFD will greatly reduce the number of product development programmes that are cancelled after wasting considerable resources.

REFERENCES

Buzzell, Robert D., and Gale, Bradley T. (1987). *The PIMS Principles*. New York: Free Press.

Hauser, John R., and Clausing, Don. (1988). 'The House of Quality.' *Harvard Business Review* (May-June), 63–73.

Hollins, Bill, and Pugh, Stuart. (1990). *Successful Product Design*. London: Butterworths.

Miles, Lawrence D. (1972). *Techniques of Value Analysis and Engineering* (2nd ed.). New York: McGraw-Hill.

Phadke, Madhav S. (1989). *Quality Engineering Using Robust Design*. Englewood Cliffs, NJ: Prentice-Hall.

Pugh, Stuart. (1977). 'Creativity in Engineering Design: Method, Myth or Magic.' *Proceedings of the SEFI Conference on Essential Elements in Engineering Education*. SEFI.

Pugh, Stuart. (1979). 'Give the Designer a Chance: Can He Contribute to Hazard Reductions?' *Product Liability International* 1(9) (October), 223–225.

Pugh, Stuart. (1981). 'Concept Selection: A Method That Works.' *Proceedings of the ICED* (pp. 497–506). Rome: ICED.

Pugh, Stuart. (1982) 'A New Design: The Ability to Compete.' *Proceedings of the Design Policy Conference* (pt. 4, pp. 12–16). London: DPC.

Pugh, Stuart. (1983). 'The Application of CAD in Relation to Dynamic/Static Product Concepts.' *Proceedings of the ICED* (pp. 564–571). Copenhagen: ICED.

Pugh, Stuart. (1984). 'Further Development of the Hypothesis of Static/Dynamic Concepts in Product Design.' *Proceedings of the International Symposium on Design and Synthesis* (pp. 216–221). Tokyo: ISDS.

Pugh, Stuart, and Smith, D.G. (1974). 'Marathon 2550: A Successful Joint Venture.' *Design Engineering* (August).

Pugh, Stuart, and Smith, D.G. (1976). 'CAD in the Context of Engineering Design: The Designer's Viewpoint.' *Proceedings of the Second International Conference on CAD* (pp. 193–198). CAD

Taguchi, G., and Clausing, Don. (1990). 'Robust Quality.' *Harvard Business Review* (January-February), 65–75.

**PART
IV**

CAD and Knowledge-Based Engineering

COMMENTS ON PART IV

After he showed the big picture and examined operational capability in Parts I, II, and III, Stuart Pugh turned in Part IV to a discussion of computers to implement design process and methods. Since the earliest use of computers in engineering and design in the 1950s, their promise for increased effectiveness and efficiency has been great. However, as Stuart pointed out, if computers are used to implement flawed design processes and methods, then the result will be rubbish.

This part emphasizes one of Stuart's favorite themes—the distinction between static and dynamic design concepts. Turbines are largely conceptually static, while biotechnology is dynamic. Computers are inherently of most value in static design. They can capture the culminations of long experience and make them available in a form that is easy and reliable to use.

However, computers are of very little value in the synthesis of dynamic designs. The total designer must be careful not to be lulled into a false sense of commercial security by proficiency with the computer. If this happens, then the competition may rush out ahead with dynamic concepts. Stuart loved to poke fun at design approaches that were very proficient at static design but largely ignored dynamic concepts. He often referred to people who promoted this approach as *hominem turbinus*. They were good at static products, such as turbines, but seemed to under value the synthesis of dynamic concepts.

Computers are very helpful with some aspects of total design but not with other aspects. As Stuart writes in Chapter 17, CAD in the Context of Engineering Design: The Designer's Viewpoint, computer 'programs will cover only but a small part of the totality of any design.'

This chapter was first presented in 1976, and computers have advanced greatly since then. This chapter reminds us of the great progress we have made with computers, and the great limitations that they still have.

In Chapter 19, The Application of CAD in Relation to the Dynamic/Static Product Concept, Stuart further developed his theory of static and dynamic concepts. It presents Stuart's assertion that the Model T Ford established the dominant concept of the automobile, which has remained conceptually static ever since. Of course, cars have improved greatly since the demise of the Model T in the mid-1920s, but this has been primarily at the subsystem level. Has the concept of the total system architecture of the automobile remained static since the Model T? Dynamism can enter at different levels of the product system (which is one of the main points of Chapter 16).

Finally, Chapter 22, Knowledge-Based Systems in Design Activity, helps to define the appropriate role of knowledge-based systems. It also presents the powerful distinction in design process between dynamic and static design. In dynamic design the product specification leads to the product concept. In static design the concept leads to the specification. This distinction is often not recognized, particularly by people who are only familiar with static industries.

In reading these chapters, look for the role of CAD (it should be responsive to the design activity model), CAD primarily for static designs, the static/dynamic continuum of product concepts, and, in the past paper, an excellent specification for knowledge-based systems.

—Don Clausing

INTRODUCTION TO PART IV

Over the years I have given serious consideration to the question of computing in design and in particular CAD. Some fifteen years ago I was pressured to swing to CAD design entirely by computer—in my opinion, at the expense of design itself or at least its understanding. I resisted the temptations as I felt we did not then have sufficient understanding of design, which we have proved beyond reasonable doubt.

The chapters in Part IV discuss the role of CAD in design education and industry and debate whether the emergence of CAD/CAM has in fact inhibited design understanding and progress.

Are we further in the dark then we thought? Specific applications apart, there are firm indications that this might be true. Computers are used today in all fields, and, with certain exceptions, we are to some extent running on the spot: we get extremely fit but do not make much progress.

The topic has evolved in sympathy with design understanding and with the issues of what type of design needs to be considered as the catalyst to improve the relationship between design and computing. The topic of product dynamism—design based on static or dynamic product concepts—is introduced. It will be seen that CAD needs to be considered and applied differently according to whether the concepts are static or dynamic. This difference in status is covered in detail in this part.

A deep, wide-ranging search of the literature on knowledge-based systems supports these differences and reveals quite a schism between design practitioners. It seems that even today, worldwide, there is still an overwhelming tendency to consider CAD as design per se. The computer dominates, particularly amongst academics, usually detached from design in practice or at best concerned with partial design. If this tendency away from practice persists, one can visualise the main work on CAD migrating back to a computer science activity. Papers with an industrial base reveal these problems, which is not as evident with those of academics.

CHAPTER

17

CAD in the Context of Engineering Design: The Designer's Viewpoint

ABSTRACT

It is the author's view that there is a widespread belief that engineering design can be reduced to that which can be put onto a computer, and that there is a need to put CAD into the context of the totality of design. The conceptual stage of design in relation to CAD is discussed at some length in this chapter, followed by a consideration of the detail stage. While computers are an extremely valuable analytical tool, there are many areas of design in which they have no application.

With Douglas G. Smith, in *Proceedings of the Second International Conference on Computers in Engineering and Building Design* (London: ICCEBD, 1976), 193–198.

Speaking as practising designers of successful machines and products over a number of years, the topic of computer-aided design is causing increasing concern to us in terms of its relative importance in the spectrum of engineering design. The problem, in our opinion, is that it is becoming simultaneously both a hindrance and a help to the designer, and to distinguish between what is in fact hindrance or help has become difficult. If the experienced designer has this difficulty, what then are the implications for the young designer who has yet to acquire experience? The main aim of this chapter is to highlight some of the practical difficulties into which the unsuspecting may unwittingly stray, if they fail to recognise and distinguish between cases where the computers can be of valuable assistance to the designer and others where they can be a positive hindrance. This distinction is most difficult to detect at the onset of any design programme—namely, at the conceptual stage, which is very much in the realms of synthesis as opposed to analysis. Computer programs claiming to synthesise a design, may, in fact, be doing just the opposite. The conceptual stage of design is indeed the most difficult, in that having taken decisions in this area, all else then follows, albeit not irrevocably.

It is our intention in this chapter to explore the conceptual stages in some depth and illustrate our thinking by reference to case material of which we have first-hand experience. By such means we hope to demonstrate our dichotomy. We then move into the areas beyond this stage and again demonstrate the use of the computer as a valuable analytic aid, without which our designs would tend towards mediocrity.

THE CONCEPTUAL STAGE

The conceptual stage of any design is or should be concerned with synthesis, which the *Concise Oxford Dictionary* defines as 'Combination, composition, putting together, (opp. analysis) building-up of separate elements, especially of conceptions or propositions or facts into a connected whole, especially a theory or systems.'

There are always two extremes of approach to design conceptualization. On the one hand, solutions are constrained by existing practice, whilst on the other, there is freedom to adopt a blank sheet of paper approach, unconstrained by practice. Reality is very rarely at either of these extremes, and depending upon such factors as the type of industry, type of problem, one is usually working somewhere between them. The implications of this in relation to CAD is we feel best illustrated and understood by reference to Figure 17.1.

By defining the conceptual stage of design, for convenience, by boundaries A and B and designating the area between them as the 'conceptual envelope,' criteria may be es-

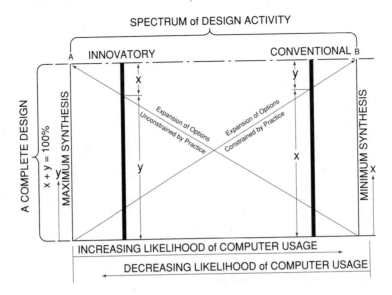

FIGURE 17.1. Spectrum of Design Activities

tablished for the relationship of CAD with this most important stage of design. In many companies, the main design emphasis is centred around developments based on well-proven concepts, represented by boundary B—that is, the reengineering of existing concepts to meet a new set of circumstances. The designer in this type of company may, for good reason, have to work within the constraints of existing practice and therefore may have little need or inclination to explore concepts beyond the immediate horizon. Such a design would demand a minimum of synthesis in broad terms; in other words, design procedures, codes of practice, regulations and analytical methods will have been developed, some of which may well be computer based. In fact, it is in such situations that CAD is likely to have been developed, some of which may well be computer based. In fact, it is in such situations that CAD is likely to have the highest utilisation. Use of established programs to the full will expand the designer's choice of options but, and this cannot be overemphasised, only within the contracts of existing practice, in the area of boundary B. It is frequently the case that designers working in such a situation think that they are working over a much broader area of the conceptual envelope and even that they are working at boundary A, completely unaware of the fact that they are actually at boundary B. Boundary A, which we shall call the 'blank sheet of paper' end, is the area of innovation or creativity (call it what you will), in that it requires maximum synthesis. We find that computer programs are of little help in this area. At this stage one has not the faintest idea of the nature of the ultimate design. How, therefore, is it possible to resolve the problem by recourse to CAD, which by definition presupposes an existing system?

We now refer to specific examples in order to validate the hypothesis of conceptual boundaries.

Thirty-Foot-Diameter Parabolic Dish Aerial for Trophospheric Scatter Communication System

Some ten years ago it had become the practice to design aerials of this type in the form of an open lattice back structure, consisting of radial rib members disposed around a central hub and interlinked by suitable circumferential members, with interbracing between ribs as shown in Figure 17.2. The backing structure was considered to carry all the load from the reflector surface, which in many cases was made up of a steel mesh. The procedure for analysis of the backing structure was therefore based on the assumption, which is quite valid, that the mesh does not contribute to the strength and stiffness of the dish. Computer programs were therefore, quite naturally and rightly, developed to analyse and optimise such structures with speed and accuracy.

Dishes had been developed and tested to underwrite the analytical work and therefore technically the design loop had been closed. However, production costing showed that dishes to this design, which we shall call design 1, would be too expensive. Concurrent with this work, the company had obtained a large contract for systems overseas, and at this cost the dishes were greatly overpriced. What was to be done? It was known that a dish to meet the specification for surface accuracy, and based upon this concept, would yield little to value analysis or cost-reduction techniques, as these had been applied during the design and development programme. It might now be said that dishes based on this concept were firmly placed at boundary B.

It quickly became obvious that not only was a new concept required but it was required quickly in order to meet contractual commitments. One was therefore forced to boundary A, the blank sheet of paper end. The computer programs, so valuable to us whilst designing within the safety and security of accepted practice, were now of little value in this context. Design 2 was produced as shown in Figures 17.3 and 17.4, based upon a system of propped cantilevers radially disposed around a central hub. These members were interlinked with circumferential members, the reflector surface being designed as a series of interlocking sectors, which had the property, when whole, of a shell. Adequate connection of this shell to the backing structure thus ensured that all the component parts of the dish would then carry load.

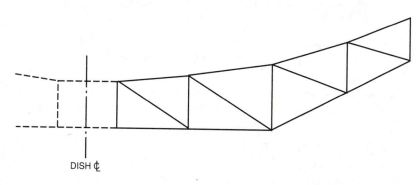

DISH ₵

FIGURE 17.2.   Design 1: Conventional Open Lattice Structure

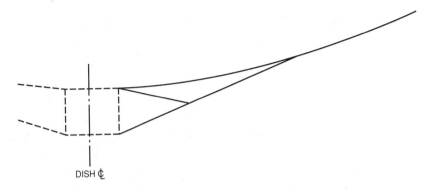

DISH ₵

FIGURE 17.3.   Design 2: Propped Cantilever Structure

FIGURE 17.4.   Photo of Parabolic Dish Aerial for
Tropospheric Scatter Communication

Within the allowable timescale for design, the analysis of the structure had to be done manually in an imprecise manner relative to design 1, with a degree of overdesign and the likelihood of the structure being suboptimum in analytical terms. These feelings were confirmed by static testing of the first dish to this design, which proved to be stiffer and stronger than required to meet the specification. However, the unit production cost of design 2 turned out to be acceptable, being about 50 percent of the cost of the established concept. The interesting point which begins to emerge here is that the generation of a new concept at boundary A will in itself give rise to computer programs based on this concept and can happen only in this order. The constraints of existing practice at boundary B had to be broken, acknowledging the fact that options could be expanded within these constraints to attain the expansion of options which the situation demanded and which all such design situations at boundary A demand.

This example is, we feel, adequately illustrative of designs based at boundaries A and B. Such is our conviction that the conceptual stage of design is almost the most important and in order to emphasise and highlight the vagaries of the conceptual boundaries, we now refer to a more recent example, taken from a postgraduate project but nevertheless a real-life situation—the chassis design of the Marathon 2550 dump truck (Pugh and Smith, 1974).

Marathon 2550 Dump Truck

The design of the vehicle started off effectively at boundary A with a blank sheet of paper approach. However, the design quickly moved towards boundary B and remained there, this being evident from the final design (see Figure 17.5). Although similar in appearance to other dump trucks, the arrangement of the subsystems allowed for considerable creative thought, which kept the design work away from boundary B.

FIGURE 17.5.   Photo of the Marathon 2550 Dump Truck

One subsystem was the design of the chassis, which from the outset was recognised as being critical, as it is highly interactive with the total vehicle concept, which in turn impinges upon reliability and ultimately the cost. As the design proceeded, it was recognised that there was considerable likelihood of using a computer program to analyse the stresses and deflections in the chassis. Time constraints did not allow writing a program from scratch, which would have been unwise in any case, due to work already in existence. Accordingly, a suitable program was found (Wardill, n.d.) which analysed grid structures but, as is not unusual, required development before it could be applied to our requirements. Before the computer program could be used at all, the general form of the chassis had to be decided. This depended upon other factors quite independent of purely analytical requirements, such as the type of steering, position of engine and driver, in fact upon the concept of the vehicle itself. The chassis in its final form, as it is now manufactured, is shown in Figure 17.6.

Having decided upon the chassis form, the computer program was used to refine the form and optimise the dimensions of the sections. However, it was necessary to validate the program by predicting the points and modes of failure of an existing dumper chassis under simulated loading conditions and to gain confidence in its use. A chassis was fatigue tested on a rig, and good correlation was achieved between the predicted and actual results.

A prototype chassis of the new design was built and tested. Trouble was in store: the prototype quickly failed. Everything looked all right 'printout-wise'! Luckily, we did not fall into the trap of producing endless printouts but asked the question, 'Why?' and after

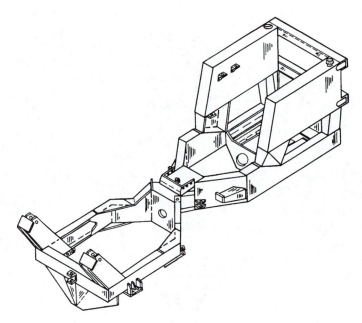

FIGURE 17.6.   Chassis of the Marathon 2550

much heart searching, an obvious factor emerged. We had failed to appreciate the limitations of the program in relation to the geometry of the new chassis design, which consisted mainly of folded box sections, giving rise to an unstable section, which the program did not identify. Needless to say, the design was modified and has since given completely satisfactory service.

Without the computer program it is unlikely that we should have achieved the target cost level for the chassis. This example demonstrates again that the use of a program is possible only after creative thought. It also provides a warning of the need to understand thoroughly the limitations of programs from an engineering point of view before using them.

The above examples are illustrative of the designer's continual predicament at the onset of any design. In other words, situations such as this are almost a daily happening, and one is continually exploring the spectrum between the conceptual boundaries.

We should like to bring the conceptual discussion to a close by reference to a current example, concerning a product which has completed prototype testing, and is now coming on the market.

The Giraffe Site Placement Vehicle

In 1973, we were associated with the design of a new type of rough terrain, materials handling vehicle for a specific market. We had little idea of the type, shape, form, or principle of the vehicle required, apart from the fact that a specification had evolved which contained a variety of requirements and constraints. To use various forms of linkage for the main superstructure, to power it hydraulically or not: these were the types of fundamental questions, to which, at the outset, we did not know the answers. Conceptually therefore, we were very much at boundary A. After the trial and iteration of many design concepts in most of which the overall principle of operation differed, the design shown in Figure 17.7 emerged, which basically consists of a pair of forks, mounted on an articulated carriage, which in turn is mounted on a telescopic boom, affixed to a four-wheel drive, four-wheel steer chassis.

Having finally arrived at this concept, computer programs were used to analyse and guide the design of the chassis and the telescopic boom. Once again, the order of things is as stated previously—conceptual stage followed by the interactive use of computer programs.

Figure 17.8 is illustrative of the intensity and direction of the activity of the projects discussed and relates to Figure 17.1 in terms of computer utilisation. It so happens that examples chosen are structural in nature and therefore form a useful set for comparison. However, the arguments propounded apply equally to any area of design.

THE DETAIL DESIGN STAGE

In publications dealing with engineering design, the impression is often given that the conceptual and detail design stages are completely independent. This is not the case in

FIGURE 17.7.   Photo of the Giraffe Site Placement Vehicle

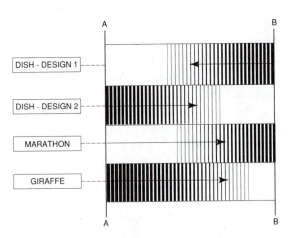

FIGURE 17.8.   Intensity and Direction of Design Activity

practice, where they are, in fact, highly interdependent, and the one phase merges into the other without clear demarcation. For example, the design of the chassis for the Marathon 2550 dump truck impinged on the design of certain details, which had to be taken into account before the chassis could be optimised by computer program.

There are many examples of computer programs which claim to 'design' such elements as cams, gears, shafts, and bearings. However, the use of the word *design* in such a context is misleading not only to some designers but also to many people outside the design field. Such programs have the advantages and limitations discussed in the previous section.

A recent example, is the design of a fly-jib for a mobile crane, which basically comprises a lattice structure, changing to a box section at the free end. One of the designs considered had the lattice members of varying length and constant cross-section, in line with normal practice. Such a design was amenable to analysis by a number of existing computer programs. However, such programs give little help in formulating the final forms of the lattice members, which, in order to minimise manufacturing costs, were made identical but had varying centroids. Without a parallel programme of empirical testing, it is impossible to determine with any degree of accuracy the criteria for failure of such members in compression. The degree of confidence in any program, modified or otherwise, to accommodate such members will be unknown. The jib was analysed manually, judgement, which cannot be programmed, being exercised in respect of failure criteria. Hence, it is an approximation to the optimum but nevertheless is being manufactured.

Situations such as this are typical of any design activity. How should the designer handle them in terms of CAD when under pressure to complete a design?

OTHER CONSIDERATIONS

We have mentioned some of the pitfalls into which the unsuspecting designer can fall in using CAD. Experience has demonstrated to us other factors which are subsidiary to the main theme of the paper, but worthy of mention since in a commercial environment they can be wasteful in the employment of personnel, in time and in money, and add to the designer's dilemma.

Identify the Problem

Designers usually think, as do most people, in terms of solutions rather than problems. In the case of a radar trailer, programs were being prepared to analyse a damper system which was giving trouble. By backtracking to ask the question, 'What is the problem we are trying to solve?' a new solution emerged, which changed the design and eliminated the need for anything but the most rudimentary analysis and simplified the design.

Mind-Shrinking

Where library programs are available, it is tempting to restrict solutions of the problem to suit the programs. For example, to use a conventional construction for the fly-jib mentioned earlier would have resulted in a structure capable of analysis by computer but not with minimised manufacturing costs. In certain situations the restriction of solutions to suit available programs may be valid, but in other circumstances such attitudes can inhibit thinking and computer programs can be 'mind-shrinkers.'

Simulation Not Synthesis

We have already discussed the relation between analysis and synthesis in relation to computer programs, but there can be equal confusion between simulation and synthesis. Computer programs represent only what has already been synthesised and are capable of allowing the exploration of options, only within the context of existing practice. Indeed, the programs themselves have restrictions which are within the limits imposed by existing practice.

Computer or Calculator?

Finally, it must be realised that the computer, which the designer accepts as an analytical tool, is only one of a range of tools from the humble slide rule through the wide selection of pocket calculators and programmable desktop calculators to digital, analogue, and hybrid computers. Designers have to choose, within the tools available to them, which is the most suitable for the task in hand. In a study carried out on the turbo-charging of diesel engines, a desktop calculator was found to be quite adequate in the context of the study for carrying out component matching. Existing computer programs were far too complex for the purpose and would have resulted in considerable extra expenditure and increased turn-round times without enhancing the end result.

The analytical aspects of design are important but are only one of the many factors that the designer has to take into account in a total design situation. This paper would be incomplete if it did not make reference to these other aspects, in order that we do not fall into the trap of shrinking design to that which can be put onto a computer.

EXTENT OF INTERACTION IN ANY DESIGN ACTIVITY

In reality, the design of anything from battleships to buildings embraces many more factors than we can possibly touch upon in this chapter. Whilst computer programs are being developed to include specific factors in particular areas, it must be recognised that in all cases such programs will only cover but a small part of the totality of any design. We made reference earlier to judgement not being programmable. Indeed, according to the type of design being undertaken, it can be said that there are as many qualitative factors as quantitative factors inherent in any design. For instance, how does one program such aspects as aesthetic appeal, ergonomic considerations, market influence, and shipping constraints? It is necessary not only to program them but also to retain an understanding of the main objectives of the design in a totally interactive sense.

We would suggest that coverage of design in such a manner, in CAD terms, is impossible, and yet this is what the practising designer is striving for in day-to-day activity. If a designer fails to obtain the best balance or compromise between all the factors, then in fact the design may be said to have failed (Pugh, 1974). A company may go out of business, lose a market, or reduce its turnover if a product does not sell. In this situation, the knowledge that the dynamic analysis of the design had been optimised to the nth degree will give little satisfaction to the designer.

We must therefore take pains to ensure that we keep CAD in context, since properly developed and applied it can be a boon. However, we consider that we are in grave danger of deluding ourselves that CAD encompasses the totality of the design activity. There is already strong evidence, and in some quarters disquiet, that young engineers with design potential truly believe that the computer can do everything. As any mathematician will tell you, to quantify infinity is an impossibility; therefore, you do not attempt it! In the same context, to attempt to quantify an infinity of opportunities is also impossible. Indeed in many instances it stops designers from thinking in an all-embracing manner, effectively switching them off.

CONCLUSIONS

If we have given the impression that we have adopted a negative attitude towards CAD, let us assure you that nothing could be further from our minds. In all humility, what we have attempted to do is probe and to highlight the sort of situation with which any designer may be confronted. In particular, we sincerely hope that this chapter will prove useful not only to designers involved with total situations and using CAD but also to CAD specialists seeking to gain a better insight into a total design activity. We must state with absolute conviction that CAD used with circumspection is a great asset. Let us try to keep it that way.

ACKNOWLEDGEMENTS

The authors wish to express their thanks to GEC-Marconi Electronics Ltd for permission to publish material relating to the 30-foot-diameter parabolic dish aerial and to the Liner Concrete Machinery Co Ltd for permission to publish information in regard to the Marathon 2550 dump truck and the Giraffe site placement vehicle.

REFERENCES

Pugh, S. (1974). 'Engineering Design: Towards a Common Understanding.' *Second International Symposium on Information Systems for Designers* (pp. D4–D6). University of Southampton: ISD.

Pugh, S., and Smith, D.G. (1974). 'Dumper Truck Design Highlights Industrial/Academic Co-operation.' *Design Engineering* (August), 27–29.

Wardill, G.A. (n.d.). 'Small Computer Procedures as Tools for Structural Designers.' Advanced School of Automobile Engineering, Cranfield Institute of Technology.

CHAPTER

18

CAD and Design Education: Should One Be Taught Without the Other?

ABSTRACT

Computer-aided design education and the place of engineering design within the spectrum of engineering education is a fairly recent innovation in educational circles. The inclusion of the former whilst the latter is still very much in embryo gives rise to certain problems which will be exacerbated unless the whole question is approached in an integrated manner. This chapter presents an integrated approach to this problem which does not offer a once-for-all solution but appears at least to minimise the problem. It also reconciles the place of computers not only in design education but also in relation to the practice of design.

From a paper presented at the International Conference on Computers and Design Education, London, 1977.

INTRODUCTION

The title of this chapter may be somewhat contradictory and controversial, but it has been chosen in order to give expression to an increasing disquiet in respect of engineering design education generally and the place therein of computer-aided design.

Is computer-aided design being treated as yet another specialism to add to the already long list of specialist subjects being taught in undergraduate engineering courses? Is it being done without an integrating framework that enables both the potential designer and specialist engineer to recognise and understand the place of his or her own activity within that framework? This chapter examines the overall question in some detail and offers a framework for integrating CAD into specialist teaching.

An approach to the teaching of computers related to design is suggested which attempts to correlate computing facilities and power with the requirements of the designer through the mechanism of actual designs. Particular attention is paid to the question of the appropriateness or inappropriateness of computer methods in order to provide a basis to enable the novice designer to better select his design aids.

ENGINEERING EDUCATION AT THE UNDERGRADUATE LEVEL

It is recognised that the teaching of undergraduate engineering courses has broadened over the years, long-established topics have been extended, and new topics introduced, but the understanding of newly graduated engineers as to just where and how the course fits them for practice remains as elusive as ever. This is particularly true of university graduates; those educated by the HNC and HND routes generally have a basis of application to which they can relate the specialist teaching.

The sandwich-type course, popular in so many U.K. universities today, goes some way to rectifying this deficiency, although recently doubts have been expressed as to whether in some instances the industrial training is integrated into a total approach or merely added on in an ad hoc manner (Smithers, 1977). Certainly, students entering the master of technology course in engineering design at Loughborough, either directly from an undergraduate course, from a sandwich course or indeed from industry, would appear not to have a sound understanding of the relationships between their taught subjects and the realities of practice. In spite of the apparent broader-based approach they are still being taught in subject compartments and the project work usually implanted to bring about the desired integration, the subjects themselves are not taught against an integrative background from the beginning. If this statement is true, is not our engineering educational structure still largely geared to the specialist engineer with only a token gesture to the engineering designer who above all else must have a substantial generalist capability? That we need specialist engineers is taken as real; we do not, however, need them to the almost complete exclusion of the generalist.

In structuring engineering education, it is our opinion that potential specialists and generalists should be taught against this background of integration, and this argument is reinforced by the fact that this is what happens in practice. With certain exceptions one cannot use, apply, or even sell the output of an aerodynamicist, stress analyst, or thermodynamicist in isolation. Their outputs require integration into a composite whole called design, and it is suggested that engineers be taught with this objective firmly in mind.

Figure 18.1 shows a structure for teaching based on the specialist topics which are absolutely essential to an engineer's education. This figure shows that any course should be taught with the distinction between the specialist and the generalist which is or should be the fundamental basis of practice. Design is always an amalgam of specialist subjects and topics, and the design or artefact does not exist that is not based upon such an amalgamation. Joselin (1970) states that 'in discussing creativity, engineering education must have substantial links in language and experience with engineering practice.' Surely this statement must apply to the whole spectrum of engineering education in addition to creativity. If, therefore, one accepts the hypothesis put forward, computers and computer-aided design fit into the same overall picture. Computing can be taught for the specialist and for the generalist. If, however, the specialist or generalist does not have an understanding based on the foregoing, the teaching of computer aids to design will at best fall upon unprepared ears and at worst add to the already existing confusion and misunderstanding of design and the place of computers in that realm.

If we now consider the progression from teaching to practice, Figure 18.2 shows practice utilising the output of specialists through the mechanism of the designer who has, as previously stated, the task of integration. Computers may be used to effect by

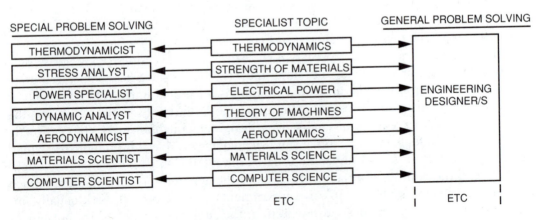

FIGURE 18.1. Teaching

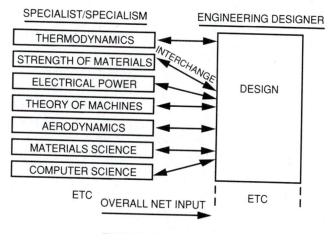

FIGURE 18.2. Practice

specialist engineers and designers, but it is also recognised that the designer may double as the specialist in many industries. Computers to most engineers fall firmly into the category of design aids unless one happens to be in the business of designing computers; they are therefore a powerful aid to design and cannot in themselves 'design' as such (Pugh and Smith, 1976).

The above reflects the authors' view on the spectrum of education related to engineering and design, with computers firmly established under the heading of aids. Engineering design students subjected to teaching against this backcloth seem to more readily assimilate not only the integrated nature of the 'practice package' but also to relate computer techniques to design in a sounder, more realistic manner. In fact, one might say they start to develop a mature discernment for the appropriateness or inappropriateness of computers in design.

This is perhaps a suitable point in this chapter to consider our approach to computer-aided design teaching by recourse to the stages present in the design activity related to the range of techniques and equipment available.

APPROACH TO TEACHING DESIGN

The proof of the benefits of CAD in any particular application is whether its use has resulted in some benefits—for example, an improved end product, a reduction in design and manufacturing costs, or a reduction in design time. It is extremely dangerous, therefore, to think of CAD in its own right rather than against the activity to which it is to be directly applied—the activity of design itself.

Engineering design is central to the operation of industry, whose function in broad terms is to supply products which meet the needs of the market. It is a complex activity, and for this reason it is difficult to comprehend and practice. However, if one is to be engaged in it, an attempt must be made to understand the component parts and their interrelations.

At its core the engineering design activity has a series of main phases—namely, market, specification, concept, detail, manufacture, and sales as shown in Figure 18.3. All products start with a market need, either real or potential; during the marketing phase the true needs should be established and other relevant information obtained leading to a written specification. The specification thereafter provides a boundary or set of criteria for the rest of the activity to which subsequent products must conform. During the conceptual stage which follows, outline schemes are prepared, culminating in a scheme which is fully designed in detail. Thereafter the detailed design is transformed into detail drawings and instructions for the purposes of manufacture, and the product is finally launched onto the market. It must be recognised that these core phases are highly interactive and iterative.

Allied to these core activities are techniques which the designer needs to make them workable, such as creativity, decision making, costing, analysis, and the like. These techniques are used by the designer as and when appropriate and add another dimension to the complexity of the activity.

What has so far been said relates to any area of design, whether a building or a consumer product, so that a further input is required—namely, technological information, which depends upon the product area in which the design is being undertaken. The technological inputs to design add yet another dimension of considerable extent and one which is rapidly expanding.

We have, during the current academic year, introduced our postgraduate students in engineering design to this approach, whereby techniques and technology are seen as inputs to the design core phases. We are certain that it has improved their ability to under-

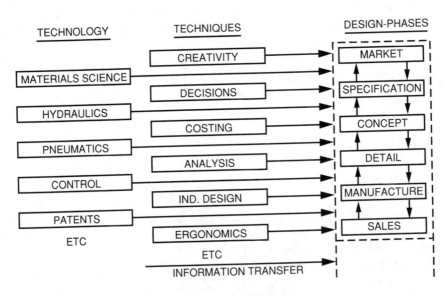

FIGURE 18.3. Relationship of Technique and Technology to Design Phases

stand the complex activity to design and also their ability to implement it. Further, we have introduced this approach in teaching design to postgraduate students in construction management and to undergraduate students in materials technology with the same effect and would suggest that it can form a basis for teaching in any area of design.

If we view the application of techniques and technology as information transfer, then we have information transfer into the design core phases as well as between them (see Figure 18.3). It is first of all necessary to consider the ways in which this information can be handled with particular reference to computing before discussing actual applications of computers to design.

LEVELS OF COMPUTING

Chambers' Twentieth Century Dictionary defines a computer as being 'a calculator, a large machine carrying out calculations of several stages automatically.' The latter part is that generally accepted, and it is certainly true that computer-aided design has traditionally been centred on large machines, whilst latterly minicomputers have entered the scene as their power has increased. However, if we think in terms of computing—the function to be carried out—then there are other means of doing so other than large or minicomputers. We may think of these as levels of computing, as depicted in Figure 18.4. While there are configurations of equipment which contravene these basic classifications, it is considered that they are the main options open to anyone wishing to compute.

The lowest level of computing is by manual means; it is not to be despised as under some circumstances it is the most practical. The lowest level of electronic aid is the hand calculator, followed by the desktop calculator, which may be supplemented with periph-

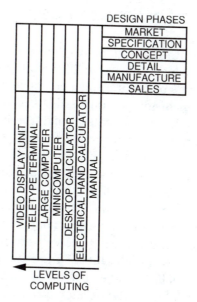

FIGURE 18.4. Computing Level: Design Phases

eral devices such as line printer and graph plotter. We then move on to the mini and large computers, accessed by punched cards or tape, and finally the supplementation of these machines by means of teletype terminals or interactive visual display units.

Developments in the field of electronics have substantially increased the power of small machines such that present-day desktop calculators can be used for operations which ten years ago would have required a fairly large computer. Thus desktop calculators fall within the scope of computer-aided design and must be included in any CAD education. However, it is our belief that CAD must be taught against a background of the function it basically carries out—that is, to compute—and this requires the progressive consideration of all computing levels from manual means onwards, or else inappropriate applications are bound to occur.

This approach has been taken for some years in teaching postgraduate students in engineering design, who are first encouraged to explore the lowest level and progress only to the higher levels where benefit will accrue. Indeed, in considering analytical techniques we encourage them to start with the simplest and increase in complexity only as and when necessary. However, even when equipped with a basic knowledge and experience of design, together with an understanding of computing levels, the decision as to whether to progress to a higher level is often not easy or obvious. We shall, by recourse to a specific project, illustrate what we have done in practice, in some cases to advantage, in others in error, indicating the reasons for the decisions made and the soundness of the applications. Calculations are only one form of information processing carried out by computers, but the above argument still holds good if 'levels of information processing' is substituted for 'levels of computing.'

COMPUTING APPLICATIONS IN DESIGN PROJECTS

If we relate the main phases of the design activity to the levels of information processing, we have potentially available to us all facilities at all phases as depicted in Figure 18.4. However, whilst the range of facilities may be potentially available, their application for a particular requirement may be impossible because of the characteristics of the equipment. Even if possible, an application might be inappropriate on account of some constraint or if appropriate may not actually be used.

Lest the argument becomes too philosophical at this point, and in order to illustrate the various categories, it seems best to consider the above-mentioned project carried out by postgraduate students. The particular project to be considered is the design of a $2\frac{1}{2}$ ton dumper which was later named the 'Marathon 2550' (shown in Figure 18.5). The project was carried out during the first eight months of 1971, during which period the large university computer, an ICL 1905, a teletype terminal with access to two time-sharing companies, and desktop calculators were immediately available to us. Furthermore, VDUs were available in other places. At that time hand calculators and minicomputers were in their infancy, and desktop calculators were generally less powerful than many current hand machines.

FIGURE 18.5.   Photo of the Marathon 2550

During the market investigation we did not find any application for computing facilities, and all our information was processed manually. The reasons for this were that we needed to have flexibility in arranging and manipulating the information, which was not possible using a computer. Indeed, it is for this reason that in all our projects so far we have found very little use for computing facilities at this phase. In our opinion for this project the manual level was the possible and appropriate one.

Moving onto the written specification, this was also prepared manually. It would have been possible to have had all the information stored on a computer. However, we considered that its use virtually as a file and typewriter did not justify the time and cost involved so that computer usage was possible but not appropriate in this particular case. As a matter of interest on another project we are currently using a desktop calculator with an extended memory and line printer for this purpose in order to assess its value. It has been chosen instead of higher-level facilities because of immediate availability and for the very practical reason that A4 paper can conveniently be used for the printout.

In the design of the dumper it was a requirement that the new design had to be as reliable as an existing design and also that the new vehicle had to be 10 percent cheaper. The chassis shown in Figure 18.6 was, therefore, a critical area and one which required a higher level of analysis than would have been practical by manual or desktop calculator methods alone. Time did not allow the development of a program from first principles, but a basic program was located at Cranfield Institute of Technology which with modifications was used for analysis of the chassis. It must be emphasised that the program did nothing to create the chassis concept; this was determined by other factors. It purely performed an analysis in terms of stresses and deflections on the configurations with which the computer was instructed. The use of a large computer was thus

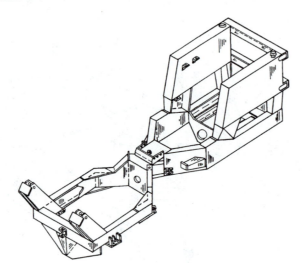

FIGURE 18.6.   Chassis of the Marathon 2550

the appropriate facility for the majority of the work. The use of the teletype terminal, although possible, was not appropriate due to the high cost of occupancy necessary. The work could also have been implemented using a VDU, but to our knowledge there was no program available, and by the time a usable program could have been developed the project would almost certainly have been completed. Furthermore, there appeared to be little to be gained by going above the computing level chosen.

An interesting example of the misapplication of computers arose at this stage in that one member of the team had the task of finding the best position of the skip lifting ram which he attempted to do by means of a computer program. After several weeks' work, no definite answer was forthcoming, and another team member with a simple cardboard model and a desktop calculator quickly arrived at an acceptable solution. In this case we consider that only levels up to the desktop calculator were appropriate.

During the detail design phase the use of the chassis program continued, and a program was also used to optimise the position and size of the hydraulic steering ram, positioned between the two halves of the articulated chassis. This program worked satisfactorily except that no account was taken of a ball joint. This resulted in the ordering of an incorrect ram on the authority of the computer printout. Fortunately this error was rectified. In this case all levels of computing were possible although purely manual methods would have been tedious, and the use of a VDU unnecessarily sophisticated. A teletype terminal or a desktop calculator could also have been used. In terms of the present day, such work would most certainly be carried out on a programmable desktop calculator.

A further example from the detail phase was the use of a program to analyse the chassis subframe for stresses and deflections, the program used for the main chassis being inappropriate due to limitations on the modes of loading. A suitable program was identified at one of the time-sharing companies and subsequently used.

Dealing briefly with the manufacturing and sales phases the use of computers was not possible in the former as the client at that time had no appropriate NC machines, and although sales data could have been stored at computer bureaux, the company did not consider it necessary as existing facilities were considered adequate.

Thus for the particular project considered, we have a pattern of computer application for the main design phases as depicted in Figure 18.7. Therefore, as with design itself, the choice of computing level is heavily dependent on the nature and structure of the problem, the availability of facilities to deal with the problem, and also the compatibility of the problem with the foregoing. For every project this will be a unique situation unless the area of design is heavily constrained by previous designs and practices, in any event each case must be judged on its merits.

We would suggest that in teaching students about CAD the system outlined in Figure 18.7 places the whole question of computing into perspective and hopefully provides a foundation of understanding for the student, leading rapidly to a sense of discernment in the use of computers. Used with circumspection the computer is a valuable tool; used indiscriminately it can lead to stifled designs and switched off designers.

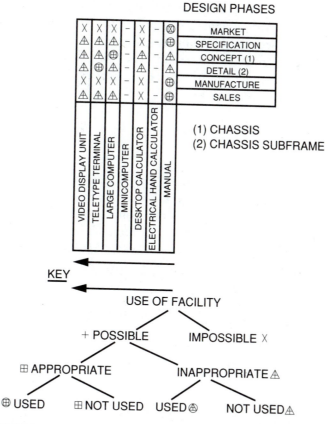

FIGURE 18.7. Computing Pattern for the Marathon 2550

CONCLUSIONS

From the foregoing, several principles emerge which we consider essential to the teaching of the specialism of computers and indeed any specialism in relation to design:

- The specialism must be taught against a background of design and not in isolation so that the relationship between the two is clearly demonstrated.
- A means of demonstrating this relationship is by building up the design activity from its core and at the same time developing techniques and technology as inputs to the core phases.
- Computing, considered as levels of information handling, should be developed progressively, considering manual means at the lowest level and working up to the most sophisticated equipment.
- Guidance should be given on the factors to be taken into account when assessing the possibility and appropriateness of computing level to a particular design application.

This approach has been adopted for several courses this year and has resulted in a much better understanding not only of the activity of design but also the relationship of the specialisms to this activity. Furthermore, this understanding has reflected to advantage in application to project work. Whilst there is still much to be done in the understanding and teaching of design, if a foundation is laid as suggested, students will be in a much better position to make valid judgements; they will also have a datum against which to relate their subsequent experience.

REFERENCES

Joselin, A.G. (1970). 'Creativity and Higher Education.' *Proceedings of the Conference on Creativity and Engineering*. Birmingham: University of Aston.

Pugh, S., and Smith, D.G. (1976). 'CAD in the Context of Engineering Design: The Designer's Viewpoint.' *Proceedings of the Second International Conference on Computers in Engineering and Building Design* (pp. 193–198). CAD '76.

Smithers, A.C. (1977). *Times Higher Educational Supplement*. 3 August.

CHAPTER

19

The Application of CAD in Relation to Dynamic/Static Product Concepts

ABSTRACT

The efficiency of the application of computer aids to design is dependent on the maturity of the product area considered. This chapter considers the question of dynamic and static product concepts and their relationship to CAD: it relates CAD usage to a model of design activity in a logical manner and concludes with examples that demonstrate the hypothesis.

From *Proceedings of the International Conference on Engineering Design* (Copenhagen 1983), 564–571.

INTRODUCTION

In a paper given in 1976 (Pugh and Smith, 1976), a tentative hypothesis was put forward stating that in any design situation there are always two extremes of approach, particularly at the conceptual stage, and that all such situations can be said to be contained within a conceptual envelope having two boundaries—(1) the boundary set by existing practice and (2) the boundary set by designs that are to come, are as yet unknown, but once evolved translate towards the boundary of existing practice. This argument was then used as the basis for considering the utilisation of computers within this spectrum of design. The two-boundary hypothesis was then used to consider the spectrum of design activity from pure innovation to conventional practice (Pugh, 1977).

Since that time, I have begun to realise that whilst the basic premise of the two boundaries is sound, it is more readily understood and explained by considering products and processes in terms of what I shall call *conceptually dynamic designs* or *conceptually static designs*, thus further extending the boundary hypothesis (Figure 19.1). Boundary A represents conceptually dynamic designs and boundary B conceptually static designs, whilst many designs lay between the two boundaries. The more the process of design becomes based on convention or product line precedents the more systematic becomes the

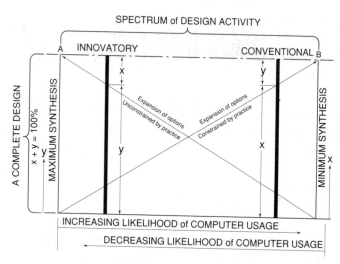

FIGURE 19.1. Spectrum of Design Activities
Pugh and Smith (1976); Pugh (1977).

design activity itself. This situation is therefore ripe for the application of computers. With designs based upon unknown concepts, the converse is also true. This chapter further explores the possibilities of the static and dynamic concept hypothesis.

CONCEPTUALLY STATIC DESIGNS (PLATEAU)

For many years now, and predominantly amongst management academics, life cycles have been exhaustively discussed. Levitt (1965) considers a product's life cycle in terms of four phases—market development, growth, maturity, and decline. His hypothesis was related to specific products from specific manufacturers creating a market, going through the phases as stated above, and then declining in sales until the point where they were no longer viable as a sales entity. De Kluyver (1977), amongst others, has taken the case further and refers to three types of product life cycle: type 1 (innovative maturity), type 2 (growth maturity), and type 3 (decline maturity). The parameters considered—sales volume versus time—again relate to the performance of a specific manufacturer's products. He states that 'the problem of how new is a new product has plagued marketing researchers for a long time' and then goes on to discuss distinct generations of engineering design, albeit in nonspecific terms. One cannot deduce the types of product design he has in mind and the concepts upon which they are based.

Lorenz (1982), in reviewing the latest MIT management pronouncements, cites Utterback in discussing innovation. The dominant design integrates within it many of the performance characteristics of early variations of the product. He then cites the Model T Ford whose front, water-cooled engine, and rear drive became the dominant design for some years.

The foregoing are all pointers which I will harness to underwrite the hypothesis of static and dynamic concepts.

I consider that the Model T Ford, for example, in setting the vogue for the modern motor car, not only became the dominant design but established a conceptual plateau for all motor cars. I suggest that automobiles are conceptually static. The Model T Ford had

- A body with windows and doors,
- Four seats and a steering wheel at the front,
- Four wheels, one at each corner,
- An engine, transmission, and brakes,
- Front and rear lights,
- Balloon tyres, and
- Front and rear bumpers (fenders).

The new Ford Sierra, seventy-five years later, has these same parts. In other words, the motor car is a prime example of a conceptually static product. Yes, designs have changed, materials have changed, and technology has advanced the product and, without doubt, improved it with the passage of time. This has taken place from the base of a fixed concept: all the changes have been in terms of rearrangement of, and advances in, the detail component designs within the fixed concept set by the Model T Ford. Has motor car design therefore reached a conceptual plateau? Is the basic concept static? The answer is probably a tentative yes, since it can be proven to be a historical fact that with the passage of time, design concepts for all products converge iteratively to a fixed static concept, after which increasing time and money is spent in refining the designs, for possibly less and less return. Many products may be said to fall into this conceptually static category and a few are given here, not in any order of preference or importance: bicycles, tractors, railway trains, steam and gas turbines, fixed-wing aircraft, internal combustion engines, houses, and ships.

Utterback and Abernethy (1975) discusses product innovation and stages of development and graphically illustrates the decline of the rate of product innovation with time (Figure 19.2a). Figure 19.2b shows the superimposition of boundaries A and B on the Utterback model.

The hypothesis put forward in this chapter defines the conceptually static product as being firmly at boundary B and, this being the case, as already stated, lends itself most readily to the efficient application of the computer. With the assumption of a static concept, future design activity and hence future designs readily lend themselves to a more systematic, mechanistic approach with increasingly efficient and effective use of the computer. Software can thus be developed upon a firm design foundation. As with the motor car, the fixed conceptual base can be expanded upon in great detail and hence is ideal for producing variations upon the fixed theme leading to greater business efficiency and better products more speedily produced. Computer-aided detail design and draughting and hence computer-aided manufacture becomes not only a reality, it becomes (more important) a viable and necessary reality.

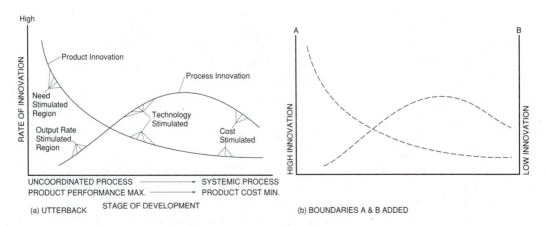

FIGURE 19.2. Innovation and Stage of Development

What, then, if conceptual maturity, the conceptual plateau, has not been reached in a particular product area and the concept is not static? Certainly our work in the Engineering Design Centre in terms of competition analysis related to new design concepts indicates that in many areas the static phase has not been attained and that many designs treated as supposedly static are in fact dynamic. Let me examine this question more closely since it vitally affects business performance and is highly interactive with efficient and effective computer usage.

CONCEPTUALLY DYNAMIC DESIGNS (NONPLATEAU)

We have found that by consistent and repeated application of a comprehensive design activity, as shown in Figure 19.3, that in many product areas concepts have *apparently* reached a plateau and in other areas they have not. Our recently introduced microscope manipulator for neurosurgery is a case in point. We have a new design concept significantly different from its competition. There is no methodology or system at present available to us in the computing field which could have efficiently aided or accelerated the process of arriving at the concept we have adopted. However, if the one we have chosen does represent the plateau for such devices (and there is no absolute guarantee of this), then we could now utilise a computer-based system to effectively and efficiently cater for variants and variations to the range based upon the assumed fixed concept.

If, however, we are wrong in our selection, then I suggest that the utilisation of computer systems will lessen our chances of discovering this, since familiarity with and the repetitive use of CAD systems can inspire false confidence, in that it is easy and often convenient to assume that a product design has reached a conceptual plateau. If in reality this is not so, then it is suggested that if the computer-oriented firm is unaware of this problem, the internal company systems, also being unaware of its existence, will be unlikely to discover the error. Competitors who do not make the same assumptions regarding static and dynamic concepts will be more likely to discover the truth of the matter since, at least being aware of the problem, they are constantly seeking viable alternatives. One should always be operating what I call the principle of controlled dissatisfaction: the designs of today can always be bettered tomorrow. The assumption is made that, until proved otherwise, existing concepts are dynamic and these by definition reside at boundary A.

To seek to avoid the static and dynamic conceptual trap, it is essential (particularly in these days of rapidly advancing technology) that comprehensive design activity as defined in Figure 19.4 be undertaken, even though ultimately the new designs that emerge from such activity are demonstrably static concepts. Will this always be the case?

In essence this means that the computer in design must be used with discretion, at what I call the front end of the design activity, where there is a translation from a qualitative approach to a more quantitative approach and thus the application of computers becomes more meaningful. However, utilisation in an information storage and retrieval role, properly conducted, will prove invaluable at this front end of design whether in boundary A or B mode. This might be expressed as the difference between the creative and the mechanistic use of computer aids to design.

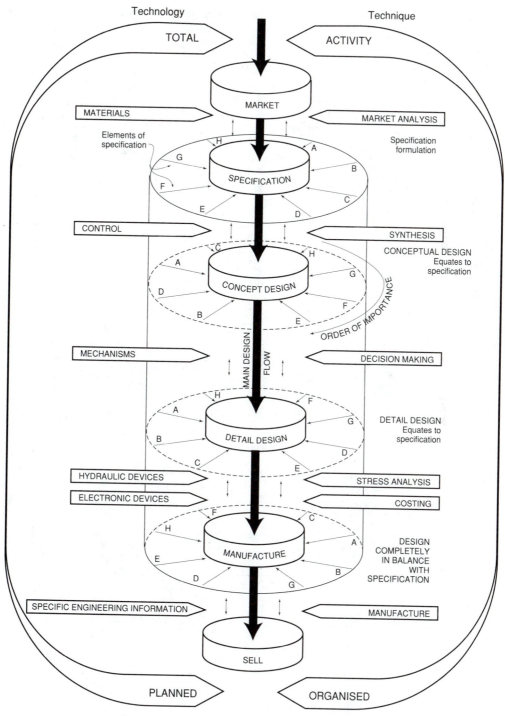

FIGURE 19.3. Design Core Bounded by Product Design
Specification: Dynamic Concept

With static concepts, the design activity model becomes as shown in Figure 19.4, and it is in this area that computer usage becomes essential to effective design and ultimately manufacture. One thing is certain, however: the assumption of static product concepts is

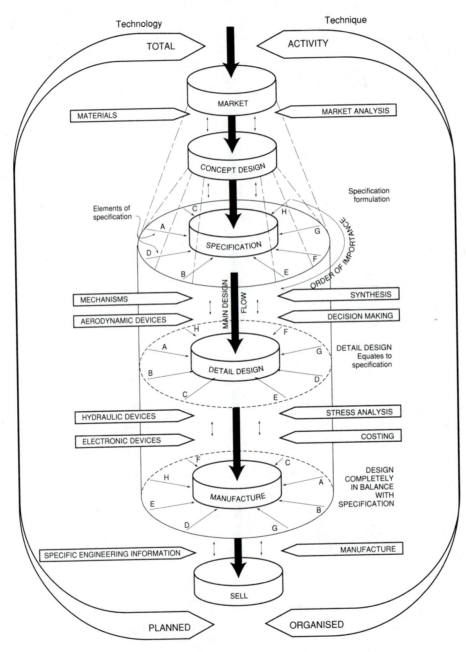

FIGURE 19.4.   Design Core Bounded by Product Design
Specification: Static Concept

in many cases open to doubt, and many companies making such assumptions are therefore vulnerable to competition. Beware of becoming frozen into a boundary B operation reinforced by sophisticated computer systems. At least keep a boundary A operation available to check out that the B operation remains soundly based and viable.

The application of and usefulness of the computer where concepts are still dynamic is little understood, although I have started work in this area. It could be said that computer usage today is almost in a total sense mechanistic and that attention must be given to the creative use of the computer at the front end of the design activity. This is considered essential to the continuation of sound product design and also innovation. One of the conclusions from the recently published EITB occasional paper (Arnold and Senker, 1982) is very salutary: 'CAD has the potential to automate drawing rather than design. The jobs of draughtsmen rather than designers are most at risk from the new technology.' The report is also concerned with 'falling innovative activity in mechanical engineering.' Could it be that increasing computer application in mechanical engineering has, on the one hand, contributed to this decline, whilst in the electronics industry, essentially a two-dimensional operation, the converse is true?

Finally, to return momentarily to de Kluyver and the product life cycle, could it be that the three types of product life cycle hypothesis is applicable only to products where designs have reached a conceptual plateau and are therefore by definition static, or do indeed concepts as opposed to products themselves have life cycles? If this is so, and certainly there is evidence to suggest that this is so, then I would put forward the proposition that only dynamic concepts have life cycles and that with the passage of time, as all designs converge to the static state, the life of a plateau concept (not in detail) becomes infinite. The concept of the differential gear was first attributed to the Chinese in 2000 B.C. (Burstall, 1963), and we have done an awful lot of development work on it since, latterly enhanced by the computer.

REFERENCES

Arnold, E., and Senker, P. (1982). 'Designing the Future: The Implications of CAD Interactive Graphics for Employment and Skills in the British Engineering Industry.' EITB Occasional Paper, ISBN 0-085083-561–5.

Burstall, A.F. (1963). *A History of Mechanical Engineering.* London: Faber & Faber.

de Kluyver, C.A. (1977). 'Innovation and Industrial Product Life Cycles.' *California Management Review*, 20 pt.1, 21–33.

Levitt, T. (1965). 'Exploit the Product Life Cycle.' *Harvard Business Review*, 43, 81–94.

Lorenz, C. (1982). 'The Terminal Risk of Failing to Innovate.' *Financial Times*, August 9.

Pugh, S. (1977). 'Creativity in Engineering Design: Method, Myth or Magic.' *Proceedings of the SEFI Conference on Essential Elements in Engineering Education* (pp. 137–146).

Pugh, S., and Smith, D.G. (1976). 'CAD in the Context of Engineering Design: The Designer's Viewpoint.' *Proceedings of CAD '76* (pp. 193–198).

Utterback, J.M., and Abernethy, W.J. (1975). 'A Dynamic Model of Process and Product Innovation.' *OMEGA, International Journal of Management Science*, 3(6), 639–657.

20

CAD/CAM:
Hindrance or Help to Design?

ABSTRACT

This chapter considers the topic of CAD/CAM in relation to design in the broader context and, in so doing, puts the computer system into the context of design. The origins of CAD are examined and found to stem from the manufacturing functions of the late 1940s with subsequent developments being production led rather than design led. This reverse evolution from the end (the product) to the beginning (the design) is seen as a major contributory factor inhibiting the design and efficient utilisation of CAD systems, particularly when allied to misconceptions about the design activity itself.

This evolution is matched to the design activity through the medium of mechanical design, where questions such as denying the emergence of new concepts are discussed.

The chapter concludes with a discussion of computer aids to design in a total sense and makes mention of work being undertaken in CAD system design which has commenced at the front end of the design activity and not at the end.

From a paper presented at the Conference on Conception et Fabrication Assistée par Ordinateur, Free University of Brussels, October 1984.

INTRODUCTION

When I was asked by Professor Jaumotte to write specifically on mechanical design, I hesitated over the word *mechanical* since my background in design has been catholic in nature and to be effective in design one has to transgress traditional discipline boundaries, whether in industry or education.

I say this with some feeling, as for the past fourteen years my own research has been concerned with establishing and understanding design in practice, and hence through practice, to translate it into a form suitable for transfer via education. This work has quite deliberately entailed working over a wide field of industry—both product and process—in order to seek out and understand, through dissecting that experience and practice, just what is this thing called design and how we can define it in a way that is understandable to others. How can we establish the general case that fits all situations and avoids, like the plague, the creation of special definitions which are divisive and yield less than effective design in a total sense?

Whilst seeking to understand design through practice, the advent of the computer and its application to design situations started to take off in the early 1970s, and one had to take into account its impact on the design activity, albeit recognising that the understanding of this activity was minimal. In fact, the rapid expansion and implementation of computer systems against this hazy background of design understanding has, and is, leading to some major rethinking in this area by leading computer experts and hardware and software companies. I return to this topic later.

I referred earlier to a hesitation over the word *mechanical* in the title of this chapter. Alas, I do not possess Professor Jaumotte's insight, and after further reflection I have realised that in mechanical engineering many claims are being made in respect of CAD and that many companies and universities are acquiring systems both to teach with and to utilise in practice—but where lie the questions of efficiency, innovation, improvement, and advance. In what context or manner does the computer enhance mechanical engineering or indeed lead to its demise?

I take my dissected design practice and research, examine these questions in detail, and, hopefully, throw some light on the whole question of computers in design.

CAD IN THE CONTEXT OF DESIGN

Several years ago, a colleague and I presented a paper (Pugh and Smith, 1976), at a CAD conference in London, which I did not think was particularly well received, since a show of hands amongst the audience revealed that possibly 1 percent of them were designers, thus revealing a communication gap of enormous proportions. In the paper a framework was established which suggested that computer usage was more likely to be effective if harnessed to existing practice and what would now be called conventional designs, rather than if they were harnessed in truly innovative situations where conventions and conventional designs have not yet been established and therefore could not be considered. This framework was in graphical form and is reproduced as Figure 20.1.

Essentially, the designs of products (all mechanical, as it so happens) were fitted into a boundary framework retrospectively in an attempt to understand the interaction with computers and also to further understand the process of the design activity. It was concluded, at that time, that 'CAD used with circumspection is a great asset; let us try to keep it that way.' Incidentally, these products are still being manufactured.

The very worrying indicator from that conference was that computer specialists, both hardware and software designers, and computer analysts were not the least bit interested in understanding design in the general case, thereby availing themselves of the opportunity to design their equipment and systems to suit designer needs. This is analogous to saying that we will design a vacuum cleaner but we will not consider the nature of the home cleaning process, sizes of dust particles, acceptable noise levels, and so on—clearly a nonsense.

Today the situation is somewhat similar, although others are becoming very worried by it, and even the CAD pundits are belatedly starting to philosophise about the nature of design and to recognise that it might assist in the speed of computer application if a few basic questions were asked about just what it is that a designers job task entails. This is of particular significance in mechanical engineering, as will be revealed later.

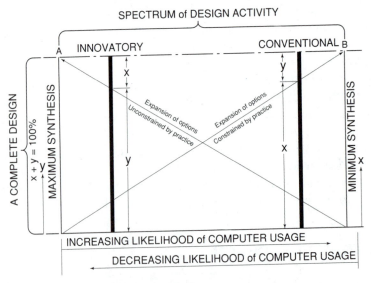

FIGURE 20.1.   Spectrum of Design Activities
Source: Pugh and Smith (1976).

Rosenbrock (1983), in discussing the interface between man and machine, states that 'What is needed is essentially simple. When new technological developments are proposed, it is only in the rarest cases that any thought is given, during the process of research and development, to the role of men and women in the system that will result. Usually it is the machines alone that are considered, and only when this development has been completed is any thought given to the essential contribution from the people, without which nothing can be produced. *This contribution, not surprisingly, is often quite unsuited to human capabilities*' (author's italics).

Since the activity of design itself is a human activity, what chance therefore is there of achieving a match between computer aids and the tasks to be performed? With certain notable exceptions, very little indeed.

Amkreutz (1984), in discussing CAD in the design practices of the future, calls for research into design itself and also into computer applications in design. It might well be suggested that this request comes a little late in the day. Will the enormous systems push developed over the past ten years allow this to happen? Has this push biased and distorted any datums for rectification of the situation? CAD specialists, by their own utterances, are already indicating quite strongly a fixed view of design and design situations which tends to match the systems in existence. Yet there are possibly one or two rays of hope. Otker (1984), in discussing integrated CAE systems comments, 'The dichotomy is not so much between "CAD" and "CAM" as well as between the "upstream" and the "downstream" part. The "upstream" part includes (1) Design analysis (product specifications), (2) Design analysis by widely applicable techniques (Finite) element analysis, optimisation, etc., (3) Drafting, (4) Process planning, (5) N/C Part Programming.' The downstream part is assumed to be manufacture and selling.

At least upstream and downstream are partially identified, although no distinction is made between differences in upstream design activity depending upon whether the product stems from existing practice and convention or is entirely novel and outside the bounds of existing practice. Possibly this situation will improve as this fact becomes recognised. Little progress will be made unless system designers, both hardware and software, talk to designers in different fields—first to establish communication and second to understand what it is that they do. According to Rosenbrock (1983), 'We do not know where we are going, but it is essential to get there as fast as we can.'

The main thrust of my work (designing apart) is to try and define just where 'there' is in relation to product and process design. Hopefully, therefore, system design will thus become not necessarily easier but better directed, with products that enable people to carry out tasks they wish to perform, and will not, as is the case with current systems, necessitate an almost traumatic redefinition of the task and the means and mechanisms with which to carry it out.

ORIGINS OF CAD

In discussing origins and datums, mention should perhaps be made of the origins of CAD. Bèzier (1984) provides a succinct description of the evolution, where the first application

of a computer was 'to define the motion of a milling cutter' in 1942. Subsequent developments into the 1950s were also concerned with manufacture. This was reiterated recently
by Fenner (1984): 'CAD grew from the needs of the automotive and aerospace industries
in the fifties.' This confirms that the drive behind the computer systems revolution was
production-led and that CAD systems that evolved were also based on that premise.
According to Fenner (1984), 'The thinking behind PDM and the CDF 5000 (Computervision equipment) is that users working for large companies manufacturing a number of
products are unlikely to be designing anything from scratch . . . , about 80% of a typical design is a modification of various parts of earlier designs.' We see yet again this dependency
on the boundary B activity being essential to support and justify the CAD system design.

This is a worrying trend, since there is increasing evidence that production-led products fare less successfully with the passage of time than design-led products. Can it be,
therefore, that the advent of CAD systems, which to be used effectively or at all, require
the fixed concept design—the ultimate goal of all production directors and, incidentally,
production academics. Scale up, scale down—the same basic parts! Perhaps the words of
Jeremy Fry (1984), in talking of valve actuators, throw some light on this: 'The reason that
the company was able to make the most of this design revelation was that it was small
and unfettered by dogma. It was also design-led and not production-led and, because we
only use subcontractors, it was not weighted by capital investment in machine tools.'

If one accepts the premise that designs of fixed concepts are an essential ingredient
of current CAD systems, then this begs a very large question over the product and
process designs of the future. In pursuing understanding of the design activity and its
relation to CAD and vice versa, the influence of CAD upon the design activity itself, it
becomes necessary to distinguish further and in a harder form the differences between
products at boundaries A and B: 'Boundary A represents conceptually dynamic designs
and Boundary B conceptually static designs, whilst many designs lie between the two
boundaries. The more the process of design based on convention or product line precedents the more systematic becomes the design activity itself. This situation is therefore
ripe for the application of computers. With designs based upon unknown concepts the
converse is also true' (Pugh, 1983). A major, almost nonvisible asset of computer systems is that in order for them to become usable and efficient, they require a systematic
approach to design activity. This requirement forces a detailed examination of the activity, and the design activity model thus assumes two forms representative of these differing conditions (see Figures 20.2 and 20.3).

Briefly, in A, we assume that designs to come have not yet reached a conceptual
plateau and thus become conventional, and in B, we assume that designs are based on
static concepts and are thus, by definition, conventional, and data bases can safety be
defined and captured. The motor car, amongst other artefacts, was given as a prime example of a likely static concept. This hypothesis is further advanced by Pugh (1984),
where it is suggested that the means of production of such products, independent of
manufacturer or country of origin, converge to become identical. This topic will be referred to later on, in the more detailed discussion of CAD in relation to products based
firmly in mechanical engineering.

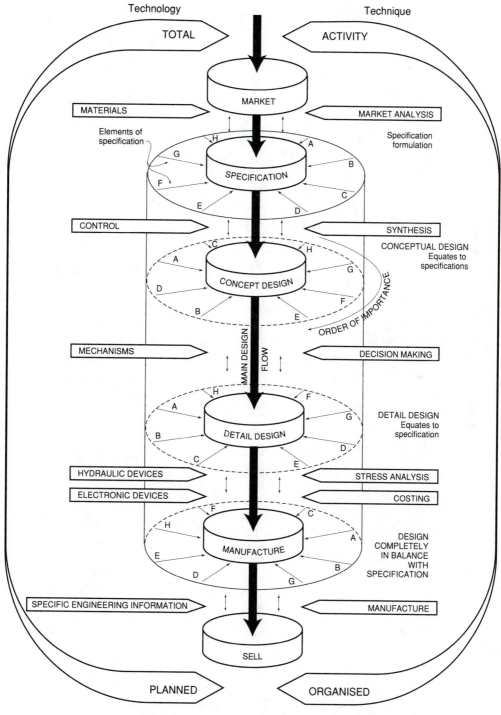

FIGURE 20.2. Design Core Bounded by Product Design
Specification: Dynamic Concept

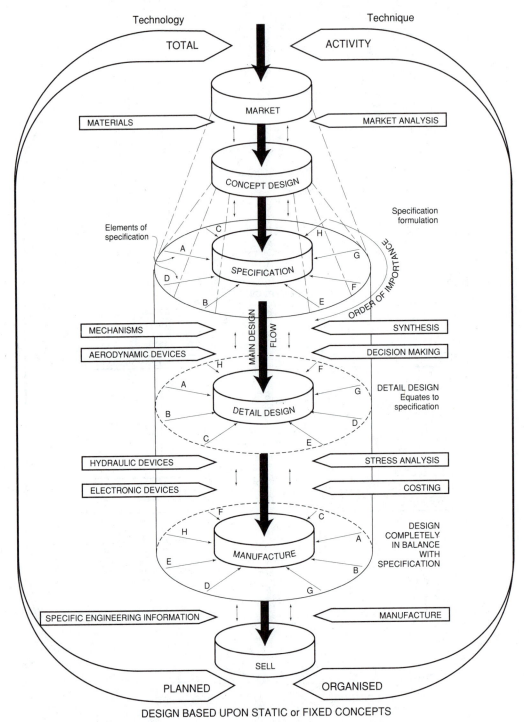

FIGURE 20.3.   Design Core Bounded by Product Design
Specification: Static Concept

So far, I have discussed the evolution and understanding of the design activity and its relationship via models to a particular product hypothesis—and its interaction with CAD, with some scant reference to the origins of CAD systems generally. Concentration upon the interactive study of design activity in relation to the whole gamut of *actual* CAD systems will throw further light on the ultimate question of hindrance or help to design.

RELATIONSHIP BETWEEN CAD SYSTEMS AND DESIGN ACTIVITY

In a detailed study (*Project Report*, 1984) of the design activity related to existing CAD systems carried out recently, some eighty-five turnkey systems were examined in great detail and correlated to the boundary A design activity model. The relative positions of the facilities are found to be as in Figure 20.4: 'The 2D draughting being mainly aimed at detail drawing, and the remaining facilities all stemming from this base, with a strong bias towards manufacture. All the facilities were really only of use somewhat after the conception stage.' These very recent conclusions thus support my earlier statements and references to the works of others actively involved in CAD but very much on the periphery of design.

In fact, the majority of systems deal only with components (not overall concepts) in mainly 2D form and, more latterly, components in 3D form. It is suggested that since efficient application of CAD systems requires a sophisticated data base and therefore frozen component forms, a frightening situation exists for mechanical engineering. Since one cannot assemble mechanical 'circuits' in the same way as electronic circuits (relatively), the data base for the former is infinite compared to the latter. This is not to deny the usefulness of CAD with mechanical systems, but to rely solely upon such systems for design, a situation which is arising in many industries, is to me confining the designs of the future to almost chance occurrence, since the rigidity of a given mechanically based data base will instil in the user the comfort of completeness. Our experience with product design over a wide range confirms this trend and that reliance on CAD systems inhibits thinking. It shouldn't but it does, and this returns one full circle to the arguments put forward by Rosenbrock (1983). Our study confirms that all available systems, to some degree, mismatch with user expectations and the tasks they wish to perform.

Why is it that it took the computer industry and no other to coin the phrase 'user friendly'? The best systems can *draw*; they cannot *design*: how can they when clearly the system designers have not studied the design activity and, therefore, must have had an inadequate product design specification to start with? We are not alone in this view.

This argument does, however, fit logically into the pattern of evolution discussed earlier; production systems utilised (and still do) 2D drawings as their input. Components drawn in 2D were translated into 3D components on machine tools working in 3D. The CAD systems therefore have continued to evolve *backwards* from the production situation. Again, this fits production-type thinking: even today many production engineers consider *design* as *drawing*, now being done faster on CAD systems. Since this is a basic

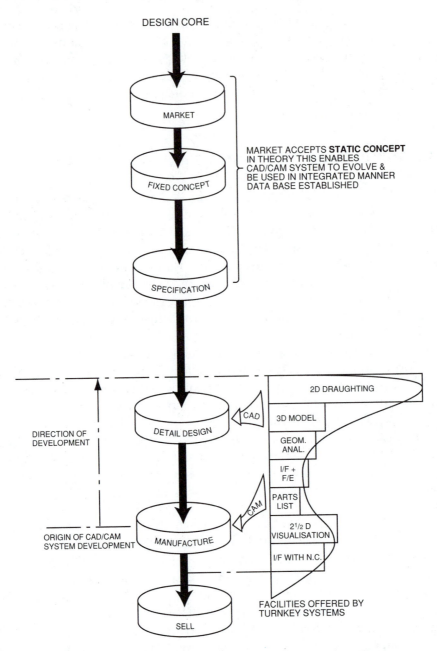

FIGURE 20.4.   CAD Facilities Matched to Design Core: Static

premise of the production function, then what better to call it than computer-aided design—now more and more vociferously being referred to as computer-aided draughting.
A Computervision executive stated very recently that his systems could only draught;
they could not design. This is not strictly true, since component detail variation based on

fixed concepts is part of the design activity, and therefore his systems make a contribution to design.

Having been a designer for over thirty years, I can confirm that backwards evolution of a product is most inefficient and wasteful, and it should not be forgotten that CAD systems are someone's product. Since current systems will only function interactively with predetermined or fixed concepts, Figure 20.5 is more representative of the truth of this situation.

Foister (1984), in discussing computer graphics in Texas industry, where he trained many people and interfaced with user companies and systems suppliers states, 'there is one area of weakness with the whole concept. I, personally, issued an open challenge to all the user companies, to all the recognised experts, to all the equipment manufacturers, to set up a board for me alongside any mainframe (or unit) and any number of operators to perform a conceptual design project from first principles—I guaranteed to beat the computer. *I had no takers!*'

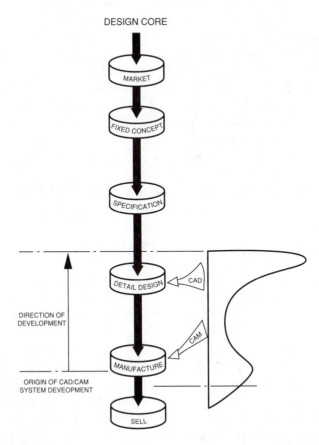

FIGURE 20.5.   CAD Facilities Matched to Design Core: Dynamic

So more and more people are saying that one cannot design using a computer. As stated earlier this, in a total sense, cannot be true. When one knows what it is that one is doing, one can use the available systems to some effect in spite of the fact that they are not *designed* for design in the total sense. A point to which I shall return at the end of my paper. Against this background, let me now attempt to close the loop on the mechanical engineering issue.

CAD AND MECHANICAL ENGINEERING

Products based essentially on mechanical engineering, from cars to turbines, bicycles to baths, are all three dimensional and have been evolving for a long period of time, and therefore many of the designs may be considered conventional, and their detail can be captured in an appropriate computer data base. This is fine if the evolution of the product is conceptually complete—the conventional motor car was referred to earlier in this context—but what if it is not complete? I, for one, am of the firm opinion that the great bulk of mechanically based designs are yet to emerge and have not yet been conceived of, that what we see today are but the first products in an infinite chain, and that the majority of existing designs are still conceptually dynamic—they must not be suppressed.

For efficiency, it is useful and convenient to manipulate conventional designs using computer-based systems that will no doubt continue to evolve. We must, however, be aware that in capturing component form and dimension in a data base, it is likely, with the exception of standard parts (which will themselves slowly evolve and change), that such components will have little if any part to play in new designs based on new concepts—that is, in conceptually dynamic situations. If we persist in using computer-based systems as our only design bank, then it is suggested that the designs of the future will be left to chance.

To give an example, the Giraffe site placement vehicle, or telescopic handler as they are now universally known, was designed in 1973. A family of vehicles now exists under the name of Markhandler. There is no way that I know of, even projecting the best of today's systems backwards in time to 1973, that a data base could have been established from which to extract this vehicle conceptually. I say this with some feeling as I happen to know that the same statement fits the next generation of such vehicles (already designed) and they too cannot be retrieved from any known store. Once the concept exists, and its viability is proven, then it can be captured in the system and there are now many companies worldwide doing this (1992).

However, in mechanical engineering and probably in all spheres of engineering, we must continue to operate people (designer)-based systems alongside machine-based systems, and this should not only become an essential aim. In my experience such a suggestion does fit the personal characteristics of designers. Some designers like developments based on fixed concepts. In fact, this statement might apply to most designers who do not like the uncertainty of truly open-ended situations. Others certainly do not. These people must not be allowed to become constrained by CAD systems, whilst the former

would probably welcome the constraints. This is not to say that computer aids should not be available and used by all types of designer. Certainly their utilisation for modelling and analysis is to be recommended, providing the users understand the analytical methods in the first place. They should be used for information retrieval generally, although what one does with large amounts of information remains an unsolved problem of increasing magnitude. Certainly current CAD systems take a lot of the drudgery out of drawing. So in mechanical engineering the use of specific programs for specific tasks is to be recommended, but for true conceptual design, these must not be linked: linkage will immediately restrict the options to what has been captured in the data base. Linkage with care and circumspection is fine for conventional fixed concepts.

Since, historically, the origins of the computer systems evolution lies in manufacture; computer-aided manufacture (CAM) therefore came first. This has led back to the interface between design and production—that is, the 2D drawing or component specification; thus we achieve computer-aided draughting (CAD). Computer-aided design (CADES) in any total meaningful form is still a long way away, if it is at all possible, since the backwards path of the systems evolution will inhibit the emergence of probable/possible systems. This situation is at last being recognised, even if slightly obliquely, and will be briefly discussed.

COMPUTER-AIDED DESIGN IN A TOTAL SENSE

The current system development backwards from production mitigates against computer-aided design (CADES) in a total sense, particularly when this statement is viewed against the background of the system design itself. As we have already seen, users are hardly even considered and the situation is rapidly getting worse. Wright (1984), discussing a microcomputer keyboard with 717 programmable keys, says 'Even in the computer world, where "user friendliness" has become an obligatory feature, there is very little equipment which even come close to the claim.'

It is submitted that, whilst hardware compaction continues, the diversion of the attention of the computer giants to software will bring only apparent relief to the problem. With the steering wheel locked in the boot (trunk) of the car, even the most efficient engine in the world is of little use. Sooner or later therefore, attention will return to the question of user needs and just what it is that people in general want to do and, in our case, designers in particular. Already there is emerging a vociferous minority which echoes these views. Sturridge (1984), discussing microcomputers says, 'Rank Xerox's Palo Alto research laboratory spent 30 work-years wrestling with the problem [of user friendliness]. At the end of the day it concluded that things are best left the way they are. The machine must mirror the way things are done now and not attempt to improve on them . . . the gadgets that are friendly are the ones that make the user feel conceptually that he is doing the same as he did before the electronics intervened.'

Having been aware of and used such systems, one is aware of the multitude designed-in inefficiencies and mismatches. We have, therefore, accelerated and widened our programme of studying 3D design through dissected practice, in order to design a system for designers. This particular programme is now well and truly launched.

The problem is recognised by Frazer, Coates, and Frazer (1983): 'Although a lot of research has been dedicated to the essential geometric algorithms for CAD systems, so far very little effort has been expended on the vital link between the programs and the human beings who will operate them. In the area of design this problem is nontrivial and demands an understanding of the nature of designing.' One must agree with these sentiments but not necessarily with the proposed solutions.

During the education and training of engineers and designers in particular, they should be made aware of the relationships between computer systems and design, and in particular that one is not a substitute for the other. There is a tendency today to envelope students with such systems, without teaching them against the background of a proper design base or model: the foregoing commentary on system evolution adding weight to this approach.

The computer, used with understanding and discretion, can be helpful to mechanical engineering design but it is not all-powerful. Used without these criteria, it is not only a hindrance, the results will ultimately be disastrous. Remember, 'Data do not yield information except with the intervention of the mind. Information does not yield meaning except with the intervention of imagination' (Levitt, 1984) (this requires people).

REFERENCES

Amkreutz, J.H.A.E. (1984). 'CAD and the Future of Design Practices.' *Proceedings of the Second European Conference on Developments in CAD/CAM*. Amersfoort: ECD.

Bèzier, P.E. (1984). 'CAD/CAM: Past, Requirements, Trends.' *Proceedings CAD 84* (pp. 1–11). Brighton: CAD.

Fenner, R. (1984). 'Mass Production.' *Computer Systems* (June), 39–46.

Foister, P. (1984). 'Computer Graphics in Texas.' *Engineering Designer* (July), 16–17.

Frazer, J.H., Coates, P.S., and Frazer, J.M. (1983). 'Software and Hardware Approaches to Improving the Man/Machine Interface.' *Computer Applications in Production and Engineering* (pp. 1083–1094). Amsterdam: North Holland.

Fry, J. (1984) 'Design Management.' *Journal Royal Society of Arts* 132(5333) (April), 304–308.

Levitt, T. (1984). 'The Globalisation of Markets.' *McKensey Quarterly* (Summer), 2–20.

Otker, T. (1984). 'The Introduction of Engineering Know-How Through CAD/CAM and the Implications to the Organisations.' *Proceedings of the Second European Conference on Developments in CAD/CAM*. Amersfoort. ECD.

Project Report. (1984). Engineering Design Centre, Loughborough University (confidential).

Pugh, S. (1983). 'The Application of CAD in Relation to Dynamic/Static Product Concepts.' *Proceedings ICED 83* (vol. 2, pp. 564–571). Copenhagen: ICED.

Pugh, S. (1984). 'Further Developments of the Hypothesis of Static/Dynamic Concepts in Product Design.' *Proceedings ISDS Conference* (pp. 216–221). Tokyo: ISDS.

Pugh, S., and Smith, D.G. (1976). 'CAD in the Context of Engineering Design: The Designer's Viewpoint.' *Proceedings CAD 76* (pp. 193–198). London: CAD.

Rosenbrock, H.H. (1983). 'Designing Automated Systems: Need Skill Be Lost?' *Science and Public Policy* 10(6) (December), 274–277.

Sturridge, H. (1984). 'Micros That Are Made for a Shout or a Caress." *Sunday Times*, 26 August, p. 49.

Wright, J. (1984). 'Not too Many Knobs on It, Please.' *Eureka* (May), p. 3.

21

CAD/CAM:
Its Effect on Design
Understanding and Progress

ABSTRACT

This chapter considers the evolution of current CAD/CAM systems against a background of the evolution of design understanding and shows that such evolutions are related to product designs which themselves may be evolutionary—which may, in fact, be conceptually dynamic as opposed to static. The static and dynamic concept situations are considered each, in turn, to establish relationships which throw some light on to CAD and design understanding. The differing interaction aspects required during designing in either static or dynamic regime are harnessed to throw more light on the topic. The whole question of CAD/CAM and design reconciled through recourse to models of design, and the suggestion made that two primary factors—one the source of CAD and the other the absence of models of design in combination—have inhibited design understanding and efficient CAD systems evolution.

From *Proceedings of the CAD/CAM, Robotics and Automation International Conference* (Tucson, AZ, 1985), 385–389.

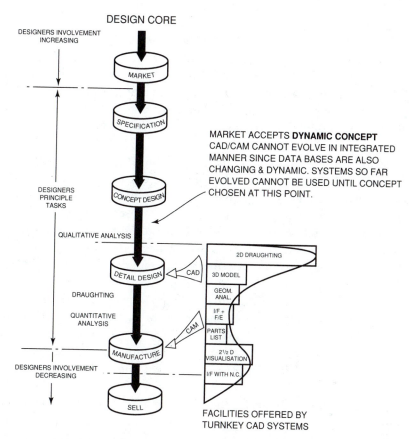

FIGURE 21.2. Facilities Offered by Turnkey CAD Systems

- The designs of the CAD equipment, hardware and software, may not in themselves lead naturally to integration and system efficiency. They work, yes. One relative to the other may be efficient or not. But are they efficient in an absolute sense—like, say, the efficiency of an internal combustion engine? Perhaps they are, perhaps they are not.

The questions raised by these three points alone could, if allowed, give rise to an enormous paper, far beyond that allowable in this instance. I will thus restrict my discourse to the second reason on the assumption that the first and third are perfect both in design and efficiency, since the one begets the other.

ITERATION BASED ON FIXED CONCEPTS

It will be generally accepted that with a truly fixed concept such as the motor car, the market accepts this until something better comes along and that a specification for a new model will be drawn up based on this assumption (Figure 21.1). Yet during design of the new vehicle many new components (detail) designs will emerge, and it is now being suggested in some quarters that CAD, because of the relative ease with which component modification may be achieved, actually proliferates variety unless adequate controls are introduced. Thus, at the detail design stage, there will be many design details emerging not just once but many times as the iteration continues to converge on to a final design. Accepting that the available data base may be massive and relevant, can it be that the data base might restrict or constrain the detail options available? I don't know for sure!

The question is, are the detail components as good, as efficient, as simple, as cheap, and so forth as would have been achieved without the constraint of the data base—albeit with much less speed, accuracy, and drawing quality? Might it be that staccato use of the data base by the designer, as opposed to its integrated usage, is more efficient and less restrictive. Does the possibility of system integration pressurise the designer into accepting the limitations of the data base again? Does one beget the other?

Certainly in the United Kingdom there is some evidence that this might be happening (Billet, 1984) possibly we are not achieving the best balance between man and machine because we are not seeking this as a primary objective. This statement applies equally to both hardware and software. The extent of the iteration in this case may be expressed typically as Figure 21.3.

The reader will have observed that I am posing many more questions than I am answering. As a designer I consider them to be very real questions which we are attempting to unscramble.

So it might be concluded that both theoretically and in practice, such systems may be used interactively and integratively but with some largely unanswered questions regarding efficiency and even suitability based on current hardware and software designs.

This, however, is considered to be the simple case. It is suggested that the dynamic concept situation is not only more difficult to grasp, but it is fraught with much more danger and many more unknowns. Let us have a look at that situation.

ITERATION BASED ON DYNAMIC CONCEPTS

In a dynamic concept situation as depicted in Figure 21.2, it is the equivalent in automobile terms of a wheel on the roof, being triangular and not round, driven by an ion motor and yet being a cheaper and more effective means of travel than the latest Ford—in other words, a design concept that has not yet been thought of, not yet conceived. How does this approach equate with the efficient application of current CAD/CAM systems?

In my opinion, until the conceptual decision is taken, until the concept is chosen, there is no equation whatsoever on an integrative or interactive basis since by definition to fulfil such criteria requires the appropriate data base. Is our data base likely to contain triangular wheels and ion motors? I suggest not. So alarmingly and significantly, not only can we not work efficiently and in an integrated fashion, but there is a distinct danger that even working with computers in a nonintegrative, staccato fashion may

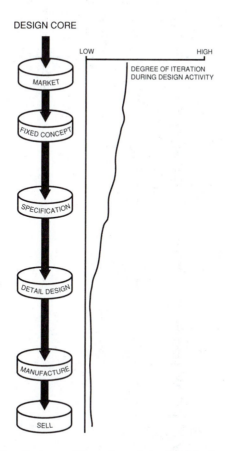

FIGURE 21.3. Degree of Iteration of the Design: Static Concept

constrain, restrain, or restrict the emergence of our conceptual designs (remember my earlier assumption that such systems are deemed to be perfect?).

Possibly, therefore, we should not be trying to achieve integrated working in this area: perhaps it is a natural mismatch, like square pegs in round holes. Certainly, if the designs of the systems are not based on an efficient specification of the correct human or machine interface, then effective utilisation in a staccato manner will leave a lot to be desired, also rendering the possibility of integrated working, even if considered a necessary and essential aim, forever elusive.

Set the foregoing against the appropriate iterative background and the likely complexity multiplies and such understanding as one thought one had rapidly evaporates to zero (see Figure 21.4). If the possibility exists that with static concepts future designs may be constrained or restricted, then that possibility becomes exacerbated in the dynamic situation.

With the randomness and variety inherent in dynamic situations, whilst it is unlikely that a particular company data base will ever be large enough and affordable enough to permit the possibility of adequate coverage, this points to the necessity (in

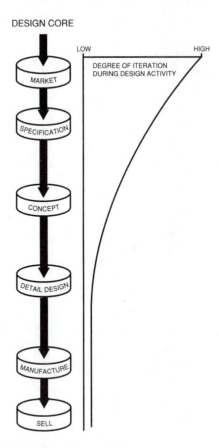

FIGURE 21.4.   Degree of Iteration of the Design: Dynamic Concept

principle) for systems of national and international data bases to be established. This in itself would give rise to many problems well beyond the scope of this chapter. Integrative working without constraints would be a true impossibility. Perhaps we are looking in the wrong areas, at the wrong problems with the wrong technologies. It is, I think, interesting to speculate in this area and to consider the words of Fuchi (1984) in discussing the Japanese approach to fifth-generation computer systems: 'They will have to be capable of electronically observing a universe of information by viewing it, reading it or listening to it and, some generations on, it is unlikely that we shall be able to predict what it is that they wish to know to arrive at their conclusions. After all, if we could we would probably not need them in the first place.'

Noting the message implied in this statement, perhaps we should not aim for integrated working with dynamic concept design activity at all. Perhaps we would be better served by applying our energies to the human and machine interaction and compatibility questions, more boldly and with more resource.

It is our own experience in product design that utilisation of CAD systems that were and are available would have seriously curtailed our options and therefore our designs. We could not have done what we have with totally integrated systems. Staccato usage, however, enhances design efficiency: you pick it up and you put it down; you don't become hooked. We are now embarked upon a systematic study of what I call the front end of design—the area between user need (market) and final conceptual choice. One specific objective is to investigate the suitability of and to design software to effectively and efficiently utilise computing systems at the various stages and to study the interactions between the stages. One big problem already evident is that not only do existing CAD system designs mismatch user expectations from the human and machine interaction viewpoint, but they are extremely cumbersome to use even in staccato mode—both from a hardware and software viewpoint. This general problem has, of course, been highlighted by Rosenbrock (1983). I would suggest that this phenomenon has a direct linkage with the problems expressed by Cooke (1984) that for the efficient design of electronic systems, 'The designer of electronic-based products has available an enormous range of basic building blocks of high and still increasing complexity. What he often does not have is either training or well-established conceptual tools for working out the levels of sophistication. Thus many engineers assume that a natural extension of their previous methods of tackling new product design will be appropriate, but this is rarely true.' He then calls for a systematic approach starting with 'defining what is wanted' and continues by formulating an ab initio model of the design activity. This brings me full circle to the title of this chapter: how have CAD systems affected understanding and progress in design?

DESIGN ACTIVITY: THE MODEL ISSUE

In evolving a model of the design activity over the years it has become increasingly apparent that since this work has commenced, our understanding of design has accelerated almost exponentially. In fact, as the model became simpler and more visible, the understanding accelerated at an alarming rate. In my opinion, the criticality of models is no longer in doubt; in fact, it is suggested that many of the current mismatches between humans and machines, and in particular CAD system design, arise because of the lack of an appropriate design model. Cooke (1984) in trying to resolve and reconcile electronic system design resorts to a model in an attempt to unscramble the situation.

It is my experience that without a systematic structured approach to product design there is no way these days that the user-need situation will ever be satisfactorily satisfied. Until such time as we adopted a model in the mid-1970s, we too carried out design on what I call a 'yahoo' basis: anything goes as long as its technology driven and utilises the latest techniques. Subsequent experience has revealed that nothing could be further from the truth.

Could it be that two primary factors have given rise to the current situation?

- The evolution of the majority of CAD systems from manufacture to design (1) took place in the United States.

- Models of design activity are generally absent from U.S. teaching and culture. At least, they do not appear, except in the most rudimentary verbalised manner, in engineering texts or research texts emanating from the United States.

This situation now appears to be changing in that modelling of the design process by computer manufacturers is becoming evident. Why is this so? I would suggest that in order to attempt to get to grips with design itself they have recognised the necessity for modelling in order to bring this about. In other words, it is required that for effective efficient CAD systems evolution, a suitable design model is not only desirable but essential, a sentiment with which I entirely agree. Why then do they not apply such systematic procedures to the design of CAD equipment itself?

So I consider that unless evolutionary models of design, leading to design understanding, become linked inextricably to computer system evolution, then design itself will become sterile and CAD systems will come only to satisfy the user-need situation in a random manner—which means inefficiently. Model evolution and design (you have to design models) is a primary concern of mine. Ehrlenspiel (1983) does, I think recognise this problem, in that his latest model attempts to break away from the rigidity of the traditional German model. To my knowledge, the Japanese, as yet, do not have a comprehensive design model, although they are working on it.

It is our experience that model evolution is inextricably linked to design practice. Could it be that the apparent randomness of many inexplicable design models is directly related to lack of such practice?

CONCLUSIONS

In this chapter I have attempted, through models of design, to relate the complex topic of product evolution to effective, efficient CAD system usage. I hope that I have succeeded. I will however resort to another diagram, Figure 21.5, which indicates how I see computer systems linking into the design core. Effective computer usage requires integrative working; the same goes for design. Are the forms of integration compatible? Figure 21.6 begs some very large questions.

REFERENCES

Billet, T. (1984). 'CAD/CAM Integration: Not Yet There.' *Draughting and Design* (September 17), 12–13.

Cooke, P. (1984). 'Electronics Design Demands a New Approach.' *Engineering* 224(4) (April), 315–317.

Ehrlenspiel, K. (1983). 'Ein Denkmodell des Konstruktion Prozesses.' *Proceedings of the ICED 83* (pp. 285–288). Copenhagen: ICED.

Fuchi, K. (1984). 'Japan Leaps Forward with the New Generation.' *Computer News* (September 13), 34–35.

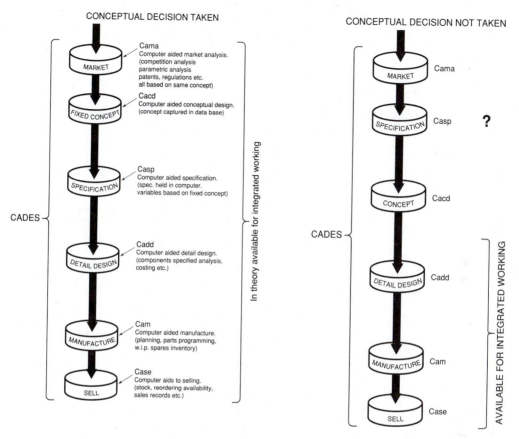

FIGURE 21.5. Software Availability: Static Concept

FIGURE 21.6. Software Availability: Dynamic Concept

Levitt, T. (1984). 'The Globalisation of Markets.' *McKinsey Quarterly* (Summer), 2–20.

Morgan, N. (1984). 'Mobilising Engineering Judgement and Imagination.' *Engineers Digest* (September), 10–13.

Pugh, S. (1983). 'The Application of CAD in Relation to Dynamic/Static Product Concepts.' *Proceedings of ICED 83* (vol. 2, pp. 564–571). Copenhagen: ICED.

Pugh, S. (1984a). 'CAD/CAM: Hindrance or Help to Design.' *Proceedings of the CFAO Conception et Fabrication Assistée par Ordinature.* Brussels: Universite Libre de Bruxelles.

Pugh, S. (1984b). 'Further Development of the Hypothese of Static/Dynamic Concepts in Product Design.' *Proceedings of the ISDS Conference* (pp. 216–221). Tokyo: ISDS.

Rosenbrock, H.H. (1983). 'Designing Automated Systems: Need Skill Be Lost?' *Science and Public Policy* 10(6) (December), 274–277.

CHAPTER

Knowledge-Based Systems in Design Activity

ABSTRACT

This chapter considers the evolution of knowledge-based systems (KBS) in relation to design activity. In order to do this effectively it commences with a definition of a core design activity in relation to products which are either conventional or novel. KBS are broken down in sympathy with the decomposition of a design—that is, whole design, partial or subsystem design, and component design. The problem of knowledge elicitation and the variability of the constraints at each level of design are considered, and conclusions are drawn in terms of conventional and novel products.

A hypothesis—to be or not to be—is proposed covering the conceptual design spectrum which suggests that the way forward is via a system of linked KMs (knowledge modules) to be designed using multidisciplinary teams, since it is considered that KBS are products in their own right and should be treated as such.

From a paper presented at the International Conference Modern Design Principles in View of Information Technology, Trondheim, Norway, June 1988.

The literature on artificial intelligence, knowledge-based systems (KBS), and design is vast and is, it seems, growing almost exponentially, and yet not before time it is all starting to focus more on the activity of design itself.

In this chapter, the evolution of KBS is considered in relation to the design activity, primarily as an adjunct to computer-aided design (CAD), since it is generally accepted that the effective design and implementation of KBS will involve extensive use of the computer. It will of necessity stay with the design activity and its ramifications, if only to retain clarity, communication, and sanity. From a consideration of the activity of design and its relationship to CAD, the chapter briefly considers the question of dynamic and static product concepts in order to provide a framework for succinctly focussing upon and harnessing worldwide KBS trends. My own views are expressed—from a design application viewpoint and an understanding of the design activity, rather than from a deep knowledge of KBS per se. In the light of this approach, conclusions will be drawn as to the lessons learned and the lessons to be learned, together with some pointers for future research likely, in my view, to yield improvements to the design activity for the practice of design.

KNOWLEDGE-BASED SYSTEMS AS A COMPONENT OF CAD

In a paper given in 1976 (Pugh and Smith, 1976), an attempt was made to relate the topic of CAD to design in practice. Consideration was given to a conceptual envelope as shown in Figure 22.1. Briefly, a boundary A product as yet nonexistent might be the Back Scratching Wogglebox with Diathermy Input (BSWWDI for short). Anyone having read this statement who manages to conjure up just what this device is to do, what its component parts are, what technology it embraces, and so on is invited to patent it! For most people, nothing appears in the mind's eye. It is truly what previously was a nonexistent product; it is novel.

On the other hand, a boundary B product such as a Porsche model 944 will not only conjure up visions of the product but will do so with envy and almost infinite vision of what it contains—especially to designers of automobiles. It is what might be described as a high-quality conventional product.

It is from this genesis that the hypothesis of dynamic and static product concepts has evolved and is being developed, where boundary A is considered to be the dynamic innovative novel boundary and boundary B the static or conventional boundary, but within which innovation can occur at subconceptual level. This is further exemplified in Pugh (1983), where two models of the design activity are given (see Figure 22.2).

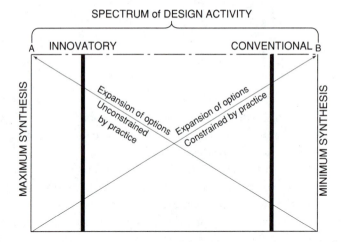

FIGURE 22.1.   Spectrum of Design Activities

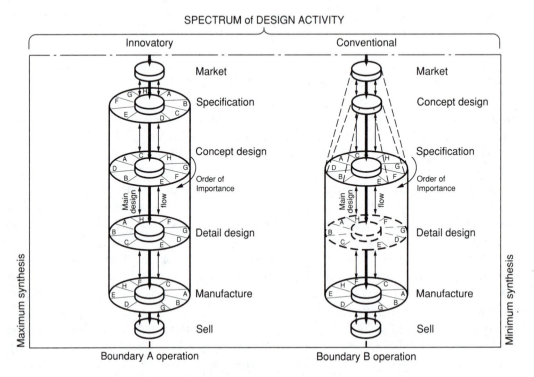

FIGURE 22.2.   Spectrum of Design Activity: Boundary A
and Boundary B Operations—Dynamic and Static

In the truly dynamic case, the expression of market or user need leads to the evolution of a product design specification (PDS) which, at least in theory, does not beg the technology or the choice of concept. Fully developed, it expresses in the context of the product to be designed all the constraints to be placed upon the design or designs yet to emerge from the activity. Design activity per se is a sterile, dormant thing until it is activated in a context and something is to be designed. In this case (the dynamic) the constraints usually have to be synthesised in relation to user need, may not have strong product line precedents, and therefore may be in a context with constraints about which very little is known or has previously been experienced.

In the static case the PDS is written with the overall conceptual choice already made—the car is a good example—and this argument is expanded upon in Pugh (1983). Most products, however, lie between the two boundaries.

The place of CAD in the context of dynamic and static concepts has been and is being agonised over as one attempts (1) better to understand the design activity and (2) to improve its efficiency (Pugh, 1984, 1985). It is worth repeating some earlier conclusions (Pugh, 1985).

In this chapter I have attempted, through models of design, to relate the complex topic of product evolution to effective, efficient CAD system usage. I will, however, resort to another diagram (Figure 22.3), which indicates how I see computer systems linking into the design core. Effective computer usage requires integrative working; the same goes for design. Are the forms of integration compatible? Figure 22.4 begs some very large questions.

In fact, Figure 22.4 begs some enormous questions which relate directly not only to CAD but also to KBS and design, since if the domain or context in which we are working has not and cannot be specified as in the truly dynamic case and the combinational experience has not been gained, then how can we specify and capture the knowledge base needed with any accuracy or certainty? I will return to the boundary A conditions later.

Since KBS require systematic knowledge acquisition from experts (by definition), it is considered essential to return to boundary B static concepts. It is also in such contexts that most KBS research seems to address itself, and this will now be considered in more detail.

RESEARCH INTO KBS AND DESIGN

The research primarily seems to fall into two categories:

- Research based upon a vision of some all-embracing design process as yet not very well defined—mainly theoretical with some application, and
- Research based upon systems design—mainly applications oriented.

Let me look into these categories a little further.

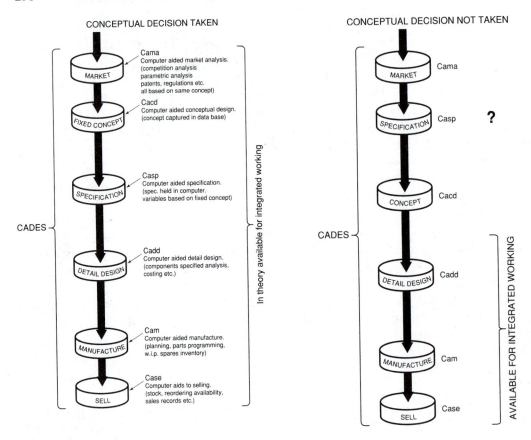

FIGURE 22.3. Software: Static Concept

FIGURE 22.4. Software: Dynamic Concept

Theoretical Orientation

Fenves (1986), in discussing the application of KBS in civil engineering, states that 'In summary, the excitement created by the emergence of KBES may be largely unfounded, as judged by the first generation KBES developed to date. Nevertheless, KBES offer a way to deal with the ill-structured aspects of civil engineering. The storage of good, practical civil engineering KBES is due not so much to the limitations of present KBES framework as to the difficulty of compiling, organising and formalising the vast body of heuristic expertise which characterises the profession.'

It is my understanding that design in civil engineering worldwide works largely to standards and codes of practice and is thus, to a great extent, highly structured when compared with the licence allowable in the design of many electromechanical or electronic products. This being the case, I find this is a somewhat distressing statement. It is, however, reinforced by Dym (1985): *'the task chosen for presentation must be reasonably narrow, performed fairly often and have a cadre of experts who agree in general terms on appropriate approaches. Expert systems are not meant for ill-defined tasks, nor are they effective where common sense and general knowledge are required'* (my italics). In my view it is more likely that working to codes and standards will result in a larger number of engineers having similar experiences, expertise, and views than is the case in electromechanical or electronics where the products are much more diverse.

Theoretical researchers are also now appreciating the fact that an intimate knowledge of the design activity itself might help in formulating KBS for design, see Gero and Maher (1987) and that they have a design problem on their hands—so they need to look to the user need end of the process and formulate a PDS for KBS design (see Lazzara et al., 1986). The discussion of Coyne et al. (1987) is interesting as KBS experts attempt to unscramble the design process, the comments of Brown being particularly poignant: 'but what worries me is that, by separating these things it's not clear that you are actually contributing much to our understanding of what design systems should look like!'

Allwood (1985) still in the field of civil engineering, foundation design, states that 'There will be many possible solutions to any design problem, but an optimal efficient solution must be sought. The measure of success of the design varies, it can be economy of material, low cost, satisfactory performance, ease of construction, elegance of appearance, etc. The expert system needs to have heuristics to test all these conditions. A generate and test control strategy seems appropriate. The logic should be deterministic, the uncertainty of the strength of materials will normally have been allowed for in the *codes of practice* that the design needs to satisfy. *Full knowledge of the conditions are required to achieve a result. It could be dangerous to design without knowing some pieces of information.'*

He is appreciative of the critical impact that the constraints have on the design problem. So a pattern is emerging: the design activity seems to be becoming the central theme, which surely must be a step in the right direction. I am deliberately avoiding any reference to the purely theoretical KBS researchers who in my view are unlikely ever to yield or design usable systems for design per se without first addressing the complex nature of the design activity, but maybe this is not their role in life!

If one accepts the premise that KBS in design is essentially a design problem which has to be addressed as such, and therefore requires the appropriate research back-up, then little progress is likely to be made in the area. This view is manifest from many areas of KBS research. As Dyer et al. (1986) states, 'Studying how people utilise their memories to create and/or adapt devices is valuable in understanding how people design and how memory is organised and applied to creative problem solving.'

Application Oriented

Applications-oriented researchers and those whose activities are probably less constrained by codes of practice tend to throw a different light on the situation and in language that I as a designer can better understand, although there are exceptions in the theoretical field (see Allwood et al., 1985). They also tend to get to closer grips with the design activity, since necessity is the mother of invention.

Edosomwan (1987) provides a very good summary of current state of the applications art. In his 'Ten Design Rules for KBES' he seems to approach the topic as a design problem. He has the user very much in mind. It is perhaps worthwhile listing his rules, which could form an *ab initio* PDS for KBS design:

1. Obtain the right knowledge base.
2. Form a knowledge-based procedure.
3. Provide an adequate structure for systems prompts and human specialisation.
4. Provide adequate KBES response time.
5. Provide adequate explanations and documentation for all system variables.
6. Provide adequate KBES time-sharing options.
7. Provide adequate user interface on KBES.
8. Provide intersystem communication ability.
9. Provide automatic programming ability and controls on KBES.
10. Provide flexibility for ongoing maintenance and update of KBES.

He focuses on the consistent repetitive task as being the basis for success, thus supporting the adherence to the boundary B design conditions adopted earlier. Heyderoff (1985) in detailing the GRASPIN approach to software design gives detailed consideration to the necessary design process in the context of software (see Figure 22.5).

Further reinforcement of the boundary B fixed-concept argument comes from Rozenblit and Zeigler (1985) considering the system entity structure for the car design problem. We know what it is, we have a lot of experience and knowledge, let's capture and use it!

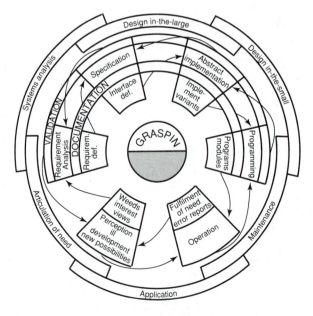

FIGURE 22.5.   GRASPIN Approach to Software Design
Source: Heyderoff (1985).

So we must have experience of the product, and presumably the top level of experience leads to the top level of expertise. The longer an overall design has been conceptually static or treated as such, the more readily we are likely to have 'shaken down' that experience and expertise into semirepetitive tasking, semirepetitive design: 'For simple repetitive tasks, a sample size of 30 humans with successful experience over five years is adequate—non-repetitive tasks are very difficult to handle!' (Edosomwan, 1987).

KNOWLEDGE ACQUISITION

The difficulty of interrogating experts and the acquisition of sound knowledge is recognised by others, such as Mittal (1984): 'Commonality of approach—all the experts followed a very similar strategy in carrying out the design in terms of how they decomposed the problem into sub-problems, worked on the sub-problems *and then related the partial designs . . . however we also learned a great deal from the differences in approaches of the experts.'*

They were dealing with a selected subsystem of copying machines and hence shared a common context, probably with variable constraints. It is suggested that variability of constraints or instability of context even with a static concept design gives rise to problems. Dankel (1986) notes that 'While it is highly desirable to employ several experts in the construction of a knowledge base, it proved to be a very difficult task—subtle incompatibilities can occur that neither the knowledge editor or experts will catch.'

Yasdi (1985) talks of a union of expertise in knowledge acquisition. This, in my view, is due to lack of definition and common view of the design activity, lack of systematic structure, and variation in the nature and type of the constraints considered (we have experienced this many times in our researches), a situation we have now overcome by formalising what design is and what it is not.

The repetitive theme of breakdown or decomposition of whole designs has led to the formulation of many partial KBS systems, partial in the sense that they deal either with subsystems or single components of whole designs. Gairola (1986) is very aware of the 'constraints on design' (see Figure 22.6): 'the main problem in using all these tools is either that they lack *entirety* in dealing with the assembly problem, or that their application is tedious, cost intensive and extremely time consuming.' To me this is symptomatic of bad systems design, stemming from loose or almost nonexistent specification of constraints within a context.

Aitchison and Wilkie (1986) recognise constraint rules but seems to limit them to the quantifiable: 'Some method should be devised to allow temporary exit from the system to access realistic design programmes, perhaps written in other languages, *to calculate the attributes of parts.*' It is suggested that parts have many attributes that cannot be calculated; they can, however, be assessed.

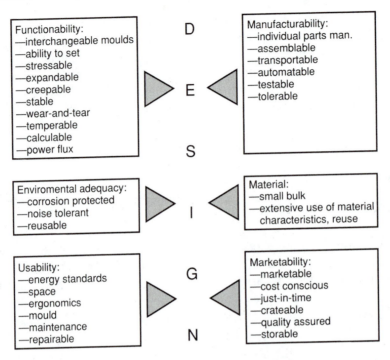

FIGURE 22.6.   Constraints on Design
(Gairola)

Components

The lowest form of decomposition of any design is to the single component and much attention is being given to this area. However, no matter what the level of the design (either whole or component), the messages are the same: context, variability of constraints within the context even for the genuine, recognisable overall static concept design. Rychener (1985) notes this difficulty: 'Emphasis here has been on solutions to components of the overall design problem, e.g. *selections of parameters* and diagnosing potential problems in partial design. *The integration of a number of such components into a complete design system is still a research problem.*'

Perhaps almost single-minded, limited KBS will be of most use to design practice, particularly at component level, as suggested by Swift, Mathews, and Syan (1986). It should be recognised that such systems must contain the appropriate prompts to trigger the designer to investigate the effects of 'change for manufacture' upon the rest of the design and its constraints. In my view this is an interminable problem with partial solutions to any part of design, but particularly within manufacturing systems refinements which tend only to accommodate those constraints of immediate import to the making function. We make it cheaper and assemble it faster, but overall product quality falls.

This partial design, partial constraint viewpoint is exemplified by Rehak and Howard (1985) in considering the design of a beam—where, incidentally, only the quantifiable constraints appear to be being addressed. However, I consider that the approach of a blackboard model, knowledge-based system where 'Knowledge is aggregated into sets of independent knowledge models (KM's) each one of which addresses a sub-problem' is fundamentally sound. I have already discussed the decomposition of a 'design' in order to handle it; here we have in parallel the decomposition of KBS design, again in order to handle it. My major worry here would hinge on the interpretation of the word *independent*, given my earlier references to discrete partial systems and the necessity for cross-linking prompts to enable smooth and safe transfer between the discrete system elements.

As stated previously, in terms of decomposition, we have nowhere to go below the component level, although atomic physicists and material scientists would no doubt take issue with this statement.

The common and strongest points emerging from the foregoing are as follows.

- Within a given context, designs and their constraints can be broken down into the following:

Total (whole) design	Total constraints
Partial (subsystem) design	Partial constraints
Component (single) (piece parts) design	Component constraints

- Acquisition of knowledge at all three levels—whole, subsystem, or component.
- The emerging treatment of KBS and design as a design problem.

These to my mind are the main themes, although arguably there have been others highlighted which are beyond the scope of a single chapter. However, there is one other almost overwhelming factor to emerge from a study of KBS and the design activity; it is so obvious as to be almost subliminal. It stems from the recognition by some researchers, such as Rychener (1983), of the difference between diagnostic and design systems, and the reconciliation of this fact with the commonest reference work of the majority of researchers in the field—Shortliffe (1976), who described a medical diagnostic KBS called MYCIN.

It is upon this latter fact that I now concentrate, bringing in the questions of context and knowledge acquisition as the hypothesis unfolds.

The Hypothesis of 'To Be or Not to Be'

The earliest applications of KBS have been in medicine and geology, both in diagnostic roles. There is no intellectual dispute or disagreement between disciplines, or in any respect, that the earth is what it is and that a human being is what he or she is. The design of KBS to address various forms of illness or natural phenomena is based upon this common premise: there is only one overall design or configuration of humans (that is, one head, two arms); there is also only design of the earth. Humans thus are in my view the *ultimate in static concepts*: the design evolved over millions of years. Humans can be firmly placed at boundary B. Their design per se will change only slowly (although genetic engineers might have us think differently); therefore, to design a KBS to deal with his ailments should theoretically carry little risk in terms of utility, accuracy, and certainty. I don't actually think it is quite as simple as this in practice, since even with medical diagnostic systems, in design terms, the subsystems and components may vary in detail. Nevertheless, as engineers we have lessons to learn from the grand design.

To return to design itself, design is not diagnostic, it is *generative*. But the interrogation of experts to acquire their knowledge and expertise is in fact a diagnostic activity; we can interview all the motor car designers and look at the designs from the various levels from the total (whole) to the components (see Oza, 1984). The experts in this case, as with the medical experts, have experimented with and understand the extent and degree of the interactions between the components of their designs—cams and valves in the case of the motor car, for example, and blood and valves in the case of the heart—although in the latter case humans didn't do the design.

So, essentially, from the diagnosis of the interactions and ailments of the system we can evolve a KBS. With humans there is the certainty that the design is fixed; can the same be said of the motor car? In designing the next car we have to operate in generative mode, and if we utilise a KBS we must recognise that the system itself has been generated from what has been treated as a static concept. This is logical since to obtain the knowledge and to be able to use it we must 'freeze the frame.' But what about the combinations we don't know about, the ones we have not yet discovered or formulated—the equivalent in medical terms of the six-headed, four-legged, three-eyed man, since in product design terms we do in fact keep coming up with this equivalence? But in product design terms, we have already experienced the problem of knowledge elicitation even with 'frame freeze' as contexts with constraints universally lacking rigour and consistency, particularly at the total (whole) level and only marginally less so at the subsystem and detail component levels. For instance, we now operate a system of constraint specification equating to the total PDS. Maher, Siram, and Fenves (1984) also discuss constraint handling to some degree.

So how certain are we of our knowledge base, obtained diagnostically by what in all probability are random methods, lacking cohesion and consistency? Not at all, according to the literature, although the closer we get to the single component level the more certain we become, according to the evidence. I return to this point later.

Let me now look at the diagnostic and generative situations in terms of the dynamic, boundary A, and static, boundary B, approach adopted earlier.

Figure 22.7 attempts to illustrate the degree of coupling between the diagnostic and the generative approach required in design, bearing in mind that in the medical situation we do not generate new designs of humans, only cures (although it may be argued that some component and subsystem redesign is underway today—with artificial joints and so on). It can be clearly argued that for the truly static concept for a subsystem akin to, say, the differential gear (see Pugh, 1983), it would be perfectly safe and reasonable to design a KBS specific to this context (given more rigour in systematic diagnosis) and to utilise it in the generative mode. Can the same be said for all car subsystems and the car itself? I think not. Therefore, even in apparently boundary B situations extreme care should be taken in the design of any KBS, since use of the system itself may ultimately inhibit progressive thinking.

The problem of constraint recognition, rigour, and formulation is becoming increasingly recognised at all levels of design activity, and Brooking (1984) 'clearly showed that an analysis of the domain affected not only MMI (man, machine interface) *but all aspects of KBS development*.' A consideration of the certainty and the stability of knowledge acquisition is shown in Figure 22.7 from a whole design viewpoint.

If this is true, it would seem that the design and use of KBS for whole design at boundary A is not only not logical but it's not possible: it is not to be. Thus in KBS terms we can differentiate between 'to be' and 'not to be' areas: boundary B is 'to be,' qualified as in the foregoing; boundary A is 'not to be'—or is it? To configure in the mind's eye the BSWWDI to which I referred earlier, one has to work with the elements of nature and those latterly provided by man. Therefore, if we design and configure a KBS in narrow but rigorous contexts, along the lines of the KMs suggested earlier but targetted initially at the lowest common denominator of a system (a single component), and provided that we *design* an adequate system of cross-prompting between them, then we can, with caution, use such systems at boundary A, provided that we retain recognition of the fact that this is what we are doing.

To design and utilise KBS at the higher levels of design at boundary A seems to me to be fraught with danger—particularly in the light of the expressed difficulties at the lower level and also in boundary B situations. Accepting the fact that any KBS stems from interrogation of experts, who are the experts in the case of the BSWWDI, considered as a whole design? It is at this point, once the product has been conceived, that we build models, prototypes, and the like in order to gain knowledge of the product, its constraints, and their interactions and our design starts to move towards boundary B. The Mk 2 BSWWDI will build upon the knowledge and the understanding gathered from the Mk 1 model; it would be considered foolish to assume that capturing all the knowledge for the whole product at this stage would auger well for the future. If we do lay down skeletal KBS for the whole design at this stage, it will grow with time and experience and will ultimately end up at boundary B. I would question whether it is wise

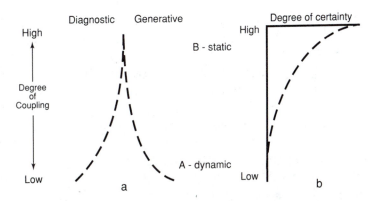

FIGURE 22.7.   (a) Relationship Between Diagnostic and
Generative Systems.(b) Knowledge Acquisition

to do this and that, perhaps for boundary A use, we will never find it possible or indeed should never go beyond the elemental approach.

So, at boundary B, all-embracing KBS are ultimately a possibility. They may be said to be truly 'to be' provided we are systematic about it. At boundary A perhaps all-embracing KBS may be truly said 'not to be,' at least for whole systems. Both ends of the spectrum may be best satisfied by a system of KMs or 'distributed knowledge bases interacting in an open communications system to support peer-level decisions' (Hamill and Stewart, 1986).

Simmons (1984) urges caution: 'The development of AI systems for engineering design is not a straight forward application of existing technology. A number of areas will involve fundamental research in AI.' In my view major steps forward in the area of KBS and design will be made only when multidisciplinary teams are put on to the problem—with a thorough understanding of the design activity—since it is a design problem; after all, this is what we do nowadays for successful design.

To conclude on a more mundane note, we are currently trying to configure and handle the context of a single shaft with *all* its constraints—a very difficult task. We should, however, always remember that 'exhaustive design rules and guidelines only serve to stifle design creativity' (Dewhurst and Boothroyd, 1986).

REFERENCES

Aitchison, T.E., and Wilkie, G.A.R. (1986). 'Expert System for the Design of Engineering Assemblies.' *Proceedings of the International Conference on Computer Aided Production Engineering* (pp. 249–257). Edinburgh: CAPE.

Allwood, R.J. et al. (1985). *Report on Expert Systems Shells Evaluation for Construction Industry Applications.* University of Technology, Loughborough U.K.

Brooking, A.G. (1984). 'Towards a Methodology for the Design of Knowledge Based Systems.' *Esprit '84* (pp. 147–158). Elsevier: North Holland.

Coyne, R.D. et al. (1987). 'Innovation and Creativity in Knowledge Based CAD.' *Proceedings of the I.F.I.P. WG 52 Conference on Expert Systems in Computer Aided Design* (pp. 435–471). Elsevier: North Holland.

Dankel, D.D. (1986). 'Expert Systems: Misconceptions and Realty.' *Proceedings of the International Congress on Artificial Intelligence SP-664, Society of Automotive Engineers* (pp. 5–9).

Dewhurst, P., and Boothroyd, G. (1986). 'Computer Analysis of Product Designs for Robot Assembly.' *Proceedings of the International Conference on Computer Aided Production Engineering* (pp. 17–23). Edinburgh: CAPE.

Dyer, M.G. et al. (1986). 'EDISON: An Engineering Design Invention System Operating Naively.' *Artificial Intelligence* 1(1), 33–45.

Dym, C.L. (1985). 'Expert Systems: New Approaches to Computer Aided Engineering.' *Engineering with Computers* (vol. 1, pp. 9–25).

Edosomwan, J.D. (1987). 'Ten Design Rules for Knowledge Based Expert Systems.' *Industrial Engineering* 19(8) (August), 78–80.

Fenves, S.J. (1986). 'What Is an Expert System.' *Proceedings of the Symposium on Expert Systems in Civil Engineering* (pp. 1–6). Seattle: SESCE.

Gairola, A. (1986). 'Design for Assembly.' *Robotics* 2(3) (September).

Gero, J.S., and Maher, M.L. (1987). 'A Future Role of Knowledge Based Systems in the Design Process.' *CAAD Futures.* Elsevier: North Holland.

Hamill, B.W., and Stewart, R.L. (1986). 'Modelling the Acquisition and Representation of Knowledge for Distributed Tactical Decision Making.' *John Hopkins APL Technical Digest* 7(1), 31–38.

Heyderoff, P. (1985). 'GRASPIN: A Coherent Methodology.' *Esprit '84 Status Report on Ongoing Work* (pp. 107–126). Elsevier: North Holland.

Lazzara, A.V. et al. (1986). 'A Framework for Extending Information Systems with Knowledge Based Processing Capabilities.' *Proceedings of the IEEE Computer Society's International Computer Software and Applications Conference Comsac* 26 (pp. 79–85).

Maher, M., Siram, D., and Fenves, S.J. (1984). 'Tools and Techniques for Knowledge Based Expert Systems for Engineering Design.' *Advanced Engineering Software* 6(4), 179–189.

Mittal, S. (1984). 'Knowledge Acquisition from Multiple Experts.' *Proceedings of the IEEE Workshop on Principles of Knowledge Based Systems* (pp. 75–81). Denver, CO: IEEE.

Oza, R.D. (1984). 'Some of the Issues in the Implementation of Artificial Intelligence in Engineering and Design.' *Proceedings of the IEEE Workshop on Principles of Knowledge Based Systems* (pp. 51–57). Denver, CO: IEEE.

Pugh, S. (1983). 'The Application of CAD in Relation to Dynamic/Static Product Concepts.' *Proceedings of ICED '83* (vol. 2, pp. 564–571). Copenhagen: ICED.

Pugh, S. (1984). 'CAD/CAM: Hindrance or Help to Design?' *Proceedings of Conception et Fabrication Assistée par Ordinateur.* Brussels: Université Libre de Bruxelles.

Pugh, S. (1985). 'CAD/CAM: Its Effect of Design Understanding and Progress.' *Proceedings of the International Conference CAD/CAM Robotics and Automation.* Tucson, AZ.

Pugh, S., and Smith, D.G. (1976). 'CAD in the Context of Engineering Design: The Designer's Viewpoint.' *Proceedings of CAD 76* (pp. 193–198). London: CAD.

Rehak, D.R., and Howard, H.C. (1985). 'Interfacing Expert Systems with Design: Data Bases in Integrated CAD System.' *Computer Aided Design* 17(9), 443–454.

Rozenblit, J.W., and Zeigler, B.F. (1985). 'Concepts for Knowledge-Based System Design' *Proceedings of the IEEE Winter Simulation Conference* (pp. 223–231). New York: IEEE.

Rychener, M.D. (1983). 'Expert Systems for Engineering Design: Experiments with Basic Techniques.' *Proceedings of Trends and Applications, Automating Intelligent Behaviour Applications and Frontiers* (pp. 21–27).

Rychener, M.D. (1985). 'Expert Systems for Engineering Design.' *Expert Systems* 2(1) (January), 30–34.

Shortliffe, E.H. (1976). 'Computer-Based Medical Consultation: MYCIN.' Elsevier: North Holland.

Simmons, M.K. (1984). 'Artificial Intelligence for Engineering Design.' *Computer Aided Engineering Journal* (April), 74–83.

Swift, K.G., Mathews, A., and Syan, C.S. (1986). 'The Potential for Intelligent Knowledge Based CAD in Manufacturing Engineering.' *Proceedings of the International Conference on Computer Aided Production Engineering* (pp. 47–53). Edinburgh: CAPE.

Yasdi, R.A. (1985). 'Conceptual Design Aid Environment for Expert: Data Base Systems.' *Data and Knowledge Engineering* (pp. 31–73). Elsevier: North Holland.

PART V

Design Teams, Management, and Creative Work

COMMENTS ON PART V

Probably the most important papers in this part are Chapter 25, The Organisation of Design: An Interdisciplinary Approach to the Study of People, Process, and Context, and Chapter 28, Organising for Design in Relation to Dynamics and Static Product Concepts, both coauthored by Stuart Pugh and Ian Morley, who is a social psychologist. This collaboration between a total designer and a social psychologist was very productive and insightful. Usually students of design teams are either engineers who know nothing about social psychology or psychologists who know nothing about total design. The work of Pugh and Morley stands out in contrast. They integrated views that approached from different directions. In particular, they emphasized the need for vigilant information processing, not mere group gropes. Many integrated product teams today could greatly benefit from a better understanding of this point: teams are necessary but not sufficient. Chapter 28 reports a study of several companies. It brings to life the concepts of the design activity model, static and dynamic design, and product design specification, as they were found in various states of capabilities in the companies. The reader might benefit by applying the assessment method to his or her organization.

This part develops the theory of static and dynamic concepts and starts with Chapter 23, Further Development of the Hypothesis of Static and Dynamic Concepts in Product Design, which I associate with my first introduction to Stuart Pugh. This paper was presented at the International Symposium on Design and Synthesis in Tokyo in 1984, and Stuart and I were introduced there by our mutual friend Ken Ragsdell. After several discussions during that conference Stuart and I quickly became friends and eventually collaborators. This chapter is marvelous, clearly making the distinction between static and dynamic design and presenting very interesting examples.

Chapter 24, Engineering Design in Practice, describes the generic product development specification and the information needs in total design. It contains a critique of the specification of a commercial product and the omissions in the specification that could have been avoided by using the generic specification. Also presented is a useful table of information needs for different levels of total design practitioners. Stuart was very keen about having the right information available. At Loughborough his student teams included information science students to help engineering students to manage the huge information requirements of a total design project.

In reading the chapters in this part look for the applications of the earlier chapters in industry, the organization of total design, the dependence of organization and the practice of total design on location in the static/dynamic continuum, and the practice of design reviews (in Chapter 29, The Design Audit: How to Use It).

—Don Clausing

INTRODUCTION TO PART V

In this part the spectrum of conceptual design is further explored with particular reference to dynamic and static concepts linked to the question of innovation. This involves concentrating on the front end of the design core. Starting prematurely—trying to innovate without sound, thorough front-end work—will render subsequent work somewhat less than effective.

The methods of Part 3 are shown in context in relation to Taguchi, QFD, and other methods in common use.

One of the greatest inhibitors to effective design performance is shown to be the poor quality of manufacturers' information. This is demonstrated using the data sheet for a variable-delivery swash plate pump in terms of its likely PDS requirements when utilised by a hydraulic system designer: note the omissions.

Design processes must be systematic and complete. It is a reasonable assumption that most random (Eureka) innovation can be bettered fairly quickly if methods are applied in a systematic as opposed to random or staccato fashion.

Analysis of the performance of a number of U.K. companies was done in terms of their products and descriptions of their design processes to a format we presented. This was very revealing since the dynamic and static variant was introduced with their position in a spectrum of design activity between two boundaries—dynamic A and static B.

Design review or design audit is discussed where it is shown to harmonise with the design core.

Working with industry and discussing design it was realised that in order to communicate effectively, it became absolutely necessary to define design and to thrash out an agreed version (with the approach taken, few problems were then experienced).

This problem became progressive over the years and Chapter 27, Balancing Discipline and Innovation, illustrates an early attempt to overcome the communication problem. This approach has stood us in good stead ever since and has enabled our research and practice to benefit. In fact, after many years trying to stabilise the SEED approach and curriculum, it was only after the establishment of a format for design, agreed on by all members, that it really took off. Prior to that there was always a degree of dissension which was not only energy sapping but extremely wasteful of time, since the discussions ultimately failed to clarify matters.

Further Development of the Hypothesis of Static and Dynamic Concepts in Product Design

---

ABSTRACT

This chapter considers the question of conventional and nonconventional products in relation to the hypothesis of static and dynamic product concepts. In the conventional case, the emergence of a generic base is highlighted and discussed in the terms of the hypothesis: randomness of approach to design in the conventional area is seen as being inefficient. In the nonconventional case it is again shown that randomness of approach, combined with rapidly advancing technology, can mask the major issues and the emergence of the generic base. It is argued that in both cases a systematic design activity needs to be adopted, and that within such activity alternative concepts to a given need situation, as defined in the product design specification (PDS), require to be synthesised. Examples are given in both conventional and nonconventional areas. Most important, attempts are made to demonstrate that, in effect, both conventional and nonconventional product areas may, in fact, be conceptually highly dynamic.

From *Proceedings of the International Symposium on Design and Synthesis* (Tokyo: Japan Society of Precision Engineering, 1984), 216–221.

INTRODUCTION

Consideration has previously been given to the dual question of dynamic and static product concepts in relation to the application of computers in design (Pugh, 1983a). A hypothesis was postulated which considered all products as fitting into a dynamic and static spectrum, and the conclusion reached was that all such products, with the passage of time, migrate towards a conceptual plateau. Until such time as the static condition or plateau is reached, all product concepts should be considered to be dynamic, which in turn demands that the design activity carried out in industry should always be structured in a dual fashion.

First, the design activity for products at the static, tending towards static concept boundary, should be clearly identified and recognised as being a valid and necessary activity in its own right. Second, a parallel, linked activity must be operated on the basis that dynamic concepts are ultimately necessary for business survival, and therefore the design activity must be structured (Pugh, 1983b) to accommodate this phenomenon and differences.

This chapter seeks to take the above hypothesis a stage further forward to gain a clearer understanding of the true nature of new product development and innovation. It will be argued that products with a long past history—long in evolution, old-fashioned, some might say—are almost always considered as conceptually static, mainly because of the passage of time. Since it has usually taken a long time for a product to reach its current state of excellence, it must perforce be the ultimate in that product sphere! It is the author's view that, whilst adopting this stance leads to a more comfortable life, an orderly product line, and, no doubt, full employment and profitability, it may be a misguided view of things. In reality, many so-called traditional (old) products are still conceptually dynamic, and to treat them as static is ultimately business suicide. This facet will be considered in some depth.

Consideration will also be given to the genuinely highly dynamic product concept situations, particularly in areas of so-called high technology where precisely the same mistakes are being made as in the past, and vulnerability to competition is rife due to lack of appreciation and understanding of the place of one's product in the dynamic and static concept spectrum.

CONVENTIONAL PRODUCTS

Without wishing to embark upon a detailed discussion of what is meant by conventional products, it is considered that cars, ships, gas turbines, bicycles, and washing machines are typical, and that electronic calculators and microcomputers are not traditional products. It is considered that this looseness of definition will not seriously affect the arguments to be put forward in the limited space available.

It is trite to suggest that once upon a time all products were novel and innovatory and thus were conceptually dynamic, and that with the passage of time they have, by an almost random process of iteration, arrived at a plateau and become conceptually static. The motor car is a good example of this, cited in Pugh (1983b), and thus all motor car design may be said to be converging on to a generic base, a single static concept, albeit with many detailed variations.

It may therefore (in the case of the passenger car) not be unrealistic to state that this convergence onto the generic product base, the ultimate static concept, will probably take 100 years or so, give or take a few years, and rightly so since it can be technically and historically justified. It is interesting to note that others concerned with innovation and technological evolution, albeit not from a design viewpoint, are indirectly recording this phenomena. William (1983) states that 'As technological evolution progresses, the product-based technological risk increases. The production functions of all surviving rivals become relatively more similar.'

Management researchers usually refer to such products as being mature: the motor car would be considered mature. However, it is suggested that there is on the one hand a great danger in assuming that mature products are conceptually static, that they have reached a plateau, and that the generic base of the traditional product has been found. The leading question is, therefore, how many conventional products have reached this zenith. It is suggested that in many product areas, where it may be convenient and comfortable to reach and also to stick to this conclusion, that such assumptions or conclusions may be ill founded and erroneous. This statement applies to many traditional products that have been assumed to be mature but conceptually are not static; they are highly dynamic, but the false assumption makes people stop looking.

"Seek and ye shall find." (Bible). It is much more comfortable not to seek but to accept the situation as it appears and then to systematise the whole business, production system, or the like. Yes, it is much more comfortable, controllable, and manageable, but it could be the ultimate in disasters. It matters not in reality whether one is in a dynamic or static concept situation; it is the *not knowing* which gives one cause for concern.

So one must seek out the generic base, which usually results in the simplest, most elegant design making the best use of the fundamental knowledge and information.

For nonbelievers it is perhaps salutary at this point to underwrite the argument so far by referring to two happenings in product areas which, worldwide to a man, would be regarded as conventional of relatively fixed design.

- The ball valve, upon which the world's sanitation has become increasingly dependent since 1748 (Wright, 1980). The first British patent for a ball valve (No. 10,837) was granted to Edward Chrimes in 1845 (Reader's Digest, 1982). This product has been conceptually static until the early 1970s with the introduction of the Torbeck valve and now is about to be overtaken by the Ve Cone valve, invented by Roberts and Associates Waste and Water Treatment (Cane, 1984) and shown in Figure 23.1. A very elegant design indeed.

- The inflatable motor car tyre on a steel rim, the first designs of which appeared in the early 1890s (*Encyclopaedia Britannica*, 1970). The design of tubed tyres and rims stabilised in the 1930s in the United States until the emergence of the tubeless tyre in 1947. Rim design changed somewhat but retained the well-base or drop centre to enable tyres to be easily mounted and dismounted. This is still in universal usage today (see Figure 23.2).

Various attempts have been made to overcome the inherent instability of the well-base rim in puncture situations. The RoSafe well-base fitment of the 1970s comes to mind as does the more widely advertised Dunlop Denovo, which, with its sophisticated equipment to lubricate and inflate the punctured tyre, also needed special rims with a reduced depth of well.

In 1984, Continental (Knight, 1984) produced a new design concept where the rim well is in effect inverted, and the tyre bead locks into grooves on the inside of the rim, as shown in Figure 23.3, a very elegant design concept.

It is significant that in each of the examples quoted the overall designs become simpler and meet the requirements demanded of the appropriate specification—the need situation—in a more cost-effective, safer, and elegant manner.

In the context of this chapter, it is suggested that with a systematic design approach based upon the assumption that the conceptual situation was still dynamic rather than static, then possibly the designs described above could have emerged earlier in time.

The two examples were chosen because they are products in universal usage, apparently simple conceptually and also of long lineage. Other examples are known to the author but in each case a large amount of text would be needed to describe the rationale behind them—well beyond the scope of this chapter.

That this suggestion is pure supposition must be acknowledged.

However, it is the author's experience in many projects selected initially to provide breadth and to avoid over specialisation with its inherent dangers (with particular regard to the effectiveness of design methods), that such breadth has revealed almost inadvertently the frailties inherent in product development based on conventional products where there are few, if any, precedents. They all have a common denominator; this is a

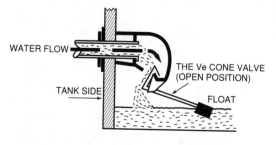

FIGURE 23.1.   Water Valve
By courtesy of *Financial Times*

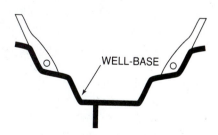

FIGURE 23.2.   Well-Based Rim

MORE SPACE FOR THE BRAKE

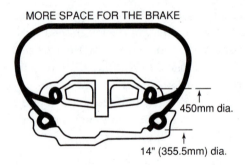

450mm dia.

14" (355.5mm) dia.

FIGURE 23.3.   New Rim
By courtesy of Eureka magazine

generic base. It is in the seeking out of such bases over a large number of wide-ranging product areas that has given rise to the hypothesis of static and dynamic product design concepts.

Finding the Generic Base: Conventional Products

The ultimate generic base concept in any product area may be defined as the concept which embraces all of the attributes of the competing concepts without attracting any of the deficiencies of those concepts. In reality this is never true and probably in the absolute sense is never attainable; however, for all practical purposes and nearer the truth, the generic base concept contains most of the attributes with very few of the deficiencies.

First, it must be assumed that all products are conceptually dynamic and that the end of the line has not been reached; otherwise the study ends here. Second, the generation of many concepts to meet a product design specification, typically between thirty and eighty, including the leading competitors, in what might be considered conventional and therefore conceptually static product areas, has in many instances led systematically (Pugh, 1981) to the emergence of new and better concepts. Two major conclusions are possible at this juncture:

1. The initially dynamic assumption has, through systematic thoroughness, proved beyond all reasonable doubt (at that time and with the technology available) to have been in error, and the product area is conceptually static, QED. The design may then be said to follow conventional conceptual lines but with detailed improvements leading to a better whole; it is still a good design.
2. This is considered much more important than (1). The initially dynamic assumption has, through systematic thoroughness, proved beyond all reasonable doubt that the product area is conceptually dynamic and the generic base for the product has been synthesised and hence discovered. The design may then be said to follow unconventional lines and thereby creates a host of competitors all assuming, possibly wrongly, that the ultimate in a new and now conventional design has been reached. The first rough terrain telescopic handler, designed at

Loughborough in 1973, is a prime example of this (Mark Handler produced by Mark (UK) Ltd.).

Suffice it to say that it is considered inefficient in any company and in any product area not to carry out design activity based on a dynamic concept basis, albeit alongside apparently static—or conventional—activity. It will be argued by many that to carry out a full dynamic activity is time-consuming and takes longer than the conventional approach, however that may be defined. This is so, but usually this only needs to be once for each product, depending on the degree of thoroughness with which the work is carried out in the first place. Thereafter, the whole design activity is accelerated and carried out faster and with more certainty than previously. Upon reflection, businesses do not really have valid alternatives, since if they pursue static approaches in dynamic situations, repetitively, they will go out of business if the competition views the situation dynamically or if they pursue a dynamic policy which always (dependent upon the quality of the people in the business) gives them the opportunity of being competitive and staying in business. It is really not a question of choice; there is ultimately no choice.

It is one thing to know that a dynamic situation exists and to find the generic base: it is quite another situation to disregard the discovery of the generic base concept and to follow convention. However, without the knowledge of the former the opportunity to exercise the latter cannot arise. There can be no excuse for not knowing.

By and large, industry worldwide is working upon the static concept assumption with conventional products. This situation is currently masked and made more difficult to unscramble by the retrofitting of more up-to-date technology, mainly microelectronics and derivatives, to conventional designs. This may prolong the life of a product; it may give it new life, but if the generic base has not been found, such new life could well be short-lived. This is not to say that experimentation with microelectronics in traditional product areas is unhealthy, per se; it is not. But in the United Kingdom it is often carried out under the heading of applied research, and if this is considered as analogous or equivalent to design, then this is unhealthy, since what is likely to emerge is the clever, functionally superior, cost-effective, conceptually vulnerable product.

The revelation of the generic product base—the ultimate static concept—should therefore always be the main aim in any design activity. New technology should always be considered in this manner and not retrofitted for short-term gain. Retrofitting, as previously stated, gives rise to conceptual vulnerability. We have many examples of this phenomena; unfortunately, these products cannot usually be named.

In a study recently carried out in a long-established conventional product area, the designs of some 250 companies—worldwide—were studied in the context of the generic base during the course of a conventional competition analysis. A generic base concept, as defined earlier, was found to cover the whole field. It was possible to fit them all into a pattern and to assess those likely to achieve the base concept and, most interestingly, added science, whilst giving product improvement, was masking the main issues.

Another recent project was carried out in cooperation with a sophisticated process company, well-endowed with high technology skills and consequently applications.

Seeking out and finding the generic base for the process area, in the end analysis, eliminated the necessity for a considerable amount of high technology equipment even though the new design itself was made possible through using high technology in a judicious manner.

Figure 23.4 shows a graph of parameters relating to vehicle-mounted aerial access platforms which come in two basic types: telescopic boom and articulated booms. This work was carried out as part of a parametric competition analysis, a normal component of any design activity.

The two concepts can be distinguished quite clearly from the graph; this revelation, combined with correlating information from many other sources, resulted in the establishment of a new concept neither telescopic nor articulated as such—not quite the generic base perhaps but possibly getting closer to it. As this equipment is under development, its form cannot be revealed in this chapter.

These examples, and others, have led to the hypothesis being put forward in this chapter. To look now at the nonconventional product—the product without precedents—should prove interesting and rewarding.

NONCONVENTIONAL PRODUCTS

There is already considerable evidence for products having a nontraditional base, particularly in areas of high technology, that the random cycle approach to design is being

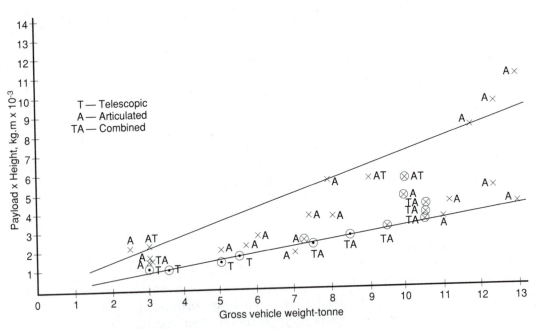

FIGURE 23.4. Parameter Relation to Aerial Access Platforms

repeated on a global basis. The author believes that the random cycle approach is triggered from two major but significantly different causes.

First, it is considered that the follow-on to fundamental research—applied research—is a major cause of confusion, particularly when it is misconstrued as design (Pugh, 1983). There are many more constituents to comprehensive design activity, and hence soundness and thoroughness of approach, than are likely to stem from applied research methodology. For example, how do the applicants of microchip research cater for the sociological, economic, aesthetic, and ergonomic considerations of the real need situation in the design of new equipment? Not very well. Lack of recognition of the design activity as being different exacerbates the confusion and feeds the random approach, leading to inadequate products. The body scanner may be a case in point here.

Second, products based upon the single idea, the eureka, which are rapidly designed, rapidly developed, and rapidly put onto the market, as in the first case, are made more difficult to unscramble by the increasing rate of technological evolution. In most cases the product or the company are likely to fail due to lack of thoroughness of approach, and there is clear evidence to this effect: within the past year some of the first robotics companies have gone out of business.

It is highly likely that this second factor is, in fact, a subset of the first, since it is from applied research or in association with the same, that eurekas are likely to arise. This phenomena—randomness of approach—is detectable: we have recently had a cursory look at robots currently on the market that indicates quite strongly in a number of areas, the drive of technology, to the ultimate dismay of the market. Figure 23.5 shows a typical robotics plot which is presented without comment except to say that models A and B should give cause for concern.

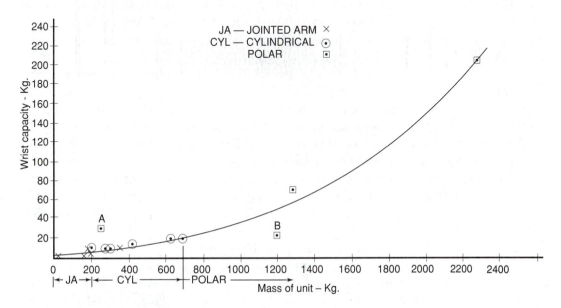

FIGURE 23.5. Parameters Related to Robotic Arms

The necessity therefore, is for a systematic design activity in the case of non-conventional products. It will be argued that without product line precedents it is difficult to carry out design in a systematic and orderly fashion—it is difficult to carry out design, per se, in any fashion!

It goes without saying that in high technology product areas, concepts are by definition going to be dynamic: there is going to be a lot of movement and change of concepts as product areas develop. It is a submission of this hypothesis that this development can take place in a systematic manner by the application of a comprehensive design activity as opposed to the random approach—which is patently inefficient. Of course it is much more difficult to minimise conceptual vulnerability in highly dynamic situations but attempting to do so will reduce the chances of it happening.

The single new product idea will be used as a vehicle to illustrate this point. In all cases the single idea (concept) in the absence of product line precedents and, as yet viable alternatives on the market, *must* be challenged by alternative concepts, synthesised on the basis that if a competitor introduced your product (conceptually) to the market, how would you, as a business, react? What concepts would you synthesise, in what manner and why, in order to ultimately produce the better competing product? In other words, the answer is overwhelmingly the forced creation of alternative concepts, the establishment of likely alternative scenarios, and the evaluation of these in a systematic manner in order to establish, beyond all reasonable doubt, that the single concept is the best that is possible (at that time), within the constraints of the product design specification. If the answer is yes, then one can be said to have minimised conceptual vulnerability in albeit a highly dynamic product area.

The dangers inherent in a lack of systematic approach with nonconventional products is much greater than with conventional products, since timescales are likely to be much shorter in the highly dynamic, volatile market. The message, therefore, is to operate in a systematically dynamic manner and not a randomly dynamic manner.

In a dynamic concept situation, the visibility of a generic product base is likely to remain obscure for some considerable time, particularly with rapidly moving technology.

Ultimately, however, the success of a product arrived at by systematic means will depend on technological balance and the skill and ability to harness the creative abilities of a multidisciplinary team within a systematically structured design activity, *without* inhibiting creativity. To some extent this has been achieved in a professional practice and, probably what is more important in the long term, it has been achieved with postgraduate students. Creativity, unconstrained and let rip, can end in disaster; conversely, creativity under control may give rise to social and cultural problems. Nevertheless, these have to be overcome if need situations are going to be satisfied more adequately in the future than they have been in the past.

I conclude with two quotes which support the hypothesis put forward in this chapter: 'With today's requirements for complex microprocessor-based projects in a wide range of industries, it is no longer good enough to produce something which seems intuitively suitable, then modify it until the result is satisfactory. Projects tackled on these lines usually start well but lead to the continual discovery of new problems. The project encounters a disastrously prolonged period when the end is in sight but never seems quite attainable. Microprocessor projects need a far more structured approach' (Cooke, 1983). Finally, at a recent conference on CAD, a speaker from a company well into efficient utilisation of such equipment was asked if he felt that the equipment was designed with both the user and his requirements in mind. He replied that 'this was the last thing you could say about the equipment, and that nothing could be further from the truth.'

REFERENCES

The Bible. Matthew Chapter 7, Verse 7.

Cane, A. (1984). 'Flush of Success: New Valve for Water Systems.' *Financial Times*, 19 January, p. 14.

Cooke, P. (1983). 'Electronic Design Warning.' *Engineering Designer* 9(6) (November), 8.

Encyclopaedia Britannica, vol. 22, pp. 13–16.

Knight, R. (1984). 'Tyre Technology Jumps into a New Era.' *Eureka* 4(1) (January), 30–31.

Pugh, S. (1981). 'Concept Selection: A Method That Works.' *Proceedings ICED* (pp. 497–506). Rome: ICED.

Pugh, S. (1983). 'Research and Development: The Missing Link—Design.' *Proceedings ICED 83* (vol. 2, pp. 500–507). Copenhagen: ICED.

Pugh, S. (1983a). 'The Application of CAD in Relation to Dynamic/Static Product Concepts.' *Proceedings ICED 83* (vol. 2, pp. 564–571). Copenhagen: ICED.

Pugh, S. (1983b). 'Engineering Out the Cost.' *Proceedings DES 83* (pp. 121–128). Birmingham: DES.

Reader's Digest. (1982). *The Inventions That Changed the World*. Reader's Digest Association.

William, J.R. (1983). 'Technical Evolution and Competitive Response.' *Strategic Management Journal* 4, 55–65.

Wright, L. (1980). *Clean and Decent*. London: Routledge and Keegan Paul.

CHAPTER

24

Engineering Design in Practice

ABSTRACT

Consideration of the information needs of engineering designers is not a simple task, since the engineering design task itself is a complex activity. Set the design task against a background of the evolution of products at varying rates and in themselves technically complex, and the size of the problem begins to become apparent. Therefore, in order to describe this situation, models of the activity of design are needed and within them the spectrum of design-type activities carried out within an engineering business. What is meant by engineering design? A possible answer is proffered which crosses the traditional discipline boundaries.

From R.A. Wall (ed.), *Product Information Finding and Using: From Trade Catalogues to Computer Systems* (Glower, 1986), ch. 12.

ENGINEERING DESIGN: TOWARDS A COMMON UNDERSTANDING

In discussion with people both in industry and the academic world, in reading articles and listening to lectures and speeches about engineering design (Pugh, 1974), a remarkable number of different approaches to the topic are discernible, conditioned no doubt by individual background, experience, and, in particular, engineering discipline. Unless a discussion is between members of the same industry or discipline, views and understanding of design will differ widely. For instance, in discussing, say, electrical engineering design, this may be considered as something quite different from mechanical, civil, electronic, and architectural design—indeed, as if there were no areas of commonality between them. The same may be said of exponents of design in respect of the latter disciplines. The reasons for this, it is suggested, are the compartmentalising and isolating of the different disciplines that prevent us from having an understanding of design common to them all and the lack of a common base for understanding, or common ground, which not only prevents or inhibits understanding but also leads to confusion.

Readers are invited to consider all design as having a boundary, to be called 'the design boundary' and defined as encompassing all the activities and factors to be taken into account by the responsible design engineer, designer, architect, and so on in pursuit of his product design. Let it be quite clear at the outset that the term *product* is used in the sense of an article to be manufactured from drawings, such as an electric razor, lawn mower, electric motor, bridge, nuclear reactor, building, and the like.

It cannot be overemphasised that design is not just a technological activity with concentration on performance and technical excellence. All designs, to be truly satisfactory and successful, must satisfy many more criteria. The criteria, activities, and factors which impinge upon and influence all product designs (factors which are not always recognised), all of which should be considered and continually borne in mind by the designers in pursuit of the best design, are listed here:

Environment
Ergonomics
Scientific disciplines (chemistry, physics, and others)
Engineering disciplines (electronics, electrical power, strength of materials, machines, and others)
Company constraints
Design and development costs
Timescale
Quality and reliability

Maintenance
Competition
Product life span
Life in service
Materials
Quantity
Aesthetics
Performance
Standards and specifications
Profitability
Research and development
Customer
Packaging
Shipping size
Weight
Market constraints
Manufacture
Product cost
Safety

This list, which is probably not exhaustive, is applicable to all design, in no matter which of the so-called fields of design may be involved. Such factors must be considered at the outset of any design activity. All designers, whatever their label, should continually monitor and bear in mind all these factors from the start to the finish of any design. It is only when this is not done that glaring omissions and oversights can be found in a design: when the product fails in service, when it costs too much to produce, when it will not go into the aircraft, when it becomes rusty ahead of time, and so the list of familiar shortcomings could go on.

It is submitted, therefore, that it is the variation in the weighting and distribution of the above factors which gives rise to the various titled forms of engineering, but that always, for a successful design, the best balance of these factors must be attained.

The designer's job (if he does it properly) could be likened to a particular circus act—namely, the plate and stick act familiar to all. Consider the designer having to keep in continuous motion a number of plates on sticks, each plate having a label taken from the above list of factors, all within the design boundary (Figure 24.1). In his day-to-day working he would have to keep each plate in motion in order to produce a satisfactory design. If he should ignore the constraints imposed by shipping regulations or, in the case of the Sydney Opera House (Yeomans, 1968), the width of the average man and other like factors, he is forgetting to keep one of his plates in motion and, like the circus performer, when a plate falls to the ground, the performance (or design) has failed.

The size and importance of the 'plates' will vary from one field of endeavour to another, but there will always be the same number of plates. Some may require spinning more regularly than others, depending upon the importance and relevance of the partic-

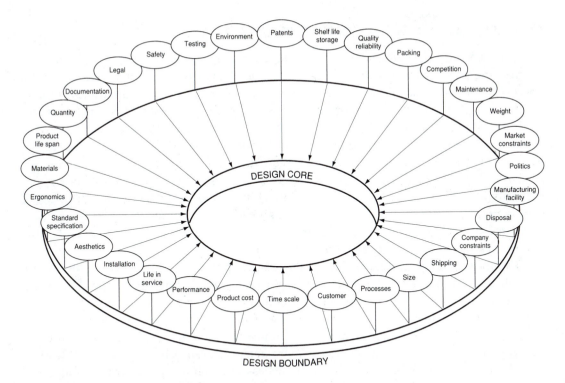

FIGURE 24.1. Elements of Product Design
Specification (PDS)

ular factor to the total design activity. The totality of these factors is known as the product design specification (PDS). The PDS is the specification prepared at the outset of any design activity. It is the specification of the product to be *designed*. It is not the specification of the *product* itself. The latter manifests itself only on completion of the design activity.

Before considering the PDS further, clearer understanding of the operational levels within a company may be useful. It is recognised that so-called designers operate at different levels in a typical engineering company from, say, a junior draughtsman up to chief designer. These differing levels of operation are usually dictated by differing levels of competence. The person responsible must keep all his plates in the air. The more junior person may operate with a reduced number of plates, but he or she must be made aware of the total activity at the earliest opportunity, in order that he or she may progress as a successful designer. Certainly, if the more junior person does not attain a total grasp, then he or she is unlikely to climb the ladder or understand the meaning of satisfactory performance.

ENGINEERING DESIGN: A COMPREHENSIVE ACTIVITY

First, it is considered that design is central to all business activities (Pugh, 1977), that in engineering-type businesses this also applies, and that through its able and successful operation, the best use is made of men, money, and materials, culminating in market satisfaction. The market or need satisfaction is becoming the most vitally important factor in engineering or product design today. With increasing competition, both at home and abroad, more and more products of a similar nature are being made available to satisfy these markets. This gives rise to a fundamental difference between engineering design in a total sense and these experienced by designers in the architectural sphere and met only in part in the civil and constructional engineering industry, although competition is rapidly increasing in the latter. In engineering companies it is becoming more and more difficult to design the right product to meet the market. Conversely, it is becoming easier and easier to design the wrong product for the market.

The model shown in Figure 24.2 will, with the passage of time, become more applicable and relevant to design in the broadest sense, irrespective of any particular discipline. Design may be construed as having a central core activity irrespective of the product. Briefly, this core consists primarily of market, specification, conceptual design, detail design, manufacture, and selling phases. Although it is recognised that other activities present in an engineering business, such as research, development, testing, and the like, are all important to overall success, one cannot proceed to a successful outcome in design without these definitive core phases.

All design starts with a need which, when satisfied, will fit into a market or create a market of its own. This statement of need must result in the formulation of a specification which then acts as the mantle or cloak which envelops all subsequent stages in the design core, becoming the frame of reference or design boundary. The specification is an evolutionary document which becomes modified and changed as the design proceeds, due to changed circumstances and interaction with the design core. Upon completion, the essential equation is a match between the product characteristics and the final specification. The specification contains a whole host of essential elements, the balance and distribution of which not only change as the design evolves but are unique in content for every design carried out. These have been given previously in the first section.

Having established the specification, one then proceeds through the conceptual, detail, and manufacturing phases to the selling stage. The main design flow is therefore from market to selling: advisedly main flow because, in practice, it is a discontinuous, iterative process which when considered in hindsight will give the impression described above. One can be in the detail design phase of a product when a new concept may be adopted. This in itself can lead to new detail designs, different from the previous ones. In flow terms, one has reversed the flow and returned to the conceptual stage: this can happen with all designs and at all stages of the design core. It will by now have been noted that, up to this point, no mention has been made of technology, science, or specific areas of design operation. This is because it has not been necessary to do so.

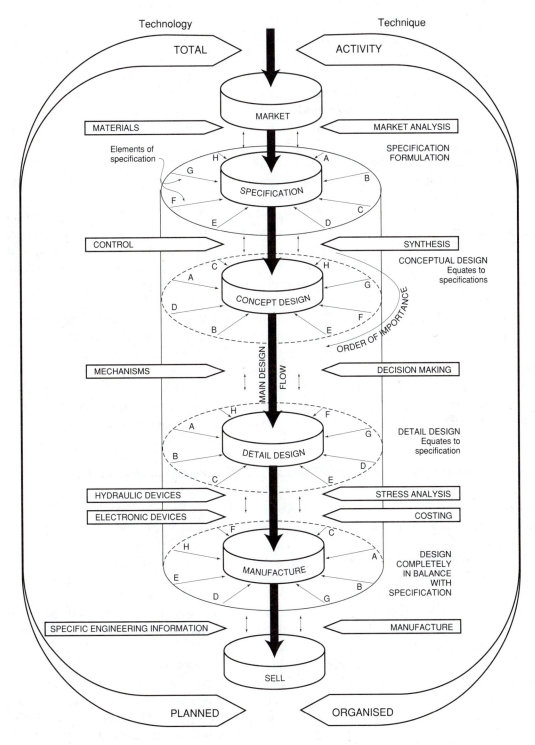

FIGURE 24.2. Design Core Bounded by Product Specification

The next consideration may be called 'techniques related to the design core,' which themselves have inputs into the core. These techniques are necessary to enable the designer to operate the core activity with increasing confidence and understanding. In other words, they can be construed as the designer's tool kit, the tools necessary to enable him to perform the basic function as delineated by the core activity. These are shown flagged in the design core model and are the techniques of analysis, synthesis, decision making, costing, and so on: the model labels can be changed to suit one's own activity.

Having arrived at a basic design core, always enveloped by a specification (PDS), and with inputs from a tool kit not only to enhance the working of the core activity but also essential to making it work at all, the stage has been reached where science and technology must be referred to and involved on an almost infinite front, in order to bring the whole activity to life. The technological inputs into the design core not only relate to specific information on materials, control, electronics, hydrodynamics, and others, but also to the specific areas of engineering in which one is working, such as electric razors, nuclear reactors, building, dustbins, and motor cars, to name but a few. However, it must never be forgotten that the whole activity is carried out within a framework of planning, organisation, and control—in other words, management.

Reference was made earlier to the finite nature of the core activity and the techniques related to it. The addition to the model of the technological inputs expands it, in effect, to an infinite situation where, at least in theory, the whole of technology is at one's disposal. There is therefore a situation which has a finite basis: that is, the design core. Add to that core the operational inputs, which in practice could become finite, and the magnitude of the problem referred to earlier may be perceived.

INFORMATION REQUIREMENTS RELATED TO OPERATIONAL LEVEL

Mention was made earlier of differing levels of the design hierarchy. People have different talents and accomplishments, and it is these differences which give rise in the main to hierarchical structures. Thus in engineering there are not only technical directors, chief engineers, senior designers, junior designers, draughtsmen, and detail draughtsmen, but also a considerable number of specialists such as aerodynamicists, physicists, stress analysts, mathematicians, production engineers, research and development engineers, and so on. The role of those operating, controlling, and participating in the design core activity is to ensure that these specialisms are coordinated and make meaningful inputs into the core activity itself.

For the sake of clarity the chart in Figure 24.3 shows the spectrum of people directly involved in the operation of the design core placed appropriately in relation to the core phases—from the top position of those encompassing the whole activity to the detail draughtsmen whose concern and output are more limited but who, nevertheless, must be made aware of the total activity or their aspirations are more likely to wither and die.

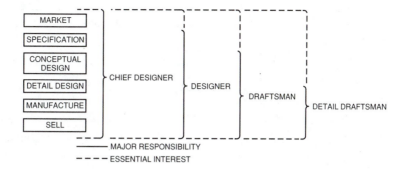

FIGURE 24.3. People and Design Core Phases

It should be noted that titles or names given to the various levels of people are those in general use throughout the engineering industry. Their involvement with the core activity is represented in two ways:

- Each has the major or prime responsibility of their position, which encompasses those core activities normally encountered in day-to-day working, and
- Each member of the team, whilst having a different level of responsibility, should have what may be called the same level of essential interest. In other words, all team members should be aware of and interested in the totality of the situation.

The types and level of information required by this spectrum of people varies enormously, and no attempt will be made to cover anything other than the general case. Likewise, no attempt will be made to delineate in detail the levels and kinds of information specific to a type of industry. The general case will be defined by considering information under the headings of technique and technology, and relating such information to the aforementioned spectrum of people, indicating in by no means rigid terms the basic necessities of each group of people (see Table 24.1).

While the listings given in Table 24.1 are by no means exhaustive, they indicate the extremely wide range of information required by operators of the design core; it should be appreciated that where there is a requirement by all concerned for, say, a particular technique of analysis, then the level of analytical ability and hence the level of information required will vary considerably. However, to illustrate this variation would probably overcomplicate the situation that the table aims to portray.

The depiction of design, information types and levels, and people in this way seeks to clarify the design position in a more coherent manner. It is fully appreciated that, to achieve a successful, competitive design, the main object of which would be to satisfy a need, the involvement of all sections within a company is essential to success. The reason for doing it this way is primarily to reinforce the argument that the design activity is central to any successful engineering enterprise and that all products or systems emanating from such enterprises are designed. The inputs and contributions of engineers of specialised disciplines, marketing and production people, and others are essential to the successful operation of the whole business function, but it is the design core operators who have to put it all together.

TABLE 24.1. Examples of Technique and Technology Information Related to Groups of People in the Engineering Industry

Technology X	Chief Designer	Designer	Draughtsman	Detail Draughtsman	Technique O
Market information	O X	O X			Market analysis
Competition details	O X	O X	X		Competition analysis
Specific engineering disciplines	O X	O X	O X	X	Stress analysis
Specific science disciplines	O X	O X	X	X	Thermodynamic analysis
Company products	O X	O X	O X	O X	Costing
OEM product data	O X	O X	X	X	Optimisation
Cost data	O X	O X	O X	X	Synthesis
Manufacturing data	O X	O X	X	X	Quality and reliability
Patents	O X	O X	O	O	Information retrieval
Materials	O X	O X	X	X	Computers
British and international standards	O X	O X	O X	O X	Manufacture
Company standards	O X	O X	O X	O X	Classification
Development and test data	O X	O X	X		Decision making
Finishes	O X	O X	X	X	Failure analysis
Fastenings	O X	O X	O X	X	Measurement
Welding	O X	O X	O X	X	Testing

PRODUCT DESIGN SPECIFICATION (PDS) AND OEM PRODUCT DATA

Accepting that all the elements discussed previously are essential to the formulation of an effective and efficient PDS, it is realised that this achievement relies heavily on (1) the attitude, ability, and awareness that information is required in these areas before the commencement of design and (2) the information itself being readily, effectively, and efficiently available.

The former will come about only from successful practice and education, while the latter will depend not so much upon the rapidity and slickness of the information retrieval systems, which of course are important, but more upon the quality and comprehensiveness of the information contained within the slick system (Pugh, 1980).

It is not proposed to deal with a PDS formulation here (the formulation stages of a new design). A fuller description of this topic is given in the Institution of Production Engineers' (1984) *A Guide to Design for Production*. It is necessary and rewarding to consider the structure of a PDS applied to Original Equipment Manufacturers' (OEM) product data, which forms a large part of the information input into the design of anything. Suffice it to say that at the start of a new design, information will be required on competitor's products, to the PDS format.

Comparisons could, however, be made on an elemental basis between the true comprehensive requirements and the availability of the OEM product data to match up these requirements. Characteristically, the situation varies from bad to awful, in that published data sheets and catalogue information leave a lot to be desired, as illustrated in Table 24.2.

TABLE 24.2.   Data Sheet for a Variable Delivery Swashplate Pump

Basic Characteristics	Information Required by a Designer for Usage in a System	Information Given in Company Data Sheet	
Performance	Flow rates litres/min/1000 rpm	OK–given as cc/rev.	
	Displacement cc/rev.	√	
	Number of pistons	√	
	Torque (input)	Only theoretical	
	Speed max. rpm	√	
	Speed min. rpm	√	
	Pressure continuous, (absolute) in bar	√	Note not related
	Maximum in bar	√	To speed? absolute
	Peak in bar	√	or gauge pressure
	@ speeds	x	
	Continuous	x	

continued

Basic Characteristics	Information Required by a Designer for Usage in a System	Information Given in Company Data Sheet
	Maximum	x
	Peak	x
	Volumetric efficiency vs speeds at varying outputs	Volumetric and overall efficiency given
	Overall efficiency vs speeds at varying outputs	Max. output only
	Bi- or unidirectional rotation to input shaft	√
	Output power KW (theoretical)	√
	Continuous output power	Only for speed range 3:1
	Casing pressure max bar	√
	Ability to run dry case	x
	Closed circuit operation	√
	Open circuit operation	x
	Vibration characteristics, amplitude vs frequency at varying pressures	x
	Ability to start on full swash (max delivery) from rest	x
	Max pressure rating of case bar (absolute)	x
	Noise levels dba	x (quiet)
	Neutral by-pass-positive neutral	x
	Feed or boost pump	
	Inbuilt/extra add on	√ (inbuilt)
	Displacement cc/rev	√
	Max inlet press (absolute) bar	√
	Max outlet press bar	√
	Internal circuit diagram	√
	Noise level dba vs press./speed	x (low noise)
	Main pump standard control requirements litres/min	x
	Provision for fitting valve block to pump body	√
	Circuit diagram of valve block	√
	Main relief valve pressure range, bar (absolute) max. min.	√
	Flow pressure characteristics for relief valve	x
	Response time for valves	x

continued

Basic Characteristics	Information Required by a Designer for Usage in a System	Information Given in Company Data Sheet
	Boost pressure RV range	√
	Controls available	
	Full servo characteristics	√ Could
	constant power	√ form a
	constant torque	√ separate
	automatic control	√ section
Materials of construction (main pump)	Types of ferrous and nonferrous specs. Finishes, paints, plating, etc.:	x (steel and bronze)
	External	x
	Internal	x
	Types of bearing:	
	Roller	√ Shown on
	Needle	√ diagram
	Ball	√ of pumps
	Pump case	x
	Input shaft	x
	Types of seal materials	x
	Types of oil or hydrocarbon,	x Fluid
	eg chlorinated hydrocarbons,	x Viscosity
	mineral oils	x Normal cST
	Phosphate testers	x
	Emulsions	√ No mention of types
Maintenance and installation	Parts likely to require maintenance,	x
	frequency	x
	Service volume required for removal in relation to pump axis	x
	A in situ remove pump,	x
	B	x
	C etc.	x
	Spares availability:	
	Components	x
	Subassemblies etc.	x
	Ability of input shaft to accept axial forces	√
	Kg	√
	radial forces	

continued

Basic Characteristics	Information Required by a Designer for Usage in a System	Information Given in Company Data Sheet
	Attitude of pumps,	
	axis vertical - shaft up/down	x
	axis horizontal	√ Assumed
	axis inclined	x
	Drainage from pump case	√
	Instruction for filling with oil	√
	Suction filtration	√
	Mainline filtration	x
	Starting procedure-direction of rotation	√
	Running-in instructions pressure/flow (speed)	√
	Range of Allen Screw/spanners etc. required for maintenance	x
Size and dimensions	Main dimensions of unit, input shaft, connections, etc.	√ Comprehensively covered
Weight	Weight of pump (bare)	x
	+ valve block	x
	+ valve block and boost pump	x
Environment (installed)	Temperature range - fluid max. min.	√
	- ambient, max. min.	√
	Humidity %	x
	Dirt, sand, dust, etc.	x
	G-loadings (shock) max. direction	x
	Corrosion - avoid	x
	Acids	x
	Alkali	x
	Etc.	x
	Abrasion	x
Shelf life (storage)	Normal shelf life -	x
	Hours, months, years	x
	Under what conditions	x
	Pumps as supplied inhibited with?	x
	Shelf life instructions	x
	Limit to shelf life, eg seals	x

continued

Basic Characteristics	Information Required by a Designer for Usage in a System	Information Given in Company Data Sheet
Standards and specifications	National/international standards with which the pump complies SAE 6000 say DTD 585 (oil), etc.	√ (reasonable cover but scant on materials and of all things, oils)
Testing	Conformance with test specs Some comment on company testing policy, is each pump tested, if yes, for how long, etc. Facilities for testing	x x
Life in service	Some idea of life—mtbfs, mttrs, etc. under idealistic conditions	x
Prices	List price—discount policy vis-à-vis OEMs Extras Price basis includes shipping and packing charges, FOB, etc.	x x x
Packing	How normally packed—cardboard box or wooden crate (shock resistant packing) Is packing environment inhibited—sizes packed, Nos. per standard ISO container, Also air container—shipping volume m³ etc.	x x
Shipping	Suitable as packed for Air UK/overseas Sea UK/overseas Road UK/overseas	x x x
Ergonomics	Lifting/handling points on pump Any protuberances to be avoided Access/servicing instructions Procedure	x x x x
Safety	Do's and don't's on start-up Do's and don't's and running procedures	x x
Availability and delivery	Timescales Quantities, etc.	x x
Servicing	Agencies, Service and spares policy Stocking policy UK/overseas	x x x
Customer	Ordering instructions Where to order from	√ Well set out √ and clear

√ = satisfactory, x = not given

Against the headings given previously as elements of the specification, a comparison of the two has been presented for a variable delivery swashplate pump. This is a data sheet which the author considers to be one of the best available. Yet the degree of shortfall to the designer's needs is quite alarming. The name of the manufacturer has been omitted for obvious reasons. The results of this comparison are not typical: it cannot be overemphasised that the majority of sources are worse.

QUALITY OF INFORMATION

From the foregoing it becomes apparent that the quality of OEM product data leaves a lot to be desired, and that there are likely to be many gaps in critical areas. It is therefore suggested that a standard for the compilation and presentation of such data is urgently required, since efficient and effective design will remain somewhat less than likely without a considerable increase in information quality. Yet is this a point readily understood by the information industry?

It is as if the information industry, driven on by information technology, is making available existing information at ever greater speed and volume without paying much attention to influencing quality and, above all else, to what one does with such information. It is suggested that the primary requirement is to improve the quality of the information available. To achieve this, the information industry must study the differing user need situations discussed earlier, before attempting to prescribe for them.

Prescription without diagnosis would not be acceptable in the medical world, and in no context should prescription based on spurious assumptions be allowed to replace true diagnosis. In the engineering design field, the quality issued has been to some extent obscured by the emergence and application of the computer at a faster rate than that of design understanding, the one masking the other (Pugh, 1984b). This has been recognised and set against a background of product design, however, and some comment is appropriate here in the context of information availability and its interaction with the design core.

COMPUTER AIDS TO DESIGN

Accepting that the primary argument from the information point of view is one of quality of information, and accepting for the sake of this argument that all information available has quality, one comes to the question of presentation and availability. This is a very complex question and in reality needs a full discussion of product evolution and product status in order to form a cogent argument and hypothesis (Pugh, 1983, 1984a). This, of course, is well beyond the scope of this book. However, some basic premises can be considered.

Type and Level of Information

Suggested requirements are as follows:

1. Designers initially need macroinformation—global information akin to abstract information—which sets out the information available and its tenets sketchily and not in great detail.
2. This macroinformation is either accepted or rejected. Rejection ends the debate. However, acceptance leads one to the next level—that is, more detailed information on the topic or subject, to feed the appetite and interest generated at the macro stage.
3. A progression of information is then required, going from macro to micro. This applies especially to OEM product data in report or paper terms: abstracts are required for such data (the macro), leading to more detailed information contained in the PDS structured equivalent (the micro).

Currently, type and level fall randomly between two stools in all areas of information technology, not just in engineering design, and only collaborative working could resolve the difficulties.

Presentation of Information

Computer-based systems can be used to store information digitally and retrieve it quickly. In the author's view it is questionable whether this is the most efficient storage means in a total sense—that is, for cost and availability. If the level and quality questions were resolved, then again the overall efficiency dilemma could also be satisfactorily resolved. It is suggested that, as with product design, the resolution of the one (the need) without the other (the means of satisfying that need), will not be achieved. Designers require a vast amount of graphical information which is probably best kept in analogue form, on fiche or optical disc. While it can be digitised, storage and redraw times are time wasting and inefficient. Optical disc storage or microfiche with computer retrieval may well prove more efficient in the long run.

The idea that information as such can be integrated into the design of computer-aided design (CAD) systems is probably misplaced, since it assumes in the ultimate a mechanistic design activity: that, because something is technically possible, it should be done. Why is it that much current ergonomics research is concerned with resolving badly designed human-machine interfaces, from both physical and psychological standpoints?

Ideally, all presentation systems should possess the following attributes:

- Rapid access,
- Macroinformation, carefully specified,
- Microinformation in comprehensive form, and
- Ability to make hard copy, either in totality or single sheet.

CURRENT INFORMATION SYSTEMS

Current abstracting systems for articles and learned papers seem to work reasonably if one has such systems within reach. They tend to work well in a university because universities are geared to information retrieval. Large industries also have access to such systems, but what about the medium to small companies and the designers within those? The situation is by no means as promising: the probable answer is for the academic institutions to forge links with local industries to encourage dual use of facilities. The politics and economics would require discussion.

The greatest difficulty would be making the industrial parties aware that they need, as a matter of some urgency, to gather information in much greater depth and breadth than hitherto. Competition from outside could cause greater attention to these matters, for otherwise the companies might ultimately go out of business. However, information selection and gathering is one thing, but what may be done with the information is quite another and beyond the scope of this chapter.

Existing OEM product information systems leave a lot to be desired and yet the faults are not all with the systems purveyors. Whilst systems package libraries such as those of Technical Indexes Limited on microfilm are excellent, they have two major drawbacks from the user's viewpoint:

- Since OEM manufacturers may have to pay to have their catalogues included in the system, it is never complete. This could cause the company or product that is best for a particular design to be overlooked.
- For the reasons already stated, since the system relies upon catalogues, their variable quality limits the useful information available. So a considerable increase in the quality of source data could make package libraries automatically more useful.

Whether storage should stay on microfilm or become captured by digital systems is debatable and will continue to be so until the alternative systems are synthesised and compared, just as one has to do in designing a competitive product, since information systems are in themselves a product in the broadest sense. Perhaps the information industry should itself adopt a structured approach to system design.

REFERENCES

Pugh, S. (1974). 'Engineering Design: Towards a Common Understanding.' *Proceedings of the Second International Symposium on Information Systems for Designers* (pp. D4–D6). Southampton: University of Southampton.

Pugh, S. (1977). 'The Engineering Designer: His Tasks and Information Needs.' *Proceedings of the Third International Symposium on Information Systems for Designers* (pp. 63–66). Southampton: University of Southampton.

Pugh, S. (1980). 'Stylised Characteristics.' Unpublished paper.

Pugh, S. (1983). 'The Application of CAD in Relation to Dynamic/Static Product Concepts.' *International Conference on Engineering Design (ICED 83).* Copenhagen: ICED.

Pugh, S. (1984a). 'Further Development of the Hypothesis of Static/Dynamic Concepts in Product Design.' *International Symposium on Design and Synthesis.* Tokyo.

Pugh, S. (1984b). 'CAD/CAM: Hindrance or Help to Design?' *Proceedings of Conception et Fabrication Assisteé par Ordinateur.* Brussels: Free University of Brussels.

Yeomans, J. (1968). *The Other Taj Mahal: What Happened to the Sydney Opera House?* London: Longman.

A Guide to Design for Production (1984). Institution of Production Engineers.

CHAPTER

25

The Organisation of Design: An Interdisciplinary Approach to the Study of People, Process, and Contexts

ABSTRACT

This chapter brings together the works of Morley and Pugh in the context of total design, considering the evolution of the understanding of leadership skills in parallel with the emerging understanding of engineering design. Models of design have brought about effective communication between the authors and industry, and they are used as the catalyst to focus upon the difference in design team characteristics required in differing design situations. These differences are elaborated through a consideration of products being either conceptually static or dynamic. It concludes by firmly establishing a sound relationship between the work in the two areas.

With Ian E. Morley, *Proceedings of the International Conference on Engineering Design* (Boston: ASME, 1987), 210–222.

INTRODUCTION

In March 1984 the Economic and Social Research Council of Great Britain sponsored a Workshop on 'The Use of Social Science to Improve the Process of Design in U.K. Industry.' A detailed report of the workshop has been given in Morley and Palmer (1984). Here we shall abstract some of the main themes.

- *Theme 1* If we are to fuse the research activities of social scientists and engineers, we need basic information about the process of design as it is implemented in different companies making different products. Surprisingly, this information seems not to be available.
- *Theme 2* Design is a vast team game. This means that it is important to find people with the right kinds of skills. It also means that design cannot be promoted in isolation from other activities within the organisation. What is needed is the creative integration of specialists from different parts of the organisation.
- *Theme 3* To achieve such integration management must be willing systematically to involve specialists from sales, engineering, marketing, manufacture, finance, and so on in the process of design. In other words, management must take corporate action aimed at structuring business activity around a central design core (Pugh, 1986).
- *Theme 4* Perhaps the major impact of the social sciences on the process of design has come from specialists on human factors or ergonomics. It is important that social scientists now attempt a wider contribution in which they integrate models of people, processes, and social contexts.

The authors were participants in the workshop (Morley was one of the organisers). They have since begun a collaboration based on the above themes. The object of the exercise is to integrate work on the social psychology of groups with work on engineering design. This chapter is a first report of the progress which has been made. It shows that engineers who build design activity models (Pugh, 1986) and psychologists who consider the performance of small groups (Janis and Mann, 1977; Morley and Hosking, 1984, 1985) have been considering similar problems, although in different contexts, and for different reasons. The authors suggest that convergence of this kind is a fruitful basis for a programme of interdisciplinary research. They will consider the kind of interdisciplinary research which needs to be done, in broad outline.

DESIGN ACTIVITY MODELS

Design activity models attempt to identify the main areas of design activity and show how they are related. Within each area they show the most important elements or subroutines to be performed. Pugh (1986) has reviewed models of design activity intended to apply to different disciplines and products. He has concluded that the key ideas may be summarised and integrated by his own model of product design activity. This model is represented graphically in Figures 25.1 and 25.2.

Figure 25.1 may be described as delineating a product design specification or design boundary model. It was itself designed to establish common ground between designers, whatever their specific title (Pugh, 1974). It represents the nature of the various constraints relevant to the design of a product—any product at all. It thus satisfies the criterion that models of design activity should have universal application, regardless of discipline, industry, or product (Pugh, 1986). It also indicates the special importance attached to product design specification in Pugh's work (Pugh, 1981, 1982, 1985).

Figure 25.1 forms part of the more complex design activity model set out in Figure 25.2. It is a first attempt to show the total activity of design, taking into account the constraints which form the design boundary.

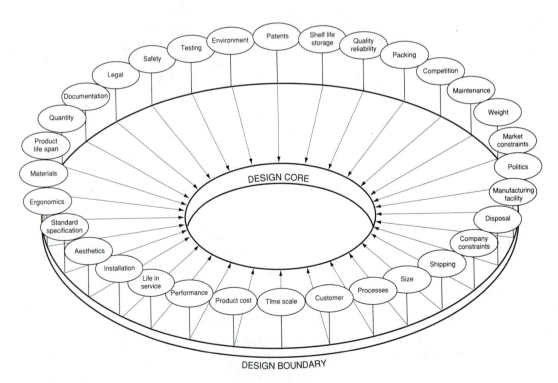

FIGURE 25.1.   Elements of Product Design Specification

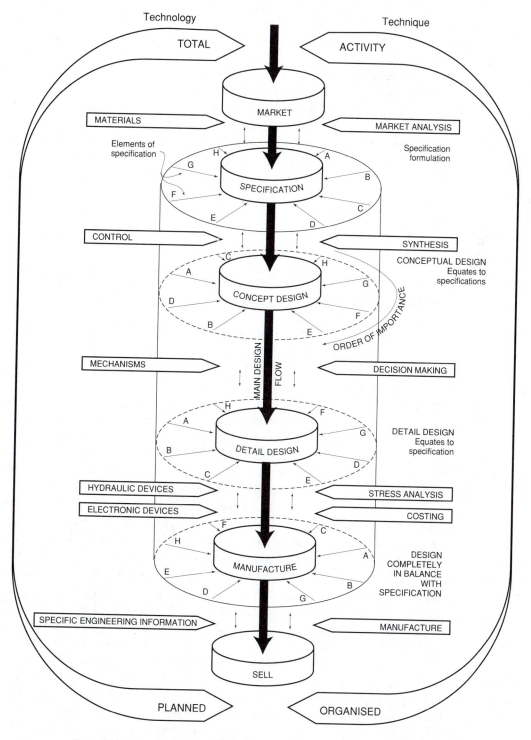

FIGURE 25.2. Design Core Bounded by Product Design Specification

The model has been described in detail elsewhere (Pugh, 1985, 1986; Kimber et al., 1985). It shows a central core of design activity, disciplined (Rawson, 1976) or restrained (Atkinson, 1972) by the nature of the agreed specification. This is the first major area of design activity, consisting of the *core phases* of market investigation, product design specification, concept design, detail design, manufacture, and sales. It is accepted that, in practice, design activity will be iterative rather than linear, so that participants rework earlier problems as interrupts occur.

It is explicitly assumed that the core phases are universal, common to all kinds of design. Other areas of design activity give designs their distinctive character. That is to say, different kinds of design may require different kinds of *information, techniques,* and *management.* Thus, much of the material included in Figure 25.2 is included for purposes of illustration. Strictly speaking, the particular inputs to the design core need to be reconsidered for each new case.

The area of *management* is of special importance because design activity requires information, resources, and support to be invested in action in the most effective way (Kanter, 1984). To quote Pugh (1983), design activity 'should be carried out under a management umbrella of planning and organisation. Place this total package within a corporate business structure and one has a picture of how design fits, or should fit, into the modern business. All facets of the business interact with the design core, and it cannot be overemphasised that since all businesses are about products to satisfy markets, to narrow or bias a business towards any one of the elements or specialisms given in the model, without balance, increases the risk of not satisfying this primary requirement.'

Pugh (1986) has recognised that more needs to be said about interactions of this kind. His most recent model, the business design activity model, attempts to locate product design activity firmly within the overall structure of a business.

A somewhat idealised version of the model is shown in Figure 25.3. Once again, it is recognised that different contexts will differ in detail. Indeed, we are currently giving designers blank templates, such as those shown in Figure 25.4, and asking them to work out models of their own activities for themselves. What is more important, however, is the idea that the design core is constrained not only by the elements of the product design specification—the product design boundary—but also by the elements of the business structure—the business design boundary. If the constraints of the business design boundary are too severe, it will be necessary to take corporate action, restructuring the business to provide designers with more information, more resources, and more support.

It is our experience that companies in diverse product fields—from lawn mowers to gas turbines—relate to the business/product design activity model. Their words may sometimes be slightly different with different emphasis on the stages, but none have difficulty in relating to and completing the blank template of Figure 25.4. Above all else such models have greatly eased communication with the companies and quickly establish a common view of the totality of design. Three aspects of Pugh's product design activity and business design activity models require further comment in the present context.

- *The personal design boundary*   It is clear that just as there is a business design boundary there is also a personal design boundary. It represents the constraints

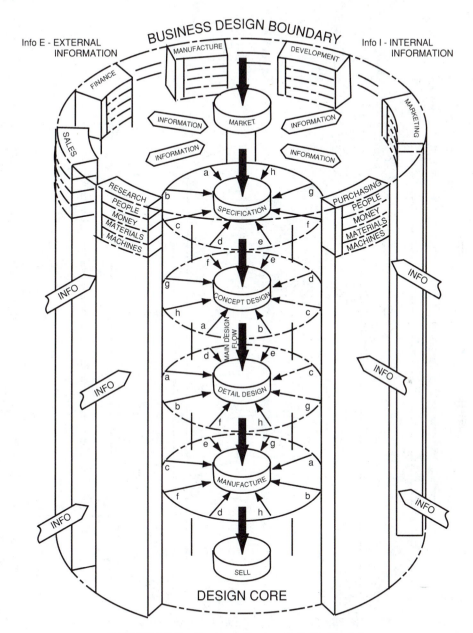

FIGURE 25.3. Business Design Activity Model

placed upon the process of total design activity (described in Figures 25.1, 25.2, and 25.3) by the personal characteristics and skills of those engaged in that activity. It is possible that there are certain personal characteristics which all creative engineering designers need. Pugh (1977) has himself produced such a list. Some of the characteristics follow directly from the activity models: motivation to ques-

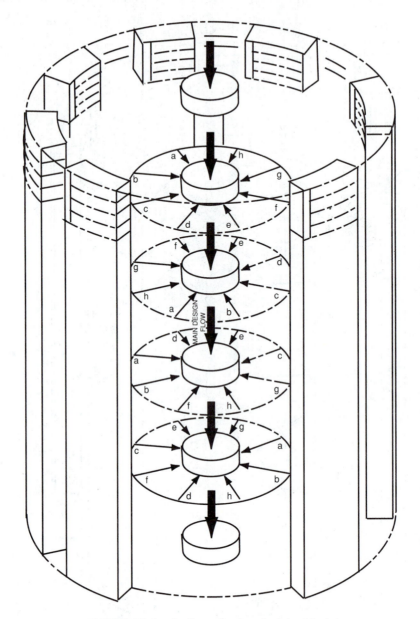

FIGURE 25.4.   Business Design Activity Model

tion the fundamentals of existing engineering practice; willingness and ability to become involved in new disciplines in pursuit of design objectives; ability to communicate with people in all walks of life, particularly engineers in specific disciplines. Other characteristics are more controversial: being equally at ease with convergent or divergent thinking; having a sense of humour.

In general it seems proper to take the position that different kinds of design, grounded in different contexts and managed in different ways, may require different kinds of skills.

- *Static and dynamic concepts in product design*   Pugh (1983, 1984) has argued that some designs, such as those of motor cars, involve incremental changes in detail around the same generic base, a single static concept. In other cases, designs have not reached a conceptual plateau; the generic base has not yet been discovered. The distinction has been marked by the preparation of two models of product design.

  Figure 25.2 shows the model appropriate to the dynamic case. Here the product design specification is generated from analysis of the market and the users' needs. At this stage there is no commitment to any predetermined concept; we do not know the nature of the final design. In terms of the motor car example, we do not yet know whether there is a wheel at each corner: the world is our oyster.

  Figure 25.5 shows the model appropriate to the static case. Here the product design specification is written on the assumption that there is little or no conceptual choice at the total system level. The generic core is assumed and specifications written on that assumption.

  The implication is that the different types of design activity may require different organisational structures, perhaps different personal characteristics and different skills. It is, however, important to note that we simply lack basic information about the process of design as it is implemented in different companies making different products. It is one thing to draw up a model of total business activity in general terms, and models of this kind are important for a variety of reasons (Pugh, 1986). It is quite another thing to be able to compare and construct models drawn up to apply to specific designs, grounded in specific contexts.

- *Successful design: the special importance of the product design specification (PDS)*   A successful design is presumably one in which the product has a secure status, is completed on time, to cost, without errors discovered at the production stage (Rzevski, 1984). Things can go wrong during any of the core phases identified in Figure 25.2. However, the PDS phase is specially important because it defines, in detail, the wide variety of constraints, technical and nontechnical, to be placed upon the design. It forms the design boundary. It is, to quote Pugh (1985, p. 5) 'the control mechanism for systematic design and the document against which emergent design must be measured and assessed.' It should itself be systematically designed. Indeed, if Pugh (1985, p. 5) is correct, 'Systematic PDS generation is . . . absolutely essential to successful product design.' It should establish the status of the product in a competitive market, and ensure that the product is worth designing in the first place.

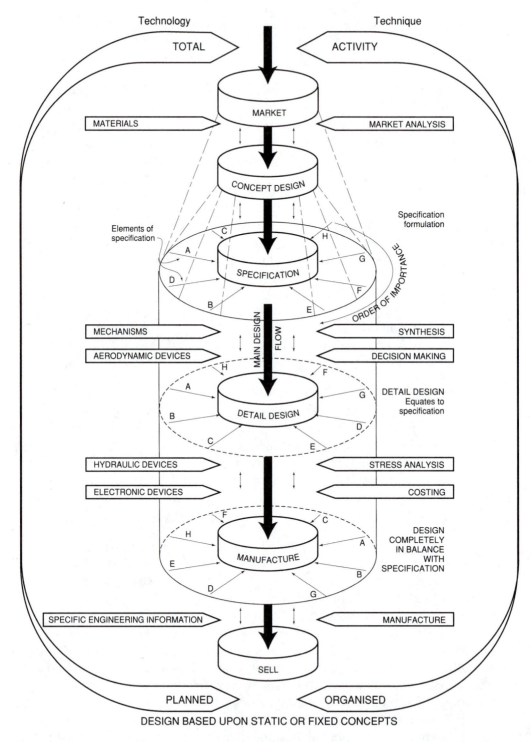

FIGURE 25.5.   Design Core Bounded by Product Design Specification: Static Concept

THE SOCIAL PSYCHOLOGY OF GROUPS

Social psychologists have paid little attention to problems of engineering design. However, they have attempted systematically to discuss the kinds of concepts needed to understand the nature of groups working within organisations (Payne and Cooper, 1981). Much of the work has been conducted from a decision-making perspective, attempting to show why some groups are successful and others are not (Janis and Mann, 1977; Belbin, 1981; Burnstein and Berbaum, 1981). In addition, attempts have been made systematically to articulate the social skills needed by leaders (Morley and Hosking, 1985; Morley, 1986) and other members of teams (Belbin, 1981; Stewart, 1979a, 1979b). Useful summaries of the literature are provided by Morley and Hosking (1984) and Golembiewski and Miller (1981).

Groups at Work

Models of product or business design activity attempt to integrate talk about people, processes, and organisations. So do models of group performance, although they recognise more explicitly (perhaps) that the relevant processes occur *within and between groups* (Morley, 1986; Morley and Hosking, 1984). There is general recognition that to understand group performance we must show how group processes are affected by individual variables, group variables, and contextual variables, although different writers make the point in different ways. Thus, in Kirkpatrick, David, and Robertson (1977), processes of search and choice are affected by individual, small group, and environmental constraints. Friedlander and Schott (1981) identify a similar (although not identical) set of individual level variables but include them under the heading of the internal dynamics of the group. Other variables are discussed under the headings of interface (between the task group and the organisation) and outputs (resulting directly or indirectly from the activity of the group).

At the level of the small group it is convenient to make a distinction between a *structural* panel of variables and a *style* panel of variables (Golembiewski, 1962). The structural panel of variables is defined in terms of differential access to resources (communication structure), differential degrees of power (status structure), differential performance of certain activities (functional role structure), differential degrees of ranking on a stratified scale of importance (prestige structure), and differential degrees of identification with the group as a whole (group cohesiveness). The style panel of variables reflects the fact that groups, like individuals, may do things in this or that idiosyncratic kind of way. There are three major components: group atmosphere, norms (shared standards defining what group members should and should not do), and leadership style.

There is considerable evidence that (1) dynamics within and between groups mean that the advantages of working in groups are often lost, (2) groups with similar structures may have very different styles and produce different outputs, (3) groups with similar styles may have quite different structures and produce different outputs, (4) group performance can be improved by social science interventions (Janis and Mann, 1977; Belbin, 1981; Morley and Hosking, 1984; Golembiewski and Miller, 1981).

Successful and Unsuccessful Teams

The design activity models outlined above imply that design teams require a combination of specialist skills. That diversity is important (Friedlander and Schott, 1981) but is not always easy to handle.

Some writers, such as Pearson and Gunz (1981, p. 152), argue that 'many of the difficulties arise because of a lack of clear definition of what is required, where particular responsibilities lie, and how the interaction should be managed.' Similar views have been stated by Heany and Vinson (1984, p. 30), who add that individuals should be selected for a team because of their skills rather than because they happen to be available. As they say, 'However elementary, the point is often ignored in practice.'

Other writers have emphasised the importance of social skills within the group (Burnstein and Berbaum 1981; Morley and Hosking, 1984, 1985), particularly those of the leader (Belbin, 1981; Morley and Hosking, 1984, 1985). Diversity implies that people have different kinds of substantive knowledge. It also implies that they will have different values, as well as different levels of motivation, commitment, and power potential. They are likely to see elements of the product design specification as having varying degrees of value and uncertainty.

Under such circumstances the process of design may be seen as having some of the characteristics of a negotiation between interdependent participants. Consequently, there are two sides to this activity and two related criteria for negotiation success. The first kind of activity is cognitive, primarily; the second is political, primarily. The former functions to help people organise their intellectual activity and think clearly about the problems they face. The latter functions to manage differences within and between groups. The cognitive processes function to make jobs no harder than they need be. The political processes function to prevent premature commitment to decisions (designs) which have not been sufficiently well appraised. In the context of design this means controlled convergent disagreement is necessary if the best choice of design is to be made. There is abundant evidence that not to disagree means not to solve the problem (Morley, 1986; Morley and Hosking, 1984). However, Burnstein and Berbaum (1981, pp. 4–5) submit that 'As members become absorbed by the task, disagreements and antagonisms are inevitable; this in turn threatens self-esteem and increases disaffection with the group, a state of affairs that, if unattended to, would degrade the decision making process. . . . Bales refers to this as the equilibrium problem . . . unless the tension is reduced, low status members will accept the influence of high status members with ever increasing hostility.' Needless to say, such activity is not conducive to task success. Psychologically speaking, this is why systematic design methods are important, and this is how they should be appraised.

It is also important to consider diversity of function (role) within a team. Belbin (1981, p. 77), for example, has argued that successful management teams contain 'individuals who balance well with one another.' The general idea is that jobs which are psychologically very different will be performed by different people. The two schemes most useful in the design context seem to be those of Clutterbuck (1979) (who looked at teams engaged in R&D) and Belbin (1981) (who looked at the performance of teams in business games). Clutterbuck's scheme is simpler, with four roles: generating, integrating, developing, and perfecting. Belbin develops the scheme by adding two leadership roles (chairman and shaper), an administrative role (company worker), and a liaison role (resource investigator). There is good evidence that effective groups pay particular importance to the latter, so that outside ideas are incorporated into the thinking of the group (Janis and Mann, 1977; Friedlander and Schott, 1981; Pearson and Gunz, 1981; Rothwell, 1977). This sort of activity is, of course, at the heart of the business design activity model outlined above. Under the heading of networking, it also forms an important element in models of social skill (Morley and Hosking, 1985; Morley, 1986; Hosking and Morley, 1985). It is clear that 'the relative importance of each role will depend upon the type of project and also the stage which it is in' (Pearson and Gunz, 1981). However, this part of the theory has not been developed.

Stages in Group Decision Making

Models of group decision making typically make the point that decision making is an integrative process involving stages of identification, development, selection, and implementation (Burnstein and Berbaum, 1981; Morley, 1986). In a design context these are the activities in the design core of Figures 25.2 and 25.4. This simply shows that design is a form of decision making. What is more important is that there are detailed analyses of the processes involved which may well be useful to those engaged in thinking about the activity of design (Burnstein and Berbaum, 1981; Morley, 1986).

Decision Making Under Stress

There is large literature dealing with the effect of decision making under stress. The general finding is that attempts to cope with stress are very often maladaptive. For example, Wohl (1981, p. 631) reports that 'individuals under stress display reduced information search, consider fewer alternatives, overreact to isolated pieces of information and generally engage in what would otherwise be suboptimal choice generation and selection.' The effect of stress is thus to make information processing less vigilant, particularly when decisions are made by certain kinds of groups (Janis and Mann, 1977; Morley and Hosking, 1984). Vigilant information processing strives towards the ideal of rational choice explicated in formal decision theory. When operating in the vigilant mode, a decision maker

- Seriously considers more than one policy or course of action (such as design strategy or product strategy);
- Carefully considers the full range of goals and constraints and the values implicated by each (such as thoroughly canvassing the elements in the PDS);
- Carefully works out the costs and benefits of each policy (such as with many concepts, many criteria, in a concept selection matrix);

- Intensively searches for new information (in the evolution of the PDS, concept selection, and so on);
- Is sensitive to new information or expert judgement, even when that information or judgement is unpalatable (concept choice made, concept changes on receipt of new information as process iterates through the design core);
- Reexamines the consequences of all known policies, including those initially discounted, before making a final choice (reestablish and rerun concept selection matrix). Makes detailed plans for implementing the policy chosen, paying special attention to contingencies.

This implies that the decision maker is seeking the best choice possible within the constraints of the task, recognising that to achieve some benefits, others may have to be sacrificed. When operating in a nonvigilant mode, a decision maker

- Considers a restricted number of policies, frequently only one (design is based on one concept, one idea);
- Considers a restricted number of consequencies, sometimes only one (ignores a formal PDS and its implied constraints);
- Evaluates each policy only once. Options are evaluated sequentially as they arise (single ideas, evaluated serially, staccato design);
- Each consequence is regarded as acceptable or unacceptable because what is required is a decision which is good enough rather than a decision which is best overall (lack of systematic design methods, staccato design, first idea is best).

Thus, it seems that the distinction between vigilant and nonvigilant information processing may provide one way of capturing differences between design based on dynamic concepts (vigilant information processing) and design based on static concepts (nonvigilant information processing). This is important because of what Pugh (1983, p. 6) has called the static/dynamic trap. As he says: 'One should always be operating . . . the principle of controlled dissatisfaction—the designs of today can always be bettered tomorrow. The assumption is made that until proved otherwise, existing concepts are dynamic.'

Consequently, it is important that some designers always operate in the vigilant mode, even if the new designs that emerge are variations on old themes.

Fortunately, Janis and Mann (1977) have detailed a set of antecedent conditions which mean that vigilant information processing is less likely to occur. They are that

- Groups have a certain structure: they are highly cohesive;
- Groups have a certain style of operation: they are given directive leadership and lack methodical procedures for search and appraisal;
- The group operates under a high degree of stress; and
- The group lacks members who perform a liaison role: it is insulated from other groups within the organisation.

Let us repeat, the potential advantages of working in groups are easily lost unless special care is taken. For further discussion the reader is referred to Janis and Mann (1977), Morley and Hosking (1984), Golembiewski and Miller (1981), and Wohl (1981).

CONCLUSIONS

Those who construct design activity models and those who construct models of group performance share many similar concerns. In each case they need to integrate information about people, process, and contexts.

Design is a team activity requiring the creative integration of specialists from different disciplines. Research and theory in the social psychology of groups may be used to complement the work of practicing designers, minimising losses from faulty process and maximising the advantages of working in groups. In particular, designers may be helped to avoid the static and dynamic conceptual trap.

Diversity within a team is crucial, but diversity needs to be carefully handled. What is needed above all else in the context of design is the use of systematic methods which provide a structure so that disagreements converge productively onto solutions all can understand and all can accept.

As Heany and Vinson (1984, p. 31) say: 'The taproot of many of the problems with new products is *not* technology. Rather, it is that the systematic nature of the product-innovation process has been ignored. Specialisation has been carried to extremes. . . . If new products are going to contribute consistently to aggressive market-share strategies, managers must pay a great deal more attention to the quality of integration and teamwork among the many specialists participating in new product development. Fortunately, there are tools at hand to help them effect a marked improvement in teamwork. One has only to use them. The rewards for doing so are substantial.' We believe that this is correct.

REFERENCES

Atkinson, J.R. (1972). 'An Integrated Approach to Design and Production.' *Phil. Trans. R. Soc. Lond.*, A273, 99–118.

Belbin, R.M. (1981). *Management Teams: Why They Succeed or Fail.* London: Heineman.

Burnstein, E., and Berbaum, M. (1981). 'Stages in Group Decision Making: The Decomposition of Historical Narratives.' *Fourth Annual Scientific Meeting, International Society of Political Psychology.* Mannheim, W. Germany: ISPP.

Clutterbuck, D. (1979). 'R&D Under Management's Microscope.' Paper presented at the Conference on International Management, February.

Friedlander, F., and Schott, B. (1981). 'The Use of Task Groups and Task Forces in Organisational Change.' In Payne, R., and Cooper, C.L. (eds.), *Groups at Work.* Chichester: Wiley.

Golembiewski, R.T. (1962). *The Small Group.* Chicago: University of Chicago Press.

Golembiewski, R.T., and Miller, G.J. (1981). 'Small Groups in Political Science: Perspectives on Significance and Stuckness.' In Long, S.L. (ed.), *The Handbook of Political Behavior* (vol. 2) New York: Plenum Press.

Heany, D.F., and Vinson, W.D. (1984). 'A Fresh Look at New Product Development.' *Journal of Business Strategy* 5(2) (Fall), 22–31.

Hosking, D.M. and Morley, I.E. (1985). 'The Skills of Leadership.' Paper presented at the Eighth Biennial Leadership Symposium, New Leadership Vistas, Texas Technical University, Lubbock, TX.

Janis, I.L., and Mann, M. (1977). *Decision Making: A Psychological Analysis of Conflict, Choice and Commitment.* London: Collier-Macmillan.

Kanter, R. (1984). *The Change Masters: Corporate Entrepreneurs at Work.* London: George Allen and Unwin.

Kimber, M.S., Coulhurst, A., Hamilton, P.H., Sharp, J.E., and Smith, D.G. (1985). 'Curriculum for Design: Engineering Undergraduate Courses.' *Proceedings of Working Party, SEED.*

Kirkpatrick, S.A., David, D.F., and Robertson, R.D. (1977). 'The Process of Political Decision Making in Groups.' *American Behavioral Scientist* 20, 33–64.

Morley, I.E. (1986). 'Negotiation and Bargaining.' In Hargie, O. (ed.), *Handbook of Communication Skills.* London: Croom Helm.

Morley, I.E., and Hosking, D.M. (1984). 'Decision Making and Negotiation: Leadership and Social Skills.' In Gruneberg, M., and Wall, T.D. (eds.), *Social Psychology and Organisational Behaviour.*

Morley, I.E., and Hosking, D.M. (1985). 'The Skills of Leadership.' In Schroiff, H.-W., and Debus, G. (eds.), *Proceedings of the West European Conference on the Psychology of Work and Organisation.* Aachen, FRG:PWO.

Morley, I.E., and Palmer, H. (1984). 'Report of the ESRC Workshop on the Process of Design at Arden House. Warwick University.

Payne, R., and Cooper, C.L. (eds). (1981). *Groups at Work.* Chichester: Wiley.

Pearson, A.W., and Gunz, H.P. (1981). '1. Project groups.' In Payne, R., and Cooper, C.L. (eds.), *Groups at Work.* Chichester: Wiley.

Pugh, S. (1974). 'Engineering Design: Towards a Common Understanding.' *Proceedings of the Second International Symposium on Information for Designers* (pp. D4–D6). Southampton: University of Southampton.

Pugh, S. (1977). 'Creativity in Engineering Design: Method, Myth or Magic' (pp. 137–146). *Proceedings of the SEFI Conference on Essential Elements in Engineering.*

Pugh, S. (1981). 'Concept Selection: A Method That Works.' *Proceedings of ICED '81* (pp. 497–506). Rome: ICED.

Pugh, S. (1982). 'A New Design: The Ability to Compete.' *Proceedings of the Design Policy Conference* (pt. 4, pp. 12–16). London: DPC.

Pugh, S. (1983a). 'The Application of CAD in Relation to Dynamic/Static Product Concepts.' *Proceedings of ICED '83* (vol. 2, pp. 564–571). Copenhagen: ICED.

Pugh, S. (1983b). 'Design Activity Model.' Engineering Design Centre, Loughborough University of Technology, June 28.

Pugh, S. (1984). 'Further Development of the Hypothesis of Static/Dynamic Concepts in Product Design.' *Proceedings of the International Symposium on Design and Synthesis* (pp. 216–221). Tokyo: ISDS.

Pugh, S. (1985). 'Systematic Design Procedures and Their Application in the Marine Field: An Outsider's View.' *Proceedings of the Second International Marine Systems Design Conference* (pt. 1–10). Copenhagen: IMSDC.

Pugh, S. (1986). 'Design Activity Models: Worldwide Emergence and Convergence.' *Design Studies* 7(3), 167–173.

Rawson, K.J. (1976). 'The Art of Ship Designing.' *Proceedings of Europort '76, International Maritime Conference.* Amsterdam: IMC.

Rothwell, R. (1977). 'The Characteristics of Successful Innovators and Technically Progressive Firms (with Some Comments on Innovation Research).' *R&D Management* 7(3), 191–206.

Rzevski, G. (1984). 'Processes within Design Teams.' Paper presented to Workshop on Design, Economic and Social Science Research Council, London, June 6.

Stewart, R. (1979a). *Choices for the Manager.* London: McGraw-Hill.

Stewart, R. (1979b). *Contracts in Management.* New York: McGraw-Hill.

Wohl, J.G. (1981). 'Force Management Decision Requirements for Air Force Tactical Command and Control.' *IEEE Transactions of Systems, Man and Cybernetics, SMC-11*, 9, 618–639.

26

Engineering Design: Towards a Common Understanding

ABSTRACT

The theme of this chapter was used to introduce the presentation of my paper 'Manufacturing Cost Information: The needs of the engineering Designer.' My main reason for so doing was not to support the paper but rather to attempt to overcome the communication gap which always seems to be present in any discussion relating to design.

In talking to delegates afterwards, it was suggested to me that it would be useful to them to have a record of this attempt to bridge the communication gap and also might serve as an apt reminder of the symposium itself.

Reproduced by courtesy University of Southampton, England. From *Proceedings of the Second International Symposium of Information Systems for Designers* (Southampton: University of Southampton, 1974), D4–D6.

In discussion with people both in industry and the academic world, in reading articles and listening to lectures and speeches about engineering design, I am always struck by the number of different approaches to the topic, conditioned no doubt by the person's particular background, experience, and, in particular, engineering discipline.

In such discussions it quickly becomes obvious that unless one is talking to someone from one's own industry or discipline, one's own view and understanding of design differs widely from the views of others.

For instance, in discussing, say, electrical engineering design, this may be considered as something quite different from mechanical, civil, electronic, and architectural design; indeed, as if there were no areas of commonality between them. The same may also be said of exponents of design in respect of the latter engineering disciplines. Why is this so? Is it not compartmentalising and isolation of the different disciplines that prevents us from having an understanding of design common to them all? Is it not the lack of a common basis for understanding, if you like lack of common ground, which not only prevents or inhibits understanding but also leads to confusion both in industry and the academic world, but also contributes in no small part to the dichotomy which exists between these two areas, when it comes to the question of an engineering education and, in particular, design education.

I have always subscribed to the view that a common basis does exist always, however, safe in the knowledge of not being able to define such a common ground. Such a statement is extremely easy to make but infinitely more difficult to quantify and express in a legible and readily understandable form.

This chapter, therefore, is an attempt to portray in very simple terms what is in reality a very complex subject. One thing is certain, as with design itself, if one starts with a complex solution to a complex problem, the chances of achieving a simple solution become very remote.

As a noted French visitor said recently, in discussing the Marathon 2550 dump truck: 'Anyone can design a nuclear power plant; to design a dump truck is very difficult; to re-design an egg is impossible.' This statement was made against a background of a lifetime in nuclear power!

To return to the subject in hand, it is only in recent months that I have evolved a means of portraying the highly manipulative activity which is design itself. It appears, at least, to be readily understood by postgraduate students in engineering design and practising designers in industry and is therefore, I suggest, also likely to be understood by people at large.

HYPOTHESIS

I would like you to consider all design as having a boundary, which we will call the design boundary, defined as encompassing all the activities and factors to be taken into account by the responsible design engineer (or designer or architect) in pursuit of his or her product design.

Let me be quite clear at the outset that the term *product* is used in the sense of the article to be manufactured from the drawings—such as electric razor, lawn mower, electric motor, bridge, nuclear reactor, building.

It cannot be overemphasised that design is not just a technological activity with concentration on performance and technical excellence; all designs to be truly satisfactory and successful must satisfy many more criteria than the aforementioned.

If we itemise all the criteria, activities, and factors which impinge upon and influence all product designs (factors which are not always recognised), all of which should be considered and continually borne in mind by the designer in pursuit of his design, we obtain the following: environment, ergonomics, scientific disciplines (chemistry, physics, et al), engineering disciplines (electronics, electrical power, strength of materials, machines), company constraints, design and development costs, timescale, quality and reliability, maintenance, competition, product life span, life in service, materials, quantity, aesthetics, performance, standards and specifications, profitability, research and development, customer, packaging, shipping, size, weight, market constraints, manufacture, product cost, safety.

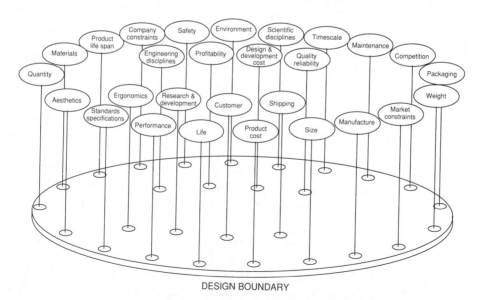

DESIGN BOUNDARY

FIGURE 26.1.   Early PDS Model: 1971

This list (which is probably not exhaustive) is applicable to all design, in no matter which of the so-called fields of design we care to consider. Such factors must be considered at the outset of any design activity.

All designers, whatever their label, should continually monitor and bear in mind all these factors from the start to the finish of any design. It is only when this is not done that we find glaring omissions and oversights in our design—when the product fails in service, when it costs too much to produce, when it will not go into the aircraft, when it goes rusty ahead of time, and so we go on listing shortcomings familiar to us all.

I submit, therefore, that it is the variation in the weighting and distribution of these factors which gives rise to the various titled forms of engineering but always for the successful design we must attain the best balance between these factors.

How, therefore, can we depict these factors in such a manner that all designers continually bear them in mind and in so doing establish the common ground?

I would liken the designer's job (if he or she does it properly) more to a particular circus act than to anything else—the plate and stick act, familiar to us all. If one considers that the designer has to keep in continuous motion a number of plates on sticks, each plate can have a label taken from our list of factors, all within the design boundary. Does the designer not in day to day working have to keep each plate in motion in order to produce a satisfactory design at the end of the day? If the designer should ignore the constraints imposed by shipping regulations or, in the case of the Sydney Opera House (Yeomans, 1973), the width of the average man, and other like factors, is he or she not forgetting to keep one of his plates in motion and, like the circus performer, when a plate falls to the ground, has not the performance, or in our case the design, failed? Looking back over events in one's career, can we not hear, loud and clear, the sound of breaking crockery? So it is with design.

The size and importance of the plates will vary from one field of engineering to another, but always we have the same number of plates. Some may require spinning more regularly than others depending upon the importance and relevance of the particular factor to the total activity.

We must, however, keep all our plates spinning all the time no matter what the design situation.

It is recognised, however, that designers operate at different levels in a company—from, say, a junior draughtsman up to chief designer, in the case of a typical engineering company. These differing levels of operation are dictated (it is hoped) by different levels of competence. The person responsible must keep all the plates in the air; the more junior person may operate with a reduced number of plates but must be made aware of the total activity at the earliest opportunity in order that he or she may progress as a successful designer. Certainly, if the more junior person does not have such a grasp, then he or she is unlikely to climb the ladder.

In asking you to accept this simple portrayal of design, do not be disarmed by the fact that the portrayal is simple; the act of keeping all one's plates in the air all the time is not only extremely difficult and demanding, but it is also very exciting and rewarding for the person who above all else enjoys situations of high tension.

I would like to conclude with a word of advice aimed at the young designer: if you do not enjoy tension situations, design is not for you. The design produced without tension is yet to be.

In all, a highly manipulative analogy to a highly manipulative process—now commonly known as 'Pugh's plates.'

REFERENCES

Yeomans, John. (1973). *The Other Taj Mahal: What Happened to the Sydney Opera House*. Camberwell: Longman Australia. This book by John Yeomans is not only fascinating reading, it also contains the largest collection of design misdemeanours under one cover known to me.

27

Balancing Discipline and Innovation

ABSTRACT

Designing and engineering the best products are important when competing in the marketplace. Unfortunately, they are not sufficient by themselves. To ensure commercial success, it needs to be coordinated with the actual users' requirements. As an engineering manager at Fuji Xerox once said, 'Western man is very good at turning unknowns into knowns. Eastern man is very good at turning knowns into commercial success.'

From *Manufacturing Breakthrough* 1(1) (January/February 1992), 9–13.

It is my contention that both objectives mentioned by the Fuji Xerox manager are achievable by Westerners, but to achieve the full cycle we must adopt new practices. What is required is a process of disciplined creativity rather than innovation.

Products do not have to be highly innovative, as long as they do the job properly: all too often, we go for high technology products that do not perform. World-class companies, including many Japanese firms, derive success from making the most reliable product.

I consider that this is a critical distinction to make, as many in the United Kingdom believe that if we invest in technology or research all will be well. It will not.

DISCIPLINED CREATIVITY

The United Kingdom (and possibly Europe) has commenced on yet another creativity and innovation cycle to try to explain why our manufacturing industries are declining. I submit that the outcome of this cycle is entirely predictable. Yet again it will be proven that we need little to stimulate our inherent creativity and innovation: it is our lack of discipline in controlling the output that has led to the aforementioned decline.

Disciplined creativity is what is required, and this is achieved by giving a good, visible structure to the design process, assisted by the appropriate methods.

TOTAL DESIGN

We need to adopt total design in all businesses in order to rectify the situation; without it we will never be competitive—even though we may momentarily appear to be innovative. It is the process of total design which will help us maintain the momentum to transfer innovation into commercial success.

Total design (see Table 27.1) is defined as the systematic activity necessary, from the identification of the market or user need through to selling, to produce competitive products for world markets.

Europeans, however, tend to practice partial design, the way design is often taught in our institutions. For a start, mechanical engineers all too often work in isolation from their counterparts in other engineering departments. This leads to mismatches between areas of the product's design and also extends cycle times.

Also, we still, in the main, cling to a rather outdated view of the design process which was traditionally completed with instructions for manufacture. All this achieves

TABLE 27.1.   Principles of Total Design

The user need, customer requirement, or voice of the customer is paramount to the success or failure of the product.

All facets of a business need to be involved in and interact with design core in parallel and not sequentially—the design team (Morley and Pugh).

To satisfy the user need, rigorous systematic working is required throughout the design core using modern methods that are both dependent and independent of the technology or product.

A product's status needs to be assessed accurately before starting any new design (Hollins and Pugh).

Within systematic working, a cyclical process of synthesis/analysis/synthesis is necessary, brought to a satisfactory conclusion by the appropriate methods.

The most up-to-date elements of engineering, based on sound engineering principles, must be used as appropriate.

Total design teams must be multidisciplinary, with sufficient expertise within the team and sufficient diversity of expertise (Friedlander and Schott).

Consideration must be given to a wide range of alternatives without prior commitment to any particular alternative.

The design team must repeatedly scrutinise and test the information and reasoning on which a design is based (Morley).

People performance is critical to total design performance.

Engineering principles are a vital subset of total design; they influence but do not necessarily relate directly to the user need.

To minimise the cycle time for completion of the design core (to minimise process losses), systematic working with modern methods and aids is required.

Total product is only achievable through total design.

Source: Pugh, S. (1990), *Total Design*. Reading, MA: Addison-Wesley.

is a design cycle which is increasingly separated from manufacturing and, thus, contributes to the division between design and manufacture and the sequential approach to the product development process.

If in using systematic engineering design processes we splendidly engineer the wrong product, the outcome of this application is zero, since the product fails to sell. The Sinclair C5 electric vehicle no doubt incorporated some splendid engineering, but its commercial success was zero.

Successful products require input from people from many disciplines, technical and nontechnical, all of which impinge on the design of the product. Unless all these inputs are coordinated and in balance, the product will fail commercially.

TECHNIQUES

Simultaneous engineering is a well-publicised phrase at present which many companies use to describe the process of simultaneously engineering a product and its means of production. If this is all the phrase means, then although this activity is essential in today's competitive world, by itself simultaneous engineering is inadequate.

The inadequacy stems from a fundamental front-end weakness which omits the market research component of marketing from the definition and, hence, the practise. For real success in the future, companies must additionally incorporate this marketing input into design and manufacturing engineering.

For this reason, total design procedures must be systematic but envelop a much wider spectrum than just traditional engineering. The design core connects the selling function (output) back to the market or user need (input)—and links together all the other activities, including specification of the need, concept design, and manufacture.

Intensive worldwide research reveals that this design core is almost universal, irrespective of product, although there might be slight semantic variations.

It is within this core that the balance of engineering and nonengineering inputs must be maintained.

In a product context, we must choose appropriate technology-dependent methods from the many available (finite element analysis, and so on) to achieve a good, sound, reliable design solution. This correct utilisation of technology is not only traditional; it is essential for success. But what has also emerged over the past twenty to thirty years is a series of techniques to assist with the efficiency and effectiveness of the design core which are completely independent of the product technology.

These include quality function deployment (QFD), Taguchi methods, and concept selection and it is in their avoidance of such techniques that many Western companies fail in the transfer of innovation into commercial success (Figure 27.1).

These techniques may be characterised as follows:

- They are independent of technology or discipline;
- Used in the design core, they flow from one to the other;
- They are reversible in that they can be used generatively or diagnostically (to analyse competition);
- The degree and success of implementation depend on visibility and understanding of the methods;
- Use of overlapping coherent methods leads to reduced product cycle times, utilises fewer people, and gives rise to more robust design;
- All methods can be used at system, subsystem, and component levels;
- Such methods have all evolved through design in practise;
- There are few mismatches between Western and Eastern methods.

My work over the past twenty years has been concerned with designing products whilst using and evaluating available methods, as well as contriving new methods to improve the processes of total design.

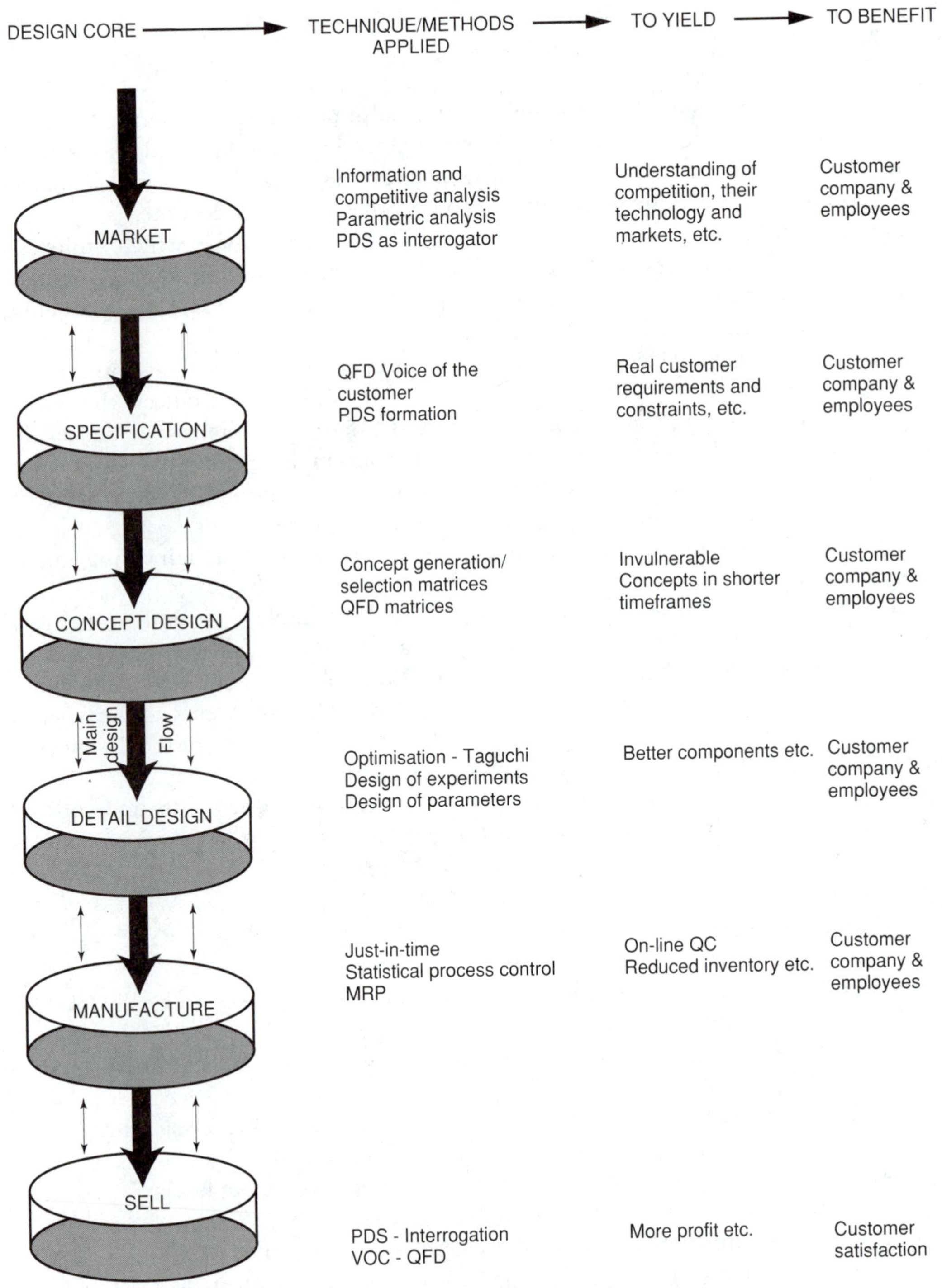

DESIGN CORE ⟶ TECHNIQUE/METHODS ⟶ TO YIELD ⟶ TO BENEFIT
APPLIED

MARKET

Information and
competitive analysis
Parametric analysis
PDS as interrogator

Understanding of
competition, their
technology and
markets, etc.

Customer
company &
employees

SPECIFICATION

QFD Voice of the
customer
PDS formation

Real customer
requirements and
constraints, etc.

Customer
company &
employees

CONCEPT DESIGN

Concept generation/
selection matrices
QFD matrices

Invulnerable
Concepts in shorter
timeframes

Customer
company &
employees

Main design   Flow

DETAIL DESIGN

Optimisation - Taguchi
Design of experiments
Design of parameters

Better components etc.

Customer
company &
employees

MANUFACTURE

Just-in-time
Statistical process control
MRP

On-line QC
Reduced inventory etc.

Customer
company &
employees

SELL

PDS - Interrogation
VOC - QFD

More profit etc.

Customer
satisfaction

FIGURE 27.1. Total Design or Product Delivery Process

What is crystal clear to me is that it is the lack of necessary input at the front end of the design core that causes most problems. All too often, the starting points for design are usually inadequate and ill-defined.

The biggest cop-out is when marketing departments recommend that the product specification be kept broad so as not to stifle design creativity. All this achieves is a project that grows until initial intents are lost and the resulting product is too expensive, takes too long to get to market, and so forth.

So, even when companies adopt techniques such as DFMA and Taguchi methods, which are very important, many people find that these techniques do not work as they should because all they do is improve rubbish due to errors made at the beginning of the design process. The only short-term solution is to include marketing in the design team. It's funny, but the Japanese do this already.

POOR PRODUCT DESIGN SPECIFICATION

The problem is that rather than address the real problem, these tools are adopted as quick fixes. The biggest weakness with many design teams is the poorly written and wholly inadequate product design specification, which is recognisable by the increasingly high rate of design changes as the product gets nearer to being launched.

THE VOICE OF THE CUSTOMER

To be effective, the design processes must start off in the right direction, otherwise the design will finish badly or at best chaotically. To achieve this, the real voice of the customer must be established and a product design specification (PDS) formulated.

The real voice of the customer becomes visible and established only if a systematic and thorough searching of information sources is carried out relating to technology and competitors—direct and indirect in terms of one's own products and also analogous products (Figure 27.2).

One powerful way of analysing the competition and acquiring insight into competitors' products and technologies is parametric analysis, where data on one's own and the competitors' products is systematically expressed and analysed. Other methods are also available, such as matrix analysis and reverse concepts selection.

This process is then followed by a brief stage where the information is used to convert the simply stated new product requirements into the PDS. The primary elements of the PDS act as triggers which lead to a detailed expression when placed in a product context (Figure 27.3).

The PDS establishes the product design boundary, which is essential. Then, within this boundary, solutions for meeting the PDS can be systemically synthesised. The method of concept selection and generation can be used to evaluate concepts against those of its competitors in a concept selection matrix or series of matrices.

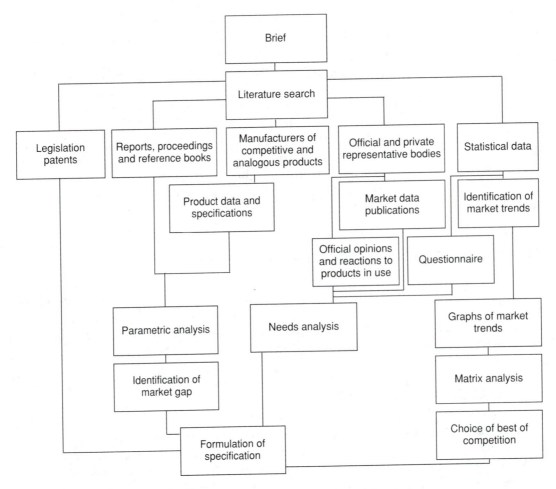

FIGURE 27.2. Establishing the Real Voice of the Customer

INTERACTION BETWEEN FUNCTIONS

Total design incorporates all the activities and has four primary boundaries relating to the process (manufacture of product), people, and the organisation, as well as the product—all of which are incorporated in the business design activity.

The design core (product delivery process) is the central theme of this business design activity, the boundary of which, encircling this core, is formed by the organisational structure of a company—its specialist activities and resources in marketing, manufacture, finance, purchasing, sales, and research and development. These include specialist people, money, materials, and machines, which need to be provided and interact with the design core to enable the business to function efficiently.

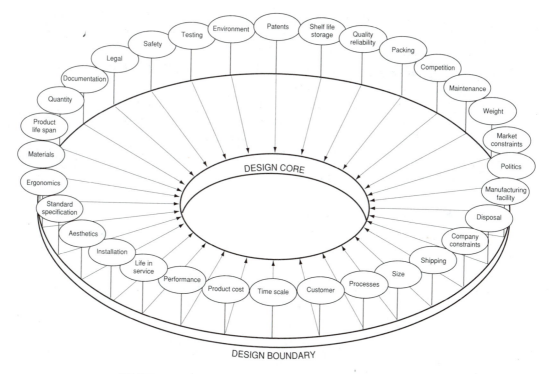

FIGURE 27.3.   Elements of Product Design Specification

To effectively operate the design core, both externally and internally generated information needs to be gathered and flowed into the core and between the various business functions. Also, for sustainable, successful products in world markets, a company needs to operate product development teams. These should be multifunctional and multidisciplinary, and the core team should consist of no more than six to eight people, including marketing.

Finally, employees need to be made aware that this is what their business is about and how they fit in. We need to specify the business design boundary and its constraints in much the same way as for the PDS, according to the type of business we are in.

CONCLUSION

What total design is about is not that much different to what many companies do now, except that it is about doing it more thoroughly and systematically and using the known techniques more effectively and in a more appropriate manner.

Adoption of such an approach requires that it is explained to everyone in the company just what total design is and what the available methods are. Show them that everything they have done in the past is relevant; all new methods do primarily is harness their collective experience and cause it to be used in a systematic manner. Above all else, the design processes must be made visible to all the staff and how they fit in and contribute to the process illustrated.

Successful implementation depends on a visible design culture whose methods are used systematically.

The Japanese do things well but not that well. World-class Western companies, such as Rank Xerox, that are practising total design—with reduced cycle times and better products—can demonstrate that they can catch and equal them. My concern is to do with betterment, which requires balanced concentration on the front end of the design core as well as sustained excellence later on.

28

Organising for Design in Relation to Dynamic and Static Product Concepts

ABSTRACT

This chapter represents part of the outcome of a pilot study carried out in U.K. industry, covering a wide and varied product field. The strategy and rationale for the study and its implementation are given. From the rationale of the study, the discussion surrounding the development of a questionnaire is given which in itself led to the construction of the interview schedule.

In five companies, members of successful design teams (teams associated with successful products on the market) were interviewed. Deductions made for the five companies as a whole are given, fitting quite clearly into a dynamic and static conceptual framework. The chapter concludes with a consideration of embodiment design as an essential part of conceptual design, to be viewed differently according to position in the dynamic and static spectrum.

With Ian E. Morley. From *Proceedings of the Institution of Mechanical Engineers International Conference, Engineering Design* (ICED 89) (Bury St. Edmunds: Mechanical Engineering Publications, 1989), 313–334. Reprinted by permission of the Council of the Institution of Mechanical Engineers.

INTRODUCTION

In June 1988, we published 'Towards a Theory of Total Design' (Pugh and Morley, 1988), which is the Part I report of our research into design practices in manufacturing industry. This chapter represents a distillation of the Part II report and describes the development of a questionnaire which may be used in a preliminary way to assess how engineering firms go about organising the process of design. We believe that feedback from such a questionnaire, comparing each organisation with others in the same sector, would be the first step in helping a firm improve its competitive performance. We shall describe how the questionnaire was developed and report some initial findings from the pilot work. The question of embodiment design in the context of dynamic and static product concepts is also addressed.

DEVELOPING THE QUESTIONNAIRE

The General Approach

We adopted an approach to designing and developing a questionnaire which had six main components:

1. A semistructured interview schedule would be constructed, initially, to be used to validate questions and to highlight strengths and weaknesses of the questionnaire approach.
2. The schedule would say what we mean by total design and establish similarities and differences between our view of design and that of the organisation concerned.
3. Respondents would then be asked to describe a project to which they had contributed, with which they were familiar, which had been completed in the recent past (or was in its final stages), and which was as typical as possible of those undertaken by their company.[1] This procedure was designed to minimise the difficulties in obtaining useful retrospective reports of cognitive and social activity (see Nisbett and Wilson, 1977; Ericcson and Simon, 1980). A similar approach was taken by Might and Fischer (1985).
4. Interviews would be carried out with people at different levels in the organisation with different roles in the project (as recommended by Maidique and Zirger, 1984). This was especially important given our emphasis upon total design. It was expected that interviews would last two to four hours.

5. Particular care would be taken to obtain descriptions of a particular project in the respondents' own words before asking questions based on theoretical models, such as the design boundary models identified in Figures 28.1 and 28.2. In effect, we followed the guidelines set out in Cohn and Turyn (1984).

6. A major review of the schedule and how it was working would be conducted after visits to three different firms. It was hoped that there would be only minor changes in the wording of questions before the review.

7. The completed interview schedule would be turned into a form suitable for use as a questionnaire.

Constructing the Initial Interview Schedule

Generating a Pool of Questions

September and October 1986 were spent generating a pool of questions which might be included in an initial interview schedule. Five main sources were used. First, the social science literature identified in the original research application made by Morley and Pugh. Consequently, particular attention was paid to the interview schedules and questionnaires used by Stewart (1976, 1979), Belbin (1981), and Kotter (1982). Second, we examined the literature concerned with the process of innovation, focussing initially on

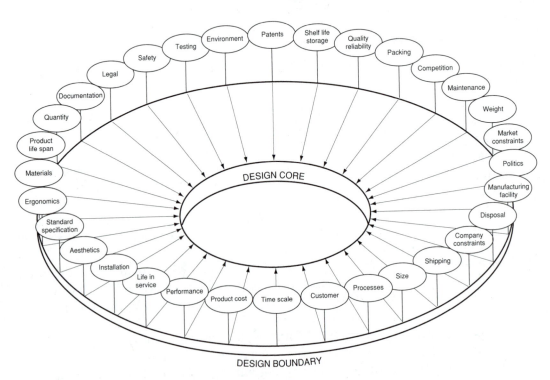

FIGURE 28.1. Elements of Product Design Specification

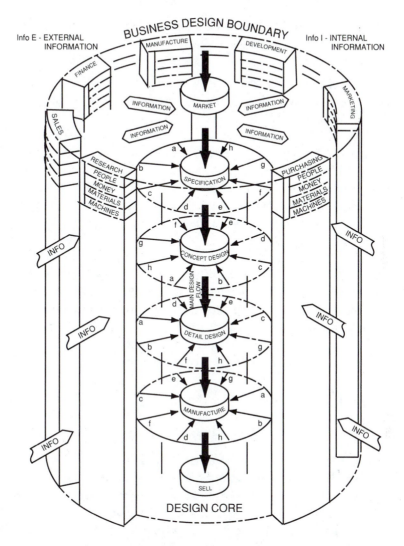

FIGURE 28.2.   Business Design Activity Model

the work of Lawrence and Lorsch (Lawrence and Lorsch, 1967; Lorsch, 1970; Lorsch and Lawrence, 1970).[2] We were particularly interested in the idea that high-performing organisations were better able to integrate key subsystems, such as sales and market research, production, and research.[3] Third, we examined some of the literature on teamwork and organisational development, such as Dyer (1977), Merry and Allerhand (1977), Helms and Wyskida (1984), and Lippitt (1982). Fourth, we examined the literature on design activity models in detail, noting especially the list of questions proposed by the Engineering Council/The Design Council in their booklet, *Managing Design for Competitive Advantage*. Finally, we generated a large number of questions of our own.

Taken together these sources produced a pool of over 300 questions which might be asked about the organisation of design.

Selecting the Questions

The first three weeks in November 1986 were spent selecting questions from the pool and ordering them into a provisional interview schedule.

Questions were selected to cover the four facets of design identified by Kimber et al. (1984)—core stages, techniques, information, and management—as detailed in the original application. Management was to be analysed in terms of the clarity of the tasks given to people, the ordering of those tasks, the choice of methods, and the assignment of resources (of various kinds). We would now say that we were attempting to establish how different firms gave *coherence* (Akin and Hopelain, 1986) or *anatomical structure* (Hubka, 1982) to the design core.

The major difficulty we experienced was in finding forms of a question which made sense given a very wide variety of organisational forms and organisational practices.

The first full version of the interview schedule contained eighty-seven questions, although thirteen of them were repeated for each design stage identified by the respondents themselves. It was anticipated that interviewers would clarify answers and ask subsidiary questions as required. Some specific prompts were identified if respondents seemed unable or reluctant to answer some of the questions.

The Structure of the Schedule

The schedule was divided into five main sections:

- *Section 1: the respondent and his job*[4] Seven questions were asked to obtain basic information about the respondent: who he was, what he did, and what he regarded as most important about the job and about the business.
- *Section 2: general characteristics of design within the company* Twenty-five questions were divided into five subsections. The first subsection of three questions established the nature and size of the design function and its relation to the business as a whole. The second subsection gave a description of what we meant by design (i.e. total design) and asked respondents to explain whether their own view differed from ours, and if so, how. From this point onwards respondents were asked to think about a particular project with which they were familiar,[5] and to which they had contributed. The third subsection contained seven questions which allowed us to find out what kind of product was being designed, how it started, and what kind of brief was involved. The fourth subsection included four questions about the management of the project. The fifth subsection included ten questions about the nature of the group of people who worked on the project.
- *Section 3: how respondents describe the process of design in their own words* Respondents were given a blank template showing the structure of Pugh's business design activity model (Figure 28.3). The following instructions were then read to them:

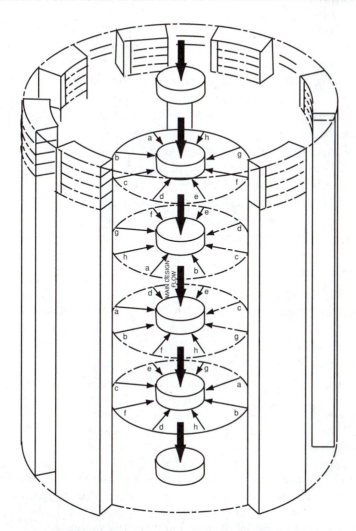

FIGURE 28.3. Business Design Activity Model

We would now like you to describe the process of design in more detail, using this template. The central column of boxes is used to indicate the main stages in the process of design from beginning (at the top) to end (at the bottom). The other parts of the template will be used to describe relationships between the project team and other parts of the organisation. We would like you to describe the same project as before. Please describe the main stages in that project, as you think they occurred. For each stage in the project please label one box in the diagram.

- It was occasionally necessary to clarify that respondents could use as many blank templates as required. If respondents showed any hesitation, it was made clear that we were simply interested in getting a description of stages in the design of the project in their own words and as they saw it. Respondents were asked to talk us through the diagrams as they were drawn. A small number of respondents preferred to draw and label a flow diagram of their own, starting with a completely blank sheet of paper. This was quite acceptable. There was never any difficulty in converting the information given into the standard format provided by the blank template.[6] When diagrams had been completed to the respondent's own satisfaction, they were asked whether these were the only stages and how long each stage lasted. They were then asked to describe each of the stages they had identified in more detail, answering the same thirteen questions for each of the stages. The questions dealt with formal responsibilities for organising the process of design (one question); how the project was organised (six questions);[7] what the major problems, challenges, dilemmas, or difficulties were (one question); what the most important things done by themselves and others were (two questions); and what proportion of their own time was devoted to the project (one question). It was expected that this section of the schedule would last for one hour, or perhaps longer.

• *Section 4: questions directly related to Pugh's business design activity models*   Respondents were only asked questions about existing theoretical models of design activity when they had described the process of design in their own words, using the blank templates (Figure 28.3), or producing flow diagrams of their own. When they had completed this task they were told the following:

We would like now to ask some questions geared specifically to one model of design activity. It is called the business design activity model. The model is represented graphically on this sheet. [Respondents were given an A4 size diagram similar to that shown in Figure 28.2.] The business design activity model portrays the business as a product-producing entity. The product design activity is represented by a design core enclosed by a business design boundary. The design core consists of a set of activities which progress iteratively in stages from analysis of market and user needs (at the top of the diagram) through to selling (at the bottom of the diagram). You will notice that the term *business design boundary* is used to describe the outer ring of the diagram. It refers to those elements of the business which contribute to and participate in the process of design. They provide resources of people, money, methods, and machines. Taken together they cause the whole enterprise to function. The design core and its relationships to the business design boundary may be described as *total design activity*.

Respondents were then asked the following:

What do you think are the major differences between the process of design, as you have described it, and the activity set out in the business design activity model?

This was followed by three questions about the nature of coordination and collaboration within the enterprise.

Respondents were then asked a number of questions about each of the activities in the design core shown in Figure 28.3, beginning with investigation of the market (five questions) and leading on to a description of the product design specification (nine questions), the conceptual design stage (nine questions),[8] and the detail design stage (one question).[9]

- *Section 5: some questions about the role of discussion and debate and some general questions about design*   Two questions were asked about the role of discussion and debate in the particular project being described. Four questions were asked about other projects within the enterprise. Two questions were asked to conclude the schedule:

Are there any aspects of design which you think are important but which have not been covered in this discussion? Are there any comments you would like to make about the questions you have been asked?

Testing the Initial Schedule in Use

The initial schedule was read by D.G. Smith of the Engineering Design Centre, Loughborough University of Technology, and by D.M. Hosking, Management Centre, University of Aston. Their comments were extremely helpful and meant that we changed some of the questions originally proposed. The basic structure of the schedule remained unchanged, however. The first interviews were conducted with A. Hayward, Product Development Director, Qualcast Garden Products Limited, Derby (24 November 1986), and G. Kirk, Chief Engineer, Engineering Research and Manufacturing Technology, Rolls Royce Limited, Derby (1 December 1986). These interviews resulted in a number of changes in wording to make questions clearer or more comprehensive.

During the period December 1986 until the middle of March 1987 interviews were conducted with members of the design teams in the following firms: Qualcast Garden Products Limited, Derby; Rolls Royce Limited, Derby; and Honeywell Control Systems Limited, Motherwell.

Revising the Initial Schedule

The next month was used to conduct a major review of work using the initial schedule. A new version of the schedule was prepared, considerably revised, and considerably shortened. The purposes of the revision were to eliminate questions which were ambiguous, misleading, produced little variance, and were cost ineffective.[10] It was also intended to cover any areas which were subsequently seen as important but omitted from the initial schedule.

It was intended that the revised schedule take no longer than two hours to administer.

The revised schedule was used to interview members of design teams at Alcatel Business Systems Limited, Romford, and Potterton International Limited, Warwick.

Producing the Questionnaire

The revised interview schedule was turned into a questionnaire by further consideration of the questions used, according to the criteria set out above. In addition, we used our experience of the previous interviews to turn questions into multiple-choice questions which could be answered simply by ticking an appropriate box.

PRELIMINARY FINDINGS

The purpose of this section is to introduce some of the things we have learned in the development of the interview schedule. We shall not identify firms by name as information was given to us strictly in confidence.

In all companies, interviews were conducted with all the members of the project teams who had worked upon the design of a successful product in the market place. What follows are our deductions, detailed comments, and conclusions upon each firm, which were fed back to the firms for their comments and approval. We structured these minireports in a coherent way in order that comparisons may be made but primarily to lead to a consideration of process losses and the stages in design.

Process Losses and Stages in Design

Previously we have written (Pugh and Morley, 1988) that order and coherence in the process of design would be more likely to lead to effective designs, other things being equal. We noted that one way of giving design order and coherence is to have a clear sequence of stages in the design core. We added that there may also be reasons for having the output of one stage formally accredited before signing it off and moving on to the next. We supposed that order and coherence of this kind would be particularly likely to minimise what Steiner (1972) has called losses from faulty process.

Firm A

Deductions   You have clearly managed the shift from a static to a dynamic context extremely well so that the process of design corresponds to that set out in Figure 28.4.

Detailed Comments

Market	This was the only company we visited in which members felt that marketing had been successfully integrated into the design process. All spoke very appreciatively of changes made in recent years, although not everyone was clear what kinds of market study had been carried out.
PDS	Very comprehensive; very well done. Everyone has a clear view that the product design specification regulates all that follows.
Concept	Only company to use systematic methods allied to systematic progress through stages. We presume this is working and would like to learn more.
Detail	Only example of decisive shift from initial to final model prototyping in our sample.
Manufacture	Seems to have been much more successfully integrated than in other firms.

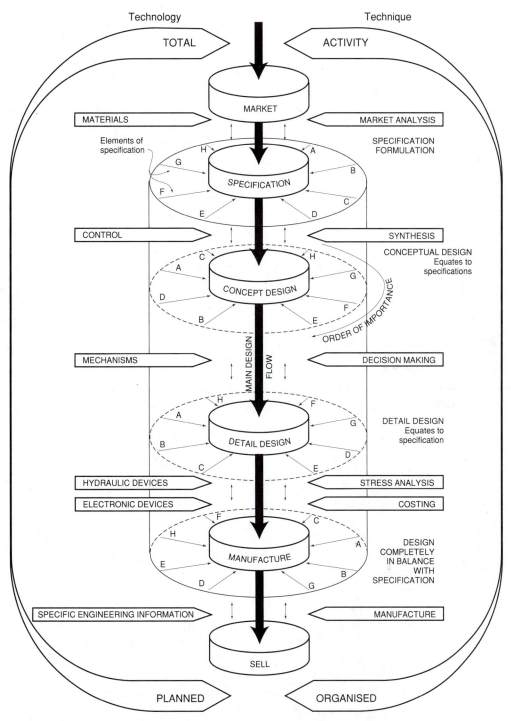

FIGURE 28.4.   Design Core Bounded by Product
Design Specification: Dynamic Concept

Conclusions

- The best example we have found so far of design being organised successfully in a dynamic context.
- This success depends very much on a shared view of the process of design, which was very evident in our interviews.
- Team members were very much aware of their roles as members of multidisciplinary teams. They understood the reasons for the shift to this form of organisation and thought the changes were very much for the better. It would have been very interesting to see whether these opinions were shared by marketing and manufacturing people within the firm.

Firm B

Deductions   You have been in business for a long time and have considerable design experience with a slowly evolving series of products. Because of this the model shown in Figure 28.5 is most appropriate. The context is, in our terms, relatively static. You know what your product is like, know in advance many of the problems you are likely to face, and know, in general terms, how they are to be solved.

Thus much conceptual design can be carried out in advance of setting down of detailed product design specification. In general your procedures have been geared to this context. However, in the project we have considered you were entering a very new area for your company, and effectively starting with a clean sheet. Thus, the model set out in Figure 28.4, setting the design process in a dynamic context, is more appropriate. In this context it is important to begin with a detailed product design specification. It is essential to lay out alternative ways of satisfying the specification and explicitly choose between them.

Detailed Comments

Market	Apparently started well, but why not integrate representatives from marketing more fully into all stages of process of design/product development? At the moment interaction seems confined to director level.
PDS	Product design specification is reasonably complete, but it is less clear this is really used to *control* the process of design. Not clear to what extent this is based on rigorous analysis of the market or competitors' product.
Concept	Need to spend much more time and effort on this part of process. Treated as static design (when dynamic). Important to educate members of design team that short cuts don't pay in the long run.
Detail	Detail design ultimately done well. Very late involvement of production meant many changes forced on product with consequent serious loss of time. Quite clear production should be involved much earlier as members of design team. Meetings with production seem too confined to director level, when fire fighting reaches its peak.
Manufacture	Involved much too little and much too late. Need to be integral part of design team.

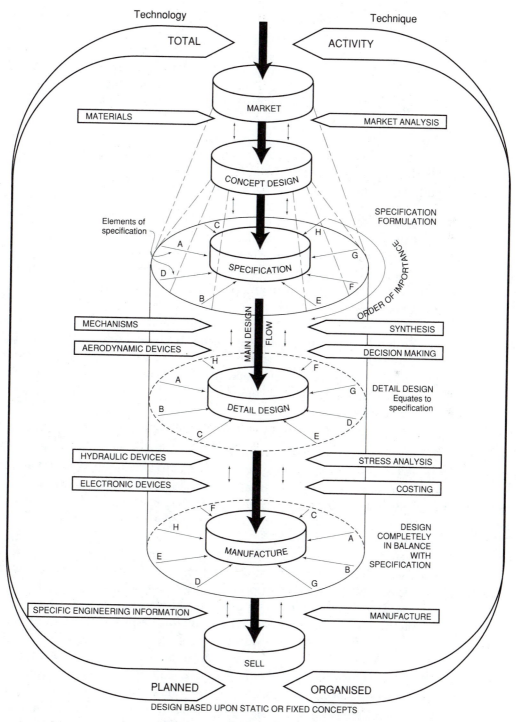

FIGURE 28.5. Design Core Bounded by Product
Design Specification: Static Concept

Conclusions   Intentions systematically to organise the process of design well understood at the director level, where liaison appears to work well. However, the process is much less well understood at lower levels, and liaison and coordination suffer accordingly. People see the process as much more ad hoc than is the intention. To turn (good) intentions into practice, participants need to be brought together and build the same view of what design means within your company.

Firm C

Deductions   The company clearly organised to operate in dynamic mode as shown in Figure 28.4. The product emerged from the application of new technology and a previously static product was made dynamic.

Detailed Comments

Market	Interviews suggest that marketing may still be a little distant from the design process. Is there a marketing member of the multidisciplinary team? We weren't clear. People were very well aware of the importance of getting the front end of the design process right.
PDS	Very thorough.
Concept	We were impressed by the systematic arrangement of the stages in design but felt that systematic *methods* could usefully be applied at this stage. Not clear whether conceptual design was a team or individual activity. Not clear people always able to generate ideas and follow these implications through.
Detail	Handled very well.
Manufacture	Seems well integrated into the overall process of design.

Conclusions   We were very impressed with the way the company was attempting to organise the process of design to show clear stages (adapted to a dynamic context) and to control the transition from one stage to another. The documentation we were shown was very thorough. However, although the procedures have been formally set out it is less clear that everyone is aware of them and of their implications.

Firm D

Deductions   You have been in business for a long time and have considerable design experience with a slowly evolving series of products. Because of this the model shown in Figure 28.5 is most appropriate. The context is in our terms relatively static. You know what your product is like, know in advance many of the problems you are likely to face, and know in general terms how they are to be solved.

Thus, much conceptual design can be carried out in advance of setting down a detailed product design specification. In general, your procedures are geared to this context. However, some aspects of the design involved very new ideas and require a more dynamic approach than the model shown in Figure 28.5. In this context it is important to begin with a *detailed* product design specification. It is then essential to lay out alternative ways of satisfying the specification and explicitly to choose between them. In some cases it may not be possible to satisfy all the elements of the PDS. It is then important to choose a design which has certain well defined and agreed strengths:

Detailed Comments

Market	Very incomplete brief from marketing compared with other companies. Marketing involved in process too little and too late. Does someone from marketing participate regularly in design meetings?
PDS	A number of important changes to existing products, showing initiative and innovative thinking, assessed intuitively against experience rather than systematically against criteria in a product design specification. Essentially, only one conceptual design considered.
Concept	Product design specification the most incomplete in any of the companies visited; *evolved backwards* through development and disagreement. Too little liaison at all stages of the process with others who should be involved in total design. This is typical of many companies in static contexts.
Detail	Detail design ultimately done well. Late involvement of marketing and production forced many changes in detail as product design specification evolved. Much time and effort could be saved by better liaison and earlier involvement of interested parties.
Manufacture	Welcome view that design extends into manufacture but representatives from production, servicing, and so on need to be consulted and involved at much earlier stage and be part of design team, as we have found in some other companies.

Conclusions   There seems comparatively little reliance on formal procedures and formal methods. The design process is characterised by informal meetings based on geographical proximity of participants. Recommend a move towards more systematic involvement of people currently outside the design team, with more meetings involving representatives of all those involved in the process of total design. There is evidence from our interviews that moves of this kind are seen to occur only when things go wrong (crises occur). More structured meetings would be welcomed and would help the process of design to converge on solutions all interested parties would understand and accept. The process would be designed constructively to move forwards rather than evolve backwards. To make this happen all participants need to be brought together and build the same view of what design means within your company.

Firm E

Deductions   You have been in business a long time and have enormous experience with the product. We consider the model set out in Figure 28.5 most appropriate to a company which has thrived on robust designs, intended to be improved incrementally with time. In simple language, you know your product and know in general terms how associated problems are to be solved. Thus, much conceptual design can be carried out in advance of setting down a detailed product design specification, and in general your procedures were geared to this context. We see the introduction of design-make teams on this product as an attempt to ensure design is as creative as possible given an essentially static (derivative) context. This means it is important to begin with a detailed product design specification. It is then essential to lay out alternative ways of satisfying the specification, and explicitly to choose between them. (If it is not possible to satisfy all of the elements of the product design specification it is important explicitly to choose a design with known and agreed strengths.) Effectively, this means moving towards the model set out in Figure 28.4.

Detailed Comments

Market	Marketing and finance still seem distant from activity of design-make teams. Otherwise what happens is very much in the spirit of our model. Product design specification was the outcome of detailed negotiations with customer.
PDS	There seems to have been a comprehensive and clear product design specification used to guide the process of design.
Concept	Basic conceptual choice established from word go.
Detail	Detail design carried out in context of fixed concepts. Robust concepts allow for many variations in detail.
Manufacture	Part of design-make team. Integration seems to have been effective because you were able strikingly to reduce the development time by approximately one year, in response to competition.

Conclusions   The organisation of the project seems to have been very successful technically. We are not able to assess whether the shift in market could have been detected earlier because we did not liaise enough to say whether this was predictable. There *may* have been a failure at the front end of the design process, or there may not. All we say is that marketing and finance were distanced from that part of the process open to our view. It is clear that those members of the design-make teams that we interviewed shared a view of what was happening, and why, and how the process should progress. What emerged also was that many processes were still very informal. We were often asked why we didn't say more about creativity. Our view is that this is not so much an individual attribute as something which follows when intelligent, motivated people operate systematically in contexts which produce a more controlled convergence on solutions everyone can understand and everyone can accept. This may mean paying more attention to formal methods, especially at the stage of conceptual design.

DISCUSSION

These reports allow us to relate order and coherence in the process of design to losses from faulty process. Our main hypothesis is that losses from faulty process will decrease as order and coherence increase. It is consistent with mini case studies such as those of Oakley (1984). It is also consistent with the data we have collected to date. In our studies firms A, C, and E obviously followed a formal series of stages in design, requiring formal procedures, tests, documents, and reports. These firms showed least sign of losses due to faulty process. In firms B and D such losses were extremely easy to find, to document and to diagnose. In firm D a faulty interface between marketing and design added at least nine months to the time to get the product ready for the market. In firm B a faulty transfer between design and manufacture meant that several hundred modifications were required after the final drawings were produced. Other examples of process loss were also evident.

It is worth commenting that losses due to faulty process seemed the most important problems in the firms in our sample. Although our evidence is by no means complete it is our very clear impression that *each* of the firms concerned ended up with a very successful, competitive product at the end of the day. Difficult technical problems were solved very efficiently. However, those firms exhibiting most process loss faced greatly extended timescales, with all the associated costs and risks.

They were characterised by lack of integration in the activities of the design core, although the design teams were often themselves extremely cohesive. They lacked a visible operational structure. They relied very heavily on informal rather than formal communications between members of the design teams.

In respect of communications, the two firms with the most obvious process losses, firms B and D, relied almost entirely on informal methods of communication.

The firms which minimised process loss (A, C, and E) formed a regular part of the process of design. For example, the project in firm E had '*project high spots* that take place once a week, and they have a set topic where an agenda will be put out, and where specific people will be asked to come along and present the status of . . . their particular part of the project to the project team. . . . This is intended to provide a project status to a wider audience than the normal project. The purpose of the project high spots is there for the whole project team to be able to comment and offer suggestion or criticism' (project leader).

There is a similar attitude to open *communication based on regular formal meetings* at firm A: 'I mentioned a quarterly meeting among section heads at operational level. We [also] have strategic planning meetings of the Department with one or two managers and all the section heads get a copy of the confidential minutes of these meetings. Because how can you expect to operate with people who are working for you as part of the team if they don't know what's between the ears? My first objective was to build a team that I could have confidence in, and trust and work with, and it was a two-way trust. Then you could open up the communications as such so that when people saw something that was confidential they could use it in a positive sense' (R&D director).

Finally, it is significant to record that since the interviews took place and the initial report was prepared, firm B has revised its procedures. It works now with a number of project teams each with a project manager, with each team comprising a mix of people from all functions in the company who stay with the project throughout its development and into production. Meetings are held regularly and formally documented, and to quote the design manager, 'This new practice has allowed projects to be better coordinated and communicated, has engendered a greater team spirit and integration between departments.'

Embodiment Design and Static and Dynamic Concepts

As previously reported, Pugh and Morley (1988) and in our reports to the firms in our sample, we have made a distinction between static and dynamic conceptual boundaries. These define a spectrum of design activity within which we may locate the firms, as shown in Figure 28.6.

We are now in a position to consider the position of embodiment design within the design core. We shall find that embodiment design has different roles depending on the location of a firm in the spectrum of design activity (whether design activity tends towards innovation or whether it tends towards convention).

The *VDI Guideline 2221* is often taken to suggest that engineering design goes through four overlapping phases: definition of the objectives, conceptual design, embodiment design, and detail design. As part of our interview schedule we had participants describe the main stages in the design of a particular product in their own words. There was no difficulty at all in matching what was said with the design boundary models discussed previously. What was very evident, however, was that no one identified a separate stage of embodiment design. *Embodiment design* was not a term universally used. When it was used, different people seemed to understand the term in different ways.

In our original interview schedule we had a question: 'How much detailed engineering or embodiment design did you carry out at the conceptual stage, before the final choice was made?' This question produced a wide range of answers, mostly using the language of detail design. If the firm was operating in a static context, answers typically pointed out that a lot of detailed work had been done on earlier designs. One of the firms was very clearly improving the performance of a previous model of the product. The project leader was asked some supplementary questions: '[Interviewer: So the conceptual design finishes with the design scheme, but where all the embodiment has been done?] Yes. [Interviewer: Then you go away and detail that?] Yes.'

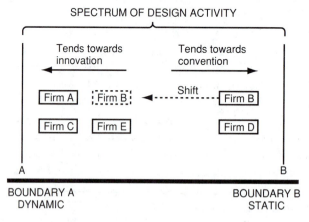

FIGURE 28.6.   Spectrum of Design Activities:
Boundary A—The Dynamic Core Model,
Boundary B—The Static Core Model
Source: Pugh (1983, 1988).

So embodiment went on throughout the process of conceptual design. This is very much in the spirit of the original *VDI Guideline 2221*, but the point is lost in simplified accounts, such as that of *Managing Design for Competitive Advantage* (1986). What is called conceptual design in *VDI Guideline 2221* results 'in a principle solution which represents the best combination of physical effects and primary embodiment features to fulfil the function structure. It may be documented as a sketch, a diagram, a circuit, or even a description.'

As we move towards boundary A (as concepts become more dynamic), it is likely that models, prototypes, and so on may be produced very early on in the process of design. This was very much the case in firm A. In this case the models and prototypes served three main functions. First, they established the technical feasibility of the product. Second, they gave the members of the team a tentative understanding of the interactions between the components embodied in the product, which collectively made up the whole. Third, they helped members of the team to gain understanding of the possible functions of the product. This becomes more and more important and essential as the design becomes more and more innovative and moves closer to the dynamic conceptual boundary, boundary A, where product precedents do not exist.

Firm A used modelling activity to gain a clear conceptualisation of the basic issues, changing some of their initial ideas and writing the changes into a thorough product design specification. The product design specification was then used strictly to regulate and control the design activity that followed. This resulted in a production prototype which required little modification and was subject to very few losses from faulty process.

It would be fair to add to this description and say that in firm A embodiment issues were explicitly addressed and resolved prior to the final conceptual choice.

As we move towards boundary B (as concepts become more static) with conventional designs which have evolved incrementally over many years, much of the embodiment engineering is taken for granted. Models and prototypes may also be built early in the design process but for reasons of a different kind. Technical feasibility, interactions between components, and understanding of function will all have been acquired over a long period of time and will be understood. In this case the reasons are more likely to be ones of cost reduction, manufacturing efficiency, style, and the like. In many companies operating towards boundary B, it might be said that prototyping precedes the start of all new design processes: the last model becomes the prototype for the next model.

Firm D is an example of this type of company, and this is illustrated by the fact that they operate the design process cyclically in reverse from the security (apparent) of the conventional product. They prototyped as if for production based upon their past experience, and each subsequent prototype variation caused the notional PDS to be revised. This notional PDS remained a fragmented set of requirements, viewed differently by different people. This fragmentation tended only to become visible in the next prototype—which brought everything together and revealed the mismatches. Each prototype was thus a manifestation of the true PDS at the time.

The multiple iterations due to lack of a PDS (comprehensive) took a long time (relatively), which in effect resulted in increased process loss, the fragmentation of both procedure and commonality of view contributed additionally to this loss.

Firm B was used to operating in the region of boundary B when they embarked upon an innovative product. They prototyped as if the product was conventional, and in this case multiple process interactions were claimed to establish technical feasibility, understanding of interactions between components and possible functions. Again the result was enhanced loss due to faulty process.

So embodiment design is clearly a subset of conceptual design whether the product is static or dynamic, but it needs to be viewed differently in each case.

This view is confirmed by the fact that we have found that all of the firms we have spoken to make a fairly clear distinction between conceptual design and detail design. In some firms, particularly those operating in static contexts, a great deal of detail design was also evident at the conceptual stage. In more dynamic contexts what was most evident was that stages which might be called conceptual design and detail design each include model building of various kinds.

In our final questionnaire we have dropped all talk of embodiment design. We have retained a distinction between conceptual design and detail design, and we have concentrated on the kinds of model building which go on at different stages in the process of design. One of the main advantages of the initial interviews was that they allowed us to find the right kind of language with which to ask these questions.

NOTES

1. One of the companies we visited had just completed what was, for them, an innovative project, moving them from a static context to a dynamic one. Interviews dealt with this project and highlighted some of the changes in attitude and organisation required.
2. This literature is reviewed in Part 1 of our report.
3. Four months later we were able to examine the questionnaire used by Benson in his studies of successful product innovation. We are grateful to Dr. Benson for his help in making his questionnaire available when it was itself at a pilot stage. We were also in receipt of a questionnaire designed by NEDO 'to be answered in two or three minutes,' which gave us a number of ideas.
4. All the respondents in this study were men.
5. Ideally, the project would have been completed in the very recent past. In one case the project was in the final stages of completion. All respondents within a firm described the same project.
6. It was never intended to duplicate these procedures in a questionnaire. We hoped, however, to learn how to interview respondents about the process of design, and we hoped to obtain valuable information to be used in developing a multiple choice questionnaire, subsequently. Both these aims have been achieved.
7. These questions were quite complex. Respondents were asked to say who were the people outside the project team it was most important for them to meet. For each person they were then asked to indicate how closely they collaborated with them using a six-point scale (1 = not at all, 5 = very closely indeed, 6 = don't know).

8. Two of the questions concerned the relationship between conceptual design and detail design. One of the questions concerned the relationship between conceptual design and embodiment design.

9. This question was based on a quotation from the Engineering Council/Design Council booklet, *Managing Design for Competitive Advantage* (p. 14). Respondents were asked, 'To what extent do you agree or disagree with the statement: . . . it is the quality of the detail design rather than differences in the concept which give one product a competitive advantage over others?' Respondents were asked to indicate their position on a seven-point scale (1 = strongly agree, 7 = strongly disagree) and to explain their answer.

10. That is, took too much time to produce very little in the way of return.

REFERENCES

Akin, G., and Hopelain, D. (1986). 'Finding the Culture of Productivity.' *Organisational Dynamics* 14(3), 19–32.

Belbin, R.M. (1981). *Management Teams: Why They Succeed or Fail*. London: Heinemann.

Cohn, S.F., and Turyn, R.M. (1984). 'Organisational Structure, Decision-making Procedures, and the Adoption of Innovations.' *IEEE Transactions on Engineering Management* EM-31, 4, 154–161.

Dyer, W.G. (1977). *Team Building: Issues and Alternatives*. Reading, MA: Addison-Wesley.

Ericcson, K., and Simon, H.A. (1980). 'Verbal Reports as Data.' *Psychological Review* 57, 271–282.

Helms, C.P., and Wyskida, R.M. (1984). 'A Study of Temporary Task Teams.' *IEEE Transactions on Engineering Management*, EM-31, 2, 55–60.

Hubka, V. (1982). *Principles of Engineering Design*. London: Butterworth Scientific.

Kotter, J.P. (1982). *The General Managers*. London: Collier Macmillan.

Lawrence, P.R., and Lorsch, J.W. (1967). *Organization and Environment: Managing Differentiation and Integration*. Boston: Division of Research, Harvard Graduate School of Business Administration.

Lippitt, G.L. (1982). *Oganizational Renewal: A Holistic Approach to Organizational Development*. Englewood Cliffs, NJ: Prentice-Hall.

Lorsch, J.W. (1970). 'Introduction to the Structural Design of Organisations.' In Dalton, G.W., and Lawrence, P.R. (eds.), *Organizational Structure and Design*. Homewood, IL: Irwin. Georgetown, Ontario: Dorsey Press.

Lorsch, J.W., and Lawrence, P.R. (1970). 'Organising for Product Innovation.' In Dalton, G.W., and Lawrence, P.R. (eds.), *Organizational Structure and Design*. Homewood, IL: Irwin. Georgetown, Ontario: Dorsey Press.

Maidique, M.A., and Zirger, B.J. (1984). 'A Study of Success and Failure in Product Innovation: The Case of the U.S. Electronics Industry.' *IEEE Transactions on Engineering Management*, EM-31, 4, 192–203.

Merry, U., and Allerhand, M.E. (1977). *Developing Teams and Organizations*. Reading, MA: Addison-Wesley.

Might, R.J., and Fischer, W.A. (1985). 'The Role of Structural Factors in Determining Project Management Success.' *IEEE Transactions on Engineering Management*, EM-32, 2, 71–77.

Morley I.E. (1981). 'Negotiation and Bargaining.' In Argyle, M. (ed.), *Social Skills and Work*. London: Methuen.

Nisbett, R.E., and Wilson, T.D. (1977). 'Telling More Than We Can Know: Verbal Reports on Mental Processes.' *Psychological Review* 84, 231–259.

Oakley, M. (1984). *Managing Product Design*. London: Weidenfeld & Nicholson.

Pugh, S. (1983). 'The Application of CAD in Relation to Dynamic/Static Product Concepts.' *Proceedings of the ICED 83* (vol. 2, pp. 564–571). Copenhagen: ICED.

Pugh, S. (1988). 'Knowledge Based Systems in the Design Activity.' *Proceedings of the Conference on Modern Design Principles* (pp. 75–89). Trondheim, Norway.

Pugh, S., and Morley, I.E. (1988). *Towards a Theory of Total Design*. Design Division, University of Strathclyde.

Steiner, I.D. (1972). *Group Process and Productivity*. New York: Academic Press.

Stewart, R. (1976). *Contrasts in Management*. New York: McGraw-Hill.

Stewart, R. (1979). *Choices for the Manager*. London: McGraw-Hill.

Winham, G.R., and Bovis, H.E (1978). 'Agreement and Breakdown in Negotiation: Report on a State Department Training Simulation.' *Journal of Peace Research* 15, 285–303.

29

The Design Audit: How to Use It

ABSTRACT

Design audit or design review forms a vital part of the strategy for a company and provides a management tool for successfully managing a design project. This chapter differentiates between the activity of reviews and the resources to enable them to be carried out. The essential structure and format of the necessary documentation are also discussed.

From *Proceedings of the Design Conference* (Birmingham, 1979), sess. 4a, paper 3, 4a/3/1–4a/3/6.

The design audit or design review forms an essential part of modern industrial practice. Properly instituted it provides a mechanism whereby designs can be improved in a balanced and best compromise manner, and thus improved designs lead to improved products. The word *design* is used to encompass the total activity of design including engineering and analysis: if this isn't the way most people see it today, then things are worse than appears.

As a practising designer, it is interesting to read the works of others, and in this respect the topic of design review is no less interesting or fascinating. Jacobs (1975), in arguing the case for design review as being a fundamental liability preventer, presents a list of functions and responsibilities to indicate how design reviews should be conducted. Napier (1979), in a paper demonstrating design review as a means to improve quality in design also, lists the activities and stages to be undertaken in a design review. There is a remarkable similarity between the two approaches and so there should be since they both represent a logical approach to design control and evolution.

However, it is suggested that each in different ways misses a vital point by not relating what to do and how to do it to anything other than a rather vague thing called a *product*. Pugh (1974) gives an insight into the areas which should be given detailed consideration prior to the commencement of any design—factors which affect a design and which are contained within what might be called the design boundary. These are the factors which form the basis for the product design specification; a further discussion of these is given in Pugh (1978).

It is suggested—nay, strongly recommended—that the product design specification, since it forms or should form the anchor point or datum for the designer of a product, should also form the basis for the design review process, since each must contain the same essential elements. The PDS is the essential reference documents for the *design* and the *design review*. To embark upon the due process of design review without an adequate PDS is almost certain to lead to frustration and wasted effort in both areas since questions will arise out of the latter which should have been asked and answered prior to design commencing. Let us explore this question further.

OPERATION OF THE DESIGN REVIEW

Before examining the mechanism of the design review, it is suggested that the following diagram (Figure 29.1) may be of assistance to the design practitioner, since he or she is the primary audience for this chapter. The PDS, since it contains or should contain all the elements appplicable to the product design, acts as the control datum or safety net

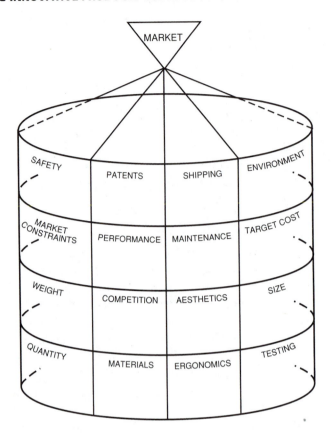

FIGURE 29.1.   Using the PDS Elements to Establish
and Control the Design Audit

against which the designer is able to measure the progress of his design from initial concepts through to production.

Figure 29.1 depicts the component parts of the PDS, and it is after this has been formulated that the first design review should take place. Is the specification realistic? Is it achievable? Are cost targets reasonable? Is the market fully understood? It is the writers' opinion that whilst many mistakes are made in design whilst it is still in progress, many more are built into a design prior to commencement due to inadequate PDS's.

Subsequently, as a guide, design reviews should take place at the conceptual stage, detail stage, prototype stage, development stage, and production stage (Figure 29.2). At each review, comparisons should be made with the specification, which in itself may have changed due to additional and more up-to-date information being available due to market shifts or a more enhanced knowledge of the product design which may also bring about specification changes. In the ultimate, hopefully, a product emerges which satisfies the specification which by definition will also satisfy the market and, by dint of the formal design review process, will be better in all aspects than without it.

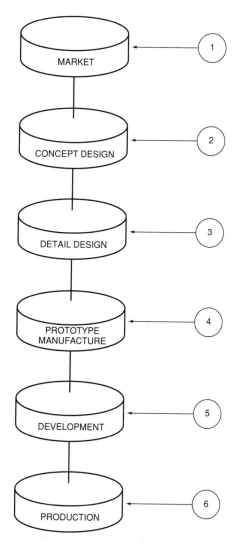

FIGURE 29.2. Suggested Stages of Design Review

The implementation of a formal review procedure also requires (unless the whole thing is to become farcical) a discipline of documentation, and hence communication, between all the parties concerned in the process. More will be said about this aspect later.

THE ACTIVITY OF DESIGN REVIEW

To the mind of the writer, a distinction must be made between the *activity* of the design review and the *resources* required in order that the actions arising from such reviews may be implemented. It is suggested that design reviews should be considered as two separate but nevertheless mutually interactive limbs, as shown in Figure 29.3. The first is the activity concerned primarily with the elements of design itself and the second with the resources necessary in order that the former may be brought to a successful conclusion—a successful design and hence product. It is necessary to make this distinction in order that difficulties of understanding a very complex activity are minimised. It is suggested that enhanced understanding will lead to smoother and more effective operation and outcome of the design review process.

THE OPERATORS OF DESIGN REVIEW

The people concerned in the formal process should be the specialists and representative of the various different resources—production, test and inspection, quality control, fi-

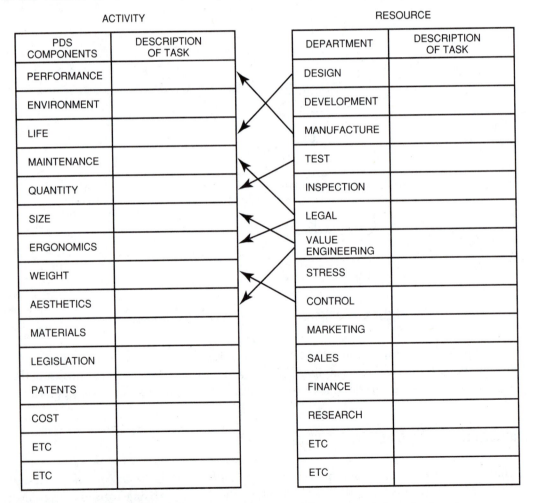

FIGURE 29.3. Activity of Design Audit: Resources Required.

nance, standards, marketing, and so on. In other words, the people who have a part to play in making the design and hence the product successful. The review team should naturally include the designers of the product. It is also essential that meetings of the team should be chaired by someone in a position and of a competence to take a balanced view of the whole interactive situation. Minutes should record decisions taken and action required, and extreme care should be exercised to ensure that the PDS is kept updated, since in day-to-day working this is the designer's reference document.

It has been suggested earlier that design reviews should be held at predetermined stages in the design activity. However, meetings of the design review team should be held when it is considered most appropriate to do so, and this timing is usually dictated by the emergence of the design itself. For example, some problem may arise which could drastically affect the total situation and require multidisciplinary discussion to resolve. Call a meeting and remember that potential problems in design do not reveal themselves to predetermined timescales; it would be nice if they did. Design reviews work and should form part of any modern engineering business.

COMPETITION ANALYSIS

Having had the opportunity over the past few years to design marketable products for industry, whilst being remote from those industries, it has become an increasing cause for concern to find that most companies do not carry out effective analysis of current and potential competition. Having done a number of such analyses, I never cease to be surprised and enlightened by the results emerging. It would appear that market researchers do not or cannot fully analyse competitive products in terms of parameters of use to the designer. That is not to say that market research information produced is not in itself useful in formulating design specifications; it is, but it never goes far enough.

Therefore, whilst the pros and cons of design reviews can be discussed at length as an activity, it is essential in the practice situation that in the future greater attention be paid to competition analysis. Why? Because inadequacy of understanding and incompleteness of the PDS can lead to inadequate products which may be well engineered and well made, but nonetheless do not satisfy the customer. To overcome this deficiency requires a greater cooperation between market researchers and designers.

REFERENCES

Jacobs, R.M. (1975). 'Design Review: A Liability Preventer.' *Mechanical Engineering* (August), 34–39.

Napier, M.A. (1979). 'Design Review: As a Measure of Achieving Quality in Design.' *Engineering* (March), 286–289.

Pugh, S. (1974). 'Engineering Design: Towards a Common Understanding.' *Proceedings of the Conference on Information Systems for Designers* (pp. D4–D6). Southampton: University of Southampton.

Pugh, S. (1978). 'QA and Design: The Problem of Cost Versus Quality.' *Journal of the Institute of Quality Assurance* 4, 3–6.

PART VI

Design for X

COMMENTS ON PART VI

The previous chapters have described broad, high-level principles and their applications. In this part Stuart Pugh presents methods for some of the more detailed concerns of design. Cost is the most important of the specific details, and several of these chapters deal with product cost.

Stuart presented some very valuable methods for basing costs on competitive benchmarks. In recent years benchmarking has become very widely used, and the results of benchmarking are incorporated into QFD. However, this emphasis has not included much specific methodology for effectively doing benchmarking of particular and important product characteristics, such as cost. In these chapters Stuart presented the concept of parametric analysis to relate the cost to capability parameters of the product so that realistic comparison with existing products could be carried out.

In the early conceptual stage of total design it is difficult to determine even moderately accurate cost estimates—yet this is the stage of design that most controls the ultimate product cost. To address this problem Stuart advocated using the equation

$$C = K \sqrt[3]{P_p P_i P_t},$$

where

C is the product cost,
P_p is the number of parts,
P_i is the number of interconnections and interfaces,

P_t is the number of different types of parts, and

K is a constant (can be determined by parametric analysis but cancels out when calculating the cost ratio between product concepts for the same industry).

This can be used in the Pugh concept selection matrix as a surrogate for cost.

Since Stuart wrote these pioneering chapters, the subject of design for X has become popular, and the number of fields covered has grown to perhaps roughly ten. However, Stuart had the view that every item in his generic product design specification (PDS) was a candidate for a 'design for' methodology. As there are thirty-four product characteristics in the PDS, the design community still has much opportunity in developing additional 'design for' methodologies. In this enterprise these papers by Stuart present a definition of the areas for future development.

In reading these chapters look for cost benchmarking methodology, cost estimating during the early conceptual phase, the PDS as template for design for X (thirty-four of them), and further emphasis on the information needs in total design.

—Don Clausing

INTRODUCTION TO PART VI

The title of this part was chosen since it represents the worldwide drift to the elucidation and combination of the attributes (or constraints) of a product which are essential to success but extremely difficult to achieve.

From the selection of chapters given, the reader will note the linking and expansion of the number and type of constraints which emerge in Chapter 30, The Engineering Designer: Tasks and Information Needs. Constraints such as cost, manufacturing, patents, materials, and so on have been expanded into the current state of the art. It must be borne in mind that these are primary constraints and can be expanded into secondaries and beyond as necessary.

Be warned, bringing all the secondaries in at the beginning of designing a product usually makes the process unwieldy and swamped with information (see Chapter 16) a common failing of basic QFD.

A good design will need all the elements given in Figure 9.5 in balance (if appropriate), if in doing a design we choose to ignore or omit some elements without good reason, we may achieve imbalance, which will automatically result in an inadequate product.

This is particularly critical and important when a company is exploring the area between a product being conceptually dynamic or static (see Chapter 19). Other elements considered in the chapter as exemplars are safety/hazards, assembly, cost, manufacture.

This part is really a sample of what is to come in future years since one of the last bastions of design method interaction to be tackled is a usable comprehensive methodology for accurately linking all the PDS elements without omission.

Take quality and assembly as two examples. Quality is achievable only by (a) a comprehensive approach and (b) a thorough approach.

The chapters presented in this part start to build this completeness, although in 1992 we have moved from twenty-four to thirty-four primary PDS elements. It can be stated after experiencing some forty projects that the PDS ties everything together particularly when it is used as the controller of the design process.

When the unknowns or uncertainties in the transition between static and dynamic concepts are taken into account, the whole interactive problem becomes enormous. We are now tackling this. In the meantime designers will have to be content with noninteractive systems where a particular element is considered singly—such as design for assembly (DFA). One has to arrive at the ground rules through individual cases, but this can be misleading; since it is not entirely unknown that having, say DFA'd product, that you render maintenance impossible or difficult, and an inadvertent 'knock-on' has been achieved.

I would recommend that design for X be considered in preparing a PDS where X can mean anything.

The Engineering Designer: Tasks and Information Needs

ABSTRACT

This chapter looks at the designer in a more traditional sense, not specifically as a team member. His or her information needs are considered according to level of operator (total system or component) detail, the difference between major responsibilities, and areas of essential interest are made whilst again reinforcing the difference between technology or technique.

Reproduced by courtesy University of Southampton, England. From *Proceedings of the Third International Symposium on Information Systems for Designers* (Southampton: University of Southampton, 1977), 63–66.

INTRODUCTION

It is becoming a matter of increasing concern to me that the very laudable efforts of information scientists to provide the information required by the people concerned with the design activity are being marred by a lack of understanding of just what is involved in the activity itself. The reasons for this situation existing are many and varied; however, there are probably two primary reasons which stand out above all else:

- The inability of the engineering designer and his associates to spell out in an understandable form just what it is that they do and what their information requirements are and
- The inadequacy of communication between information scientists and designers compounded by the fact that many such scientists have no first-hand knowledge of engineering design and therefore cannot by themselves correct the deficiencies.

At the 1974 Conference it was apparent that a problem may exist in this area and preliminary attempts were then made towards bridging the communication gap (Pugh, 1974). Since the Conference I have become even more aware of the problem and, based upon the premise that 'if one is going to service a motor car adequately and efficiently and also give the owner complete satisfaction, it cannot but help to have a better understanding of motor cars,' this chapter attempts to take another step in this direction.

The opportunity, therefore, to set down further thinking in this area is one for which I am extremely grateful, and it is the intention of this chapter to set out in a logical and more readily understandable way the type and levels of information required by the people concerned with the design activity in practice. It is of particular importance that the level of information required should be matched to the level at which the person working within the design hierarchy is situated. This is to be done against a new model of the activity itself which hopefully will make the situation more understandable.

ENGINEERING DESIGN: THE ACTIVITY

First, I do not propose to launch into a consideration of the relative correctness or values of the many definitions of engineering design that have come about over the years. To do so would probably lead to endless debate without securing the main objective, which is the understanding of the design activity and its place in the overall spectrum of an engineering business or artefact producing organisation.

Second, I would ask you to accept the premise that engineering design is central to all such activities and that, through its able and successful operation, the best use is made of men, money and materials culminating in market satisfaction. The market or need satisfaction is becoming the most vitally important factor in engineering or product design today. With increasing competition both home and abroad resulting in more and more products of a similar nature being available to satisfy these markets, it is essential that the right products or artefacts are designed to meet these markets. This factor does, I think, give rise to a fundamental difference between engineering design in a total sense and that experienced by designers in the architectural sphere and met only in part in the civil and constructional engineering industry, although competition is rapidly increasing in the latter. In engineering companies, it is becoming more and more difficult to design the right product to meet the market; conversely, it is becoming easier and easier to design the wrong product for the market.

The model shown in Figure 30.1 will, therefore, with the passage of time, become more applicable and relevant to design in the broadest sense irrespective of any particular discipline. Design may be construed as having a central core activity which is present in all design work. Briefly, this core consists primarily of market, specification, conceptual design, detail design, manufacture, and selling phases. Although it is recognised that other activities present in an engineering business such as research, development, testing, and the like are all important to overall success, one cannot proceed to a successful outcome in design without these definitive core phases.

All design starts with a need which when satisfied will fit into a market or create a market of its own. This statement of need must result in the formulation of a specification which when established then acts as the mantle or cloak which envelops all subsequent stages in the design core; it becomes the frame of reference or design boundary. The specification is an evolutionary requirement which becomes modified and changed as the design proceeds due to changed circumstances and interaction with the design core. However, upon completion, the essential equation is a match between the product characteristics and the final specification. The specification contains a whole host of essential elements, the balance and distribution of which not only change as the design evolves, but which are in content unique for every design carried out (Pugh, 1974).

Having established the specification we then proceed through the conceptual, detail, and manufacturing phases to the selling stage. The *main design flow* is therefore from market to selling; I say main flow as in practice it is a discontinuous, iterative process which when considered in hindsight will give the impression described above. To expand upon this point, you can be in the detail design phase of a product, and a new concept emerges, in which case, conditions being suitable, the new concept may be adopted which in itself will lead to new detail designs, different from those previously. In flow terms you have reversed the flow and returned to the conceptual stage; this can happen with all designs and at all stages of the design core.

You will by now have noted that to this point in the chapter, no mention has been made of technology or science or specific areas of design operation; this is because it is not necessary to do so at this stage and in fact can lead to confusion and misunder-

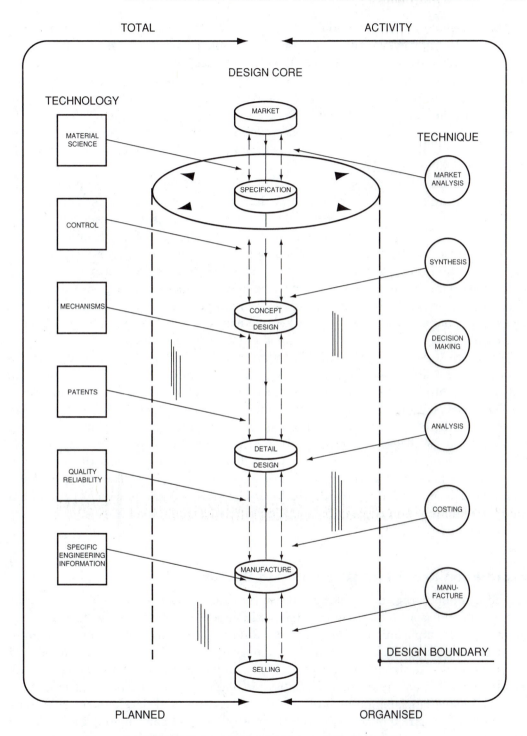

FIGURE 30.1.   Design Activity as a Planned and
Organised Activity

standing of what design is all about. This statement is made without any disrespect for science and technology, which are themselves essential to the success of any design.

I would like next to consider what I call techniques related to the design core and having inputs into the core. These techniques are necessary in order to enable the designer to operate the core activity with increasing confidence and understanding. In other words, they can be construed as the designer's tool kit, the tools necessary to enable him to perform the basic function as delineated by the core activity. These are shown in Figure 30.1 as being techniques of analysis, synthesis, decision making, costing, and modelling.

So we have arrived at a basic design core, enveloped always by a specification and with inputs from a tool kit in order not only to enhance the working of the core activity but also essential to making it work at all. We have now reached the stage where we must refer to and involve science and technology on an almost infinite front in order to bring the whole activity to life. The technological inputs into the design core relate not only to specific information on materials, control, electronics, hydrodynamics, and so on but also to the specific areas of engineering in which one is working, such as electric razors, nuclear reactors, buildings, dustbins, and motor cars to name but a few. It must never be forgotten, however, that the whole activity is carried out within a framework of planning, organisation, and control—in other words, management.

Reference was made earlier to the finite nature of the core activity and the techniques related to it; the addition to the model of the technological inputs expands it in reality to an infinite situation where at least in theory the whole of technology is at one's disposal. We therefore have a situation the basis of which is finite—the design core. Add to the core the operational inputs which in practice can be infinite, and we begin to grasp the magnitude of the problem. It is upon this model of practice that we base our teaching of design to postgraduate students in the Engineering Design Centre; they seem to understand it and in some measure start to practise it. Having defined and detailed the design activity in depth and breadth, I now propose to consider the extent and type of information required by the activity related to the level of person operating within the activity itself.

Information Requirements Related to Operational Level

I made mention earlier of the design hierarchy without wishing to imply that hierarchies exist for reasons other than on a basis of competence levels. People have different talents and accomplishments, and it is these differences which give rise in the main to hierarchical structures. Thus we have in engineering not only technical directors, chief designers, senior designers, junior designers, draughtsmen, and detail draughtsmen, but also a considerable number of specialists such as aerodynamicists, physicists, stress analysis, mathematicians, production engineers, research and development engineers, and so on. It is the role of those operating, controlling, and participating in the design core activity to ensure that such specialisms as mentioned above are coordinated and made meaningful as inputs into the core activity itself.

For the sake of clarity Figure 30.2 shows the spectrum of people involved directly in the operation of the design core placed appropriately to the core phases; from the top people encompassing the whole activity to the detail draughtsmen whose concern and output are more limited but who nevertheless must be made aware of the total activity otherwise their aspirations are more likely to wither and die.

It should be noted that titles or names given to the various levels of people are those generally in use throughout the U.K. engineering industry. I have represented their involvement with the core activity in two ways: (1) their major or prime responsibility is their position, which encompasses those core activities normally encountered in day-to-day working, and (2) each member of the team, whilst having a different level of responsibility should have what I call the same level of essential interest. In other words, all team members should be aware of and interested in the totality of the situation.

The types and level of information required by this spectrum of people varies enormously, and I do not attempt in this chapter to cover anything other than the general case. No attempt is made to delineate in detail, levels, and types of information specific to a type of industry.

I propose to define the general case by considering information under the headings of technique and technology and relate such information to the aforementioned spectrum of people indicating in by no means rigid terms the basic necessities of each group of people. This is shown in Table 30.1.

Whilst the listings given are by no means exhaustive, they indicate the extremely wide range of information required by the operators of the design core. In a situation where there is a requirement by all concerned for, say, a particular technique of analysis, then the level of analytical ability and hence the level of information required will vary considerably. However, to illustrate this variation on the diagram would, I feel, overcomplicate the situation I am trying to portray.

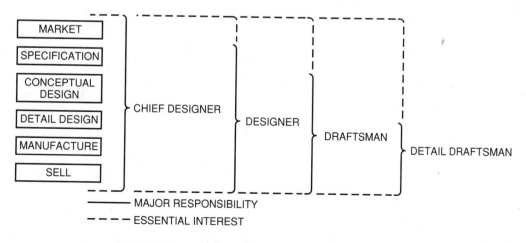

FIGURE 30.2. Differentiation Between Design Team Members' Responsibilities and Areas of Interest

TABLE 30.1. Interaction Between People and Techniques and Technology

Technology X	Chief Designer	Designer	Draughtsman	Detail Draughtsman	Technique O
Market information	O X	O X			Market analysis
Details of competition	O X	O X	X		Competition analysis
Specific engineering disciplines	O X	O X	O X	X	Stress analysis
Specific science disciplines	O X	O X	X	X	Thermodynamic analysis
Company products	O X	O X	O X	O X	Costing
Cost data	O X	O X	O' X	X	Synthesis
Manufacturing data	O X	O X	X	X	Quality and reliability
Patents	O X	O X	O	O	Information retrieval
Materials	O X	O X	X	X	Computers
BS. and international standards	O X	O X	O X	O X	Manufacture
Company standards	O X	O X	O X	O X	Classification
Development and test data	O X	O X	X		Decision making
Finishes	O X	O X	X	X	Failure analysis
Fastenings	O X	O X	O X	X	Measurement
Welding	O X	O X	O X	X	Testing

In depicting design, information types and level, and people in this way, I have sought to clarify the design position in a more coherent manner. I fully appreciate that in order to achieve a successful, competitive design whose main objective is to satisfy a need, the involvement of all sections within a company is essential. My reasons for doing it this way are primarily to reinforce the argument that the design activity is central to any successful engineering enterprise and that all products or systems emanating from such enterprises are designed. The inputs and contributions of engineers of specialised disciplines, marketing and production people, and others are essential prerequisites to successful operation of the whole business function. It is, however, the design core people who have to put it all together.

I only hope that the foregoing will lead to a better understanding of not only the design activity but also of design information needs. If this is the case, then information scientists may gain a clearer insight into the designers' needs and therefore be better able to perform their function in the overall spectrum.

REFERENCE

Pugh, S. (1974). 'Engineering Design: Towards a Common Understanding.' *Proceedings of the Second International Symposium on Information Systems for Designers* (pp. D4–D6). Southampton: University of Southampton.

31

Design Is the Biggest Exposure

ABSTRACT

This chapter considers safety as being a critical design for X for any product. It deals with thoroughness and completeness applied systematically to the design of anything and establishes the linkage between the activity of design review and the resources to achieve the desired quality.

From Stuart Pugh's lecture: *The Improvement of Product Safety* given at the International Press Centre, London, 29–30 October 1981.

The improvement of product safety is the outcome of sounder approaches to, and sounder practice in, product design. Improved product safety is best and most economically achieved by seeking to obtain the best mix of many factors in a particular design situation. It is *not* achieved by concentration on a single factor called *safety*.

As a practising designer I should take issue with the statement made by Howard Abbott that 'Probably design is responsible for more product failures than anything else.' In a sense he is absolutely right in that current design practices and the products emanating from design offices provide some evidence to support this view. On the other hand, whilst there are good and bad practitioners in any profession, I would suggest that poor product design, in a total sense, is mainly the responsibility of management in its various strata. This statement in no way absolves the designer from responsibility within his or her ambit.

The key to the whole question of sound product design, and hence sound products for the marketplace, is basic and fundamental: it is complexly a matter of thoroughness. I say complexly since the whole activity of product design is a complex matter whether viewed from the standpoint of safety, quality, cost, or any other single factor; it is never *simply* a matter of thoroughness

Let me expand on this point by stating that thoroughness is the key to wholesome products and, by definition, the safest attainable within the set of constraints given in the product design specification (a topic to which I shall give attention later). I mean much more than meticulous attention to technical detail and the latest technology. My hypothesis is that 'thoroughness of both approach and of practise leads to more thorough products' in an absolute sense.

The bright idea, the eureka quickly engineered and put into production (emotion rules), are the exception and may be extremely risky; most successful design, in my experience, is more a question of meticulous slog. Beware the singleton solution of applying the latest technology; it is extremely risky in my experience. Unfortunately, there still persists, in some areas of industry, what I would call the 'grown-ups Lego approach to design' based upon a 'let's do something at all costs philosophy.' This approach is just not good enough today—too risky, too unreliable, and likely to mismatch the market and user expectations. Once such mismatches occur, then safe products may be rendered less safe by demands being placed upon them which, in the absolute sense, are not unreasonable, but nevertheless were not seen prior to design commencement.

Thoroughness of approach? Thoroughness of practice? What do I mean?

COMPETITION ANALYSIS

If you are in a competitive field, you cannot afford not to analyse and fully understand such competitors in terms of other than just knowing of and about them. Such analysis, if soundly structured and carried through, yields immense amounts of information and forces approaches to and viewpoints of products which otherwise would be impossible to trigger by any other means.

This part of the activity extends way beyond the traditionally accepted market research norms. Whilst such research is vital, in practice the results yielded from such studies are invariably of little tangible use to the design practitioner.

A thorough parametric analysis should be carried out and conclusions drawn as appropriate, including the qualitative and quantitative aspects of safety, both deliberate and inherent in the products. This leads straight to the formulation of the product design specification (see Chapter 12).

PRODUCT DESIGN SPECIFICATION (PDS)

From the foregoing competition analysis and company inputs, a product design specification should be drawn up which embodies many primary factors (listed below and graphically illustrated in Figure 31.1).

Safety	Environment	Competition
Shelf life (storage)	Maintenance	Packing
Weight	Market constraints	Size
Manufacturing facility	Customer	Product Cost
Life (in service)	Ergonomics	Aesthetics
Processes	Shipping	Quantity
Materials	Product life span	Politics
Company constraints	Testing	Patents
Quality and reliability	Performance	Timescale
Standards and codes of practice		

These in themselves act as a mental trigger to yield a whole host of secondary factors, the outcome being a thorough and comprehensive understanding of the product design upon which you or your designers are about to embark. This, then, should become the ab initio control document for the total design activity. Even today, I know that many products have and are being evolved without any formal PDS.

No design should be embarked upon without a formal PDS; it is the foundation upon which product well being and safety are built. Interestingly and logically, the primary headings form the basis not only for the PDS itself but also for many other stages in the design activity which in themselves constitute bricks in the wall of thoroughness. For instance, the initial basis for any competition analysis should be the primary de-

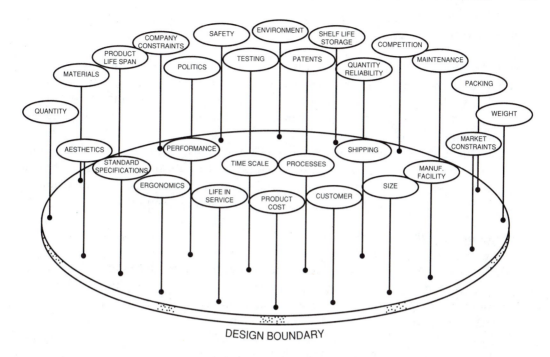

FIGURE 31.1. PDS Boundary

scriptors outlined previously: they are the logical starting points for such analysis since they are applicable to all products. Other bricks will become apparent as the chapter proceeds (see Chapter 16).

SOLUTIONS TO MEET THE PDS

Perhaps a little should be said about the respective viability of design solutions to meet the product design specifications. Let me say right away that obtaining adequate solutions in quantity is not a particular problem; it is sorting out the solutions with a degree of certainty that leads not only to enhanced products but also products with enhanced safety is the fundamental problem.

Methods have been developed, based upon practice, which enable such selections to be made with a greater certainty of success in all areas, including *safety*. The method of controlled convergence (Pugh, 1981) which we are developing in the Engineering Design Centre produces the results in practice. It is a method based firmly on the PDS and, not surprisingly, uses as its basis the primary elements of the PDS. Again, this is logical hypothesis, since if one is in practice to produce successful products, is it not obvious that the designers should attempt to meet the PDS to the best of their ability. The linkage with the PDS will be expanded upon in this chapter.

THEREAFTER

During the subsequent design activity, following the selection of the design or designs with which to take forward to the detail design, drawing, prototyping, and ultimately, manufacture and selling, as well as from the commencement of the activity—control of the activity itself is paramount. Such control is best achieved by the effective and efficient procedure of design review.

Now, much has been said and written about design reviews, much of it, in my opinion, leaving the reader bemused, confused, and, worst of all, bewildered as to how to implement such a procedure. In principle, implementation is, at least on paper, simplicity in itself; in practice, however, like most things, it's a little more difficult.

DESIGN REVIEWS

The design audit or design review forms an essential part of modern industrial practice; properly instituted, it provides a mechanism whereby designs can be improved in a balanced and best compromise manner, and thus improved designs ultimately lead to improved products and improved product safety.

The operative elements of the design review procedure are, again, logically, the primary elements of the PDS (Pugh, 1979).

The first design review takes place after the formulation of the PDS and thereafter at stages in sympathy with the evolution of the design (see Figures 31.2 and 31.3). A notional six stages are suggested.

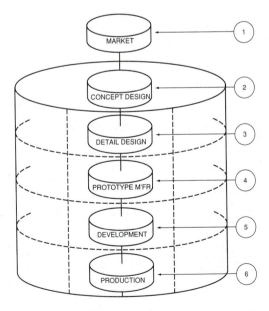

FIGURE 31.2. Linkage of PDS Elements to the Design
Core and Stages of Design

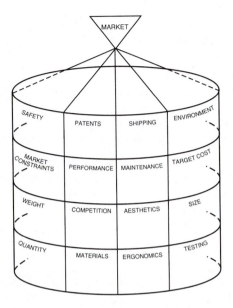

FIGURE 31.3. Linkage of PDS Elements to the Design
Core and Stages of Design

The distinction should be made between the *activity* of the review procedure itself and the *resources* required to implement it (see Figure 31.4).

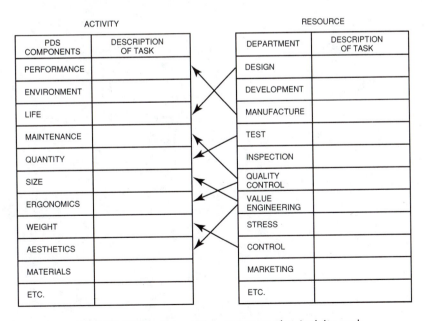

FIGURE 31.4. Interaction Between the Activity and
Resources Required for Design Review

Design review teams are essentially multidisciplinary and should be chaired by someone with the gifts of impartiality and total design competence. This then is the recommended control of those stages of activity that have the greatest bearing on product success and safety.

Granted that defective detail design of components can render products less than acceptable and, in some cases, dangerous. Efforts are still needed in this area to produce usable tools to enable this to happen (Pugh, 1977). Tears and impassioned pleas have lost their novelty and were never effective.

I have tried to no more than outline the key elements which lead to enhanced product soundness, hence safety. It is complexly a question of thoroughness of procedures, backed up by sound technical competence and understanding.

REFERENCES

Pugh, S. (1977). 'Load Lines: An Approach to Detail Design.' *Production Engineer* (January/February), 15–18.

Pugh, S. (1979). 'The Design Audit: How to Use It.' *Proceedings of the Design Engineering Conference* (sess. 4a, paper 3, pp. 4a-3-1–4a-3-6). NEC.

Pugh, S. (1981). 'Design Decision: How to Succeed and Know Why.' *Proceedings of the Design Engineering Conference* (sess. 5a, paper 3, pp. 5a-3-1–5a-3-10). NEC.

CHAPTER

32

Manufacturing Cost Data for the Designer

ABSTRACT

This chapter is primarily concerned with the costing of designs at both the component and total (global) system levels with particular emphasis on speed and accuracy of costing systems to be used by the designer. Speed of utilisation of any system is considered essential.

From *Proceedings of the Engineering Design Conference* (London, 1977), 17-1–17-16.

INTRODUCTION

The engineering designer's knowledge of total product costs and indeed the cost of manufacture of the component parts that collectively make up the whole is a cause for concern to many people. Industry itself still seems to be primarily concerned and preoccupied with production in that the 'let's build it and get out of the door' mentality still prevails in many quarters, with only a gradually awakening recognition of the role and status of the design function within the overall production system. Whilst there is an increasing awareness of this role, unfortunately, it manifests itself mainly through verbalisation rather than through definitive courses of action.

What then has this to do with manufacturing cost data? In fact, a very great deal. First, it must be recognised that in manufacturing industry, in the end analysis, the culmination of any product design and manufacturing activity is reduced to, and assessed on, a cost basis. Granted there are many other functions which together make up a successful operation; cost, however, is present in all such activities and therefore is a common factor. Second, it must be recognised and understood that the major cost in any product is initiated by a person different from the one who incurs the costs (Pugh, 1974). Everyone in the design chain from the technical director to the detail draughtsman has a part to play in the control and understanding of costs. Yet one finds that the availability of costing data or methods of costing on a formal basis seems to be inversely proportional to their position in the design chain. In other words the technical director may be very knowledgeable on costs whilst the detail draughtsman knows little of anything about them.

Cost information is not by and large provided for design engineers or detail draughtsmen as a matter of formal policy, yet they build the costs into the product. Is it not a situation analogous to saying to a blind man 'we recognise that you are blind, we sympathise with your predicament; however, you must successfully negotiate the obstacle course of life without white sticks and guide dogs, but heaven help you if you bump into anything'?

This lack of knowledge of costs has been demonstrated over the years with postgraduate students and hundreds of practising designers from industry. On being asked to assess the cost of simple, ordinary components they exhibit a cost variation of 20:1! This fact in itself is frightening enough. Is it therefore surprising that design to a target cost is too difficult to achieve without recourse to a continual design/estimate/redesign/reestimate programme with its consequent consumption of time, money, and temper? This situation should not exist to such a great extent as methods and means are available to enable the designer to assess product costs as the design proceeds to much better than 20 percent of true cost, which indeed makes a nonsense of a 20:1 variation.

413

It is the intention of this chapter to examine these methods and to comment upon their facility and effectiveness, bearing in mind that any costing system for use by designers, who, after all, are employed as designers and not estimators, should preferably satisfy the following criteria:

- It should primarily be simple, quick, and easy to use.
- It should be sufficiently accurate to enable the best manufacturing route to be selected at the design stage.
- It should preferably be based on factual rather than on comparative costs (Pugh, 1974; Henry, 1975).

TARGET COSTING

The establishment and underwriting of the target production cost at the onset of the design phase is the most important and essential objective as far as the designer is concerned, and it is perhaps ironical that in practice the target cost is likely to be in error and is usually low. In the author's experience, many given target costs are not only low, but it is impossible to achieve the specification requirements within the target cost. If the latter is the case, the whole design exercise may prove to be a frustrating waste of time, and in this situation the costing, however accurate, of the component parts which make up the whole becomes irrelevant.

It is at this point that the designer has for his own well being to take an interest in the marketing aspects of the product and in particular its likely competition. A combined design and marketing approach is required in the best interests of the company as both disciplines look upon the potential new product from very differing viewpoints—the marketing person with knowledge and understanding of the market and competition, the designer with intimate knowledge of what can and cannot be achieved technically. Creative, unconstrained thinking, which should be inherent in the designer's make-up, should be applied to the market and competition through the mechanism of a parametric analysis of what already exists. The competition should be investigated in all its aspects, the primary objective being to establish a correlation with one's own specification. The parameters used are always peculiar to the product area for such situations leading ultimately to the establishment of target costs (Pugh, 1974) ranging from printing machinery, construction equipment, through to nuclear components. It always works!

Recourse to an example to illustrate a typical parametric approach is probably the best way of conveying the basic approach to the reader of this chapter.

ROUGH TERRAIN SITE PLACEMENT VEHICLE

Some few years ago we were concerned with the design of a vehicle to an excellent and all enveloping specification prepared by marketing people—a target selling price being specified. With a knowledge of the relationship between manufacturing cost and selling price for the company the selling price could be translated into a manufacturing cost. The question therefore in all such situations is whether the specification is achievable at the target cost specified.

Briefly, the specification called for a vehicle which would handle materials in the manner of a conventional fork lift truck with one important overriding difference—a substantial forward reach capability was called for. It is a simple fact of life that if one holds a weight in one's hand and stretches forward and the weight is large enough, then one falls over. In engineering terms our forward reach capability would have to be counterbalanced. Net effect, counterbalance adds weight to the vehicle.

Another factor also affects the equation—namely, the lift height. To raise a load to a height of 1 foot will require a certain weight of structure to support it; to lift the same load to a height of, say, 40 feet will require a heavier structure. Again, the net effect is to increase the weight of the vehicle. In this instance, therefore, in addition to cost, we have isolated four important parameters—payload, lift height, forward reach, and vehicle weight. How can we utilise these parameters to underwrite and check our target cost? Remember we have as yet no overall vehicle concept in mind and design work had not started.

In the United Kingdom there are manufacturers of conventional rough terrain fork lift vehicles with negligible forward reach capability, and in the United States there are some very large vehicles with forward reach capabilities far in excess of our specification. On the basis of the parameters already isolated a function (payload x lift height) was plotted against vehicle unladen weight for both types of vehicle (see Figure 32.1). It

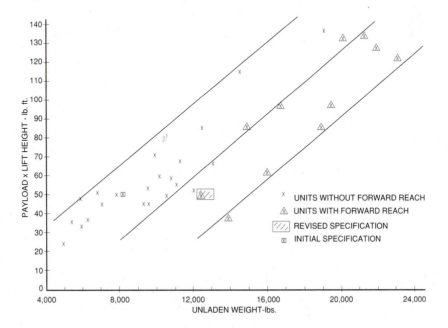

FIGURE 32.1. Parametric Analysis: Payload Versus Lift Height for Rough Terrain Forklight Trucks

can readily be seen that the two types of vehicle fall into two distinct bands, the vehicles having forward reach being heavier for a given (payload x lift height) than the conventional type. The specified function for the new vehicle was and still is 50,000 pound-feet, which from Figure 32.2 predicts a vehicle unladen weight of 12,000 pounds.

Refer to Figure 32.2, which shows the cost pattern for conventional vehicles averaging at that time 45 pence per pound of ULW. The target cost given in the specification represented on this basis a vehicle ULW of 8,000 pounds, which clearly from Figure 32.1 is unattainable if the forward reach capability specified is to be retained. The decision was therefore quite clear, there being two alternatives: (1) to proceed with the design to specification with an increased target cost or (2) to remove the forward reach capability from the specification and leave the target cost as specified. In fact, the vehicle was designed (see Figure 32.3) on the basis of the former criteria and its final ULW is 13,000 pounds! In other words the specification as it stood originally was unattainable.

The procedure for the establishment and underwriting of target costs may be summarised as follows:

- Use the information that exists on competitors' products to the full.
- Establish cost patterns for competitors and one's own products.
- Do *not* as a matter of course become necessarily pure and logical in one's approach to parametric analysis; an apparently illogical approach can lead to a logical outcome.

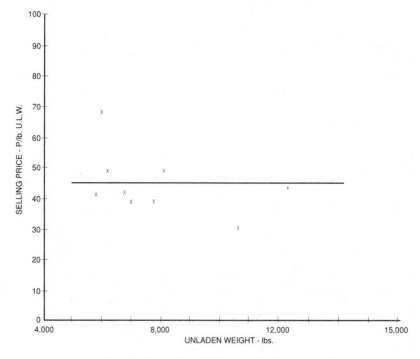

FIGURE 32.2. Weight Versus Unladen Weight

FIGURE 32.3.   Photo of Giraffe Site Placement Vehicle

Perhaps the most recent and publicised example of a large-scale parametric approach of this nature has been carried out in designing the new Ford Fiesta small car (Wilkins, 1976). Certainly industry could adopt this approach with more of a vengeance, it being almost certain to lead to better and more cost-conscious design.

COST OF COMPONENTS

Reference has already been made to cost patterns. The previous example used what may be termed a global all-enveloping cost pattern to establish the target cost. It is also a fact of engineering life that cost patterns exist for the component parts that together make up a complete product design. Why therefore are they not determined in increasing numbers and over an increasing spectrum of the engineering industry? Obviously, the ultimate number of cost patterns is probably infinite; this fact, however, should not deter anyone from making a start with one's own products and component parts thereof. Cost patterns would seem to be an approach to costing which satisfies the essential criteria referred to earlier in this chapter. Other approaches to the problem have and are being developed and will be referred to later.

The essential feature of cost patterns is that they are based on real costs and a number of these evolved out of necessity are to be found in Pugh (1974). The cry, often heard, that the expression of costs in real terms is illusory and that escalation quickly makes them unrealistic and inaccurate is unjust; the 20:1 variation referred to earlier is a very salutary reminder of this fact. Having established patterns in the first place, it is a relatively simple matter to update them. Figures 32.4, 32.5, and 32.6 show the patterns for injection moulded components in various materials and are illustrative of this point.

Mouldings are to commercial standards for a minimum quantity of 5,000, and the cost boundaries A and B represent respectively complex and simple mouldings.

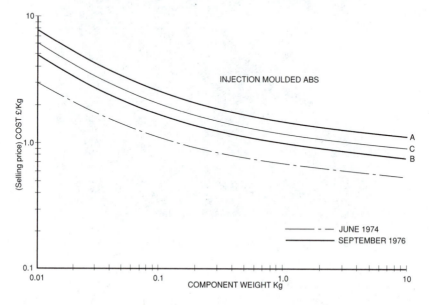

FIGURE 32.4. Polymer Component Costs

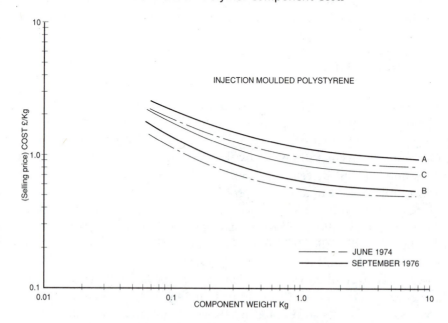

FIGURE 32.5. Polymer Component Costs

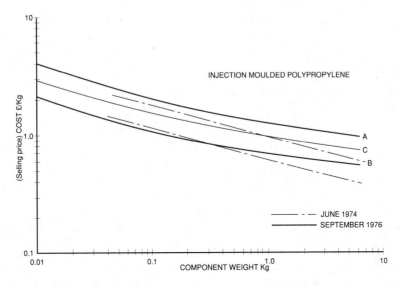

FIGURE 32.6. Polymer Component Costs

Typically A would represent a 5,000 off quantity of a complex moulding, B a simple moulding of 250,000 off, and C a simple moulding of 25,000 off.

One problem facing any designer is always the choice of material and manufacturing route for the component in question; if, for example, we assume that a number of materials are technically equivalent for our component, how does the designer decide which route to specify? Usually he relies on his experience, which is probably limited, or he may shop around seeking quotations from suppliers and from within his own organisation; to generate a number of alternatives is extremely time consuming and costly and has one certain drawback: the design is held up.

If the designer were issued with cost pattern information covering a wide spectrum of processes and updated on a regular basis, decision making would not only be improved but the designer would also gain a better understanding of actual costs. If for our mythical component an ABS moulding might be considered suitable, then the cost could be rapidly assessed by reference to Figure 32.4—probably two minutes' work. The designer would have virtually an instant answer to compare with other instant answers obtained from the cost patterns for other suitable materials. Application of this method to all components as the design proceeds at least keeps one in the cost ball park to fairly close limits and reduces the risk of the design exceeding the target cost.

Whilst cost pattern coverage is slowly increasing, the main impetus in this area must come from industrial organisations and designers themselves; this is where the bulk of the data lies. The following procedure is to be recommended for the preparation of cost patterns:

1. Obtain data from within the company and without, with as wide a size of weight range as possible. These components must have like classification and

quantities, such as steel castings to a material specification, fabricated components in stainless steel, or turned components in brass to BS 249.

2. Check that the cost information is to a consistent base, as the data will primarily have been recorded for production, estimating, and cost control reasons and is unlikely to be specifically for design use.

3. Establish the likeliest parameters (usually weight, area, enveloping volume, and the like), and plot on a cost-rate basis (cost/parameter versus parameter basis). This is to be preferred as almost all components exhibit a strong rate variation not always apparent with, say, a straight plot of cost versus parameter.

4. Only shift to less readily obtainable parameters if the simple parameters do not yield results.

5. If tight and definitive patterns do not emerge, this usually means one of two things—either wrong choice of parameters or correct choice of parameter with excessive scatter, making the establishment of the pattern elusive. In the case of a strong pattern with excessive scatter, the likelihood is that the scatter components represent exceptions to the cost pattern and subsequent investigation of the scatter data will usually reveal that the components are poorly designed for manufacture with excessive processing content.

A word of warning and encouragement here: do not be put off by colleagues who tell you it is a waste of time; in my experience the end result always justifies the effort involved.

Figure 32.7 shows diagrammatically the process from start to finish of a product design. At commencement the initial components package, plus assembly cost and other peripherals, usually exceeds the target cost; as the iterative process of design proceeds the component numbers, from the methods of manufacture change until ultimately, on completion, the final summation of component costs plus peripherals,

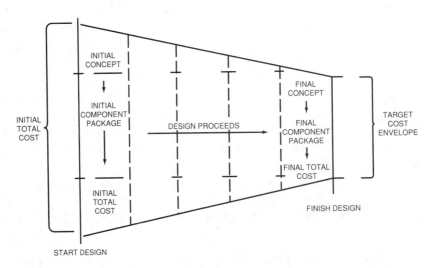

FIGURE 32.7. Approaching the Target Cost

matches the target cost. It is in the detail build up of the package that component cost patterns give the most assistance.

OTHER METHODS OF COSTING AT THE DESIGN STAGE

In reviewing the methods of costing available to the designer and discounting the traditional methods of the estimator which are much too time consuming, one finds that very few shorthand methods have emerged over the years. One or two may be construed as being applicable to complete products in a general sense, whilst others deal specifically with particular processes or components.

Of the former, Starkey (1976) describes a method of product costing based upon a Pareto distribution, which in engineering cost terms means that the bulk of the cost of a product lies with a small percentage of the component parts. However, the bicycle example used to demonstrate the Pareto technique is doubtful, not so much for an established design but certainly for a design which does not yet exist. He states that 'starting from the very broadest concepts with no details yet established' (in other words, no design), he rapidly deduces the cost of a bicycle and its component parts in great detail by assuming that the new design of a bicycle will have the same component parts as those which already exist. This means, does it not, that he is using someone else's design as the basis for costing: therefore, a design does exist, which conflicts with the earlier statement. Even so, the method is based on relative or comparative costs, and if in the example quoted the costs of tyres and tubes in relation to the other components vary significantly, then the whole costing exercise becomes less accurate.

Henry (1975) outlines a system he has developed and used in the teaching of design. Based primarily on a series of cost factors which themselves are based on ISO tolerance grades, the system tends towards the traditional methods of estimating in that each stage of, say, the machining of a component is considered, and hourly rates for labour figure large in the equation. The example given would indicate that the system is tolerably accurate and useful. The distinction between manufacturing cost and selling price is not, however, clear. In the example given of the costing of a pressure vessel, the estimated costs arrived at by using the system are presumably at manufacturing cost level, and this cost is compared directly with three quotations for the same vessel, which no doubt represent someone's selling prices; therefore, the two are not directly comparable.

Some recent work was carried out at UMIST (Brunn, 1976) that concerned a technique for rapidly evaluating the production times of turned components on various types of lathes and is extremely useful. Cost curves are given for capstan, centre, and NC lathes. This system of costing, evolved with the production estimator rather than the designer in mind, is based upon a system of complexity factors established according to the degree of difficulty of the machining operation. The UMIST claim of fifteen minutes for a typical hand calculation certainly makes it a system of great usefulness, although from the designer's viewpoint it is probably less useful as in order to carry out the above calculation accurately, the component has to be fully defined.

Experience of the German system of costing—VDI Richtlinien (1969), which is a comparative costing system, and fairly easy to use—would indicate that it is not particularly accurate for the simple reason that it relates everything to the cost of mild steel, and therefore if the relative costs of other materials vary, the system error increases.

The *Materials Optimiser* (Waterman, 1974), published by the Fulmer Research Institute, contains a section on the costing of components and also a section on the costs of materials which is updated on a yearly basis. The section on component costing gives some useful information on various processes and several cost patterns, although rather regretfully they are for specific components, which limits their usage.

CONCLUSIONS

It is hoped that the foregoing adequately portrays the current state of the art in respect of component and product costing, although doubtless other techniques do exist and are used behind closed doors of industry. If the author has thrown new light on the methods available or indeed confirmed views already in existence then this chapter will have succeeded in its objective.

As previously mentioned, costing has infinite boundaries and in order to broaden the picture already created a table is included which contains references to other publications having relevance in specific topic areas.

ACKNOWLEDGEMENT

The author would like to acknowledge the assistance given by R.G. Rhodes of the Loughborough University of Technology Library, particularly in respect of the publications listed in Table 32.1.

TABLE 32.1.

Bockstiegel, G. (1974)	'Technical and Economic Aspects of P/M: Hotforming. Part 2. Comparative Cost Analysis.' *Powder Metallurgy International.* 6(4) (compares alternative production process costs for a ring gear).
Butler, M.J. (1976)	'Mould Costs and How to Estimate Accurately.' *Plastics and Polymers* 41(152) (April) (determination of mould costs by regression analysis, based on ten parameters inherent in all mouldings, stresses the accuracy of the system).
Coates, J.B. (1976)	'Tool Costs and Tool Estimating.' *Production Engineer* (April) (parametric estimating based on a regression approach, press tools and injection moulding tools). *continued*

Denholm, D.H. (1974)	'Cost Estimating for Job Shop Work.' *Industrial Engineering* 6(3) (March) (structural job shop estimations show wide variations, a specially designed worksheet narrows the differences).
Flick, E.W. (1973)	'Use These Two Charts to Determine the Cost of Coatings.' *Material Engineering* (November) (cost of paint finishes on basis of solids content, thickness of deposit, cost per mil thickness).
Kay, S.R. (1974)	'Factorial Estimating System: Factest.' Paper presented at the Third International Cost Engineering Symposium, London, October, Association of Cost Engineers (factors relating undefined elements of a chemical plant to main plant items).
Mathews, L.M. (1974)	'Want to Cut Costs to the Bone?' *Electronic Design,* 22(6) (March 15) (the learning curve, effect of design change on learning curve).
Reynolds, D.E.H. (1974)	'Decreasing Welding Costs in Heavy Fabrication.' *Welding and Metal Fabrication* 42(3) (March) (cost pattern depicting weld costs versus size of weld).
Roberts, C.H. (1975)	'Cost Reduction Starts with the Designer.' *Engineering* (October) (relative cost of surface finishes, comparative costs of tooling and labour for production of holes by various methods, relative cost of tolerances).
Watson, G.E. (1973)	'How Much Will That New Product Cost?' *Machine Design* 45(5) (19 April) (predicting the cost of a new product before design, parametric analysis based on a regression equation).

REFERENCES

Brunn, P.J. (1976). 'The Selection Output and Utilsation of Lathes.' Ph.D. thesis, UMIST.

Henry, T.A. (1975). 'A Costing System Suitable for Use in Teaching Design.' *Proceedings of the Institution of Mechanical Engineers.* Education and Training Group. vol. 189 (June).

Pugh, S. (1974). 'Manufacturing Cost Information: The Needs of the Engineering Designer.' *Proceedings of the Second International Symposium on Information Systems for Designers.* Southampton: University of Southampton.

Starkey, C.V. (1976). 'Costs for Designers: Cost Estimates.' *Engineering Designer* 2(3) (May/June).

VDI Richtlinien (1969). 2225, sec. 2, June.

Waterman, N.A. (1974). *Fulmer Materials Optimiser.* Fulmer Research Institute.

Wilkins, G. (1976). 'Billion Dollar Baby.' *Daily Telegraph Colour Supplement* (September).

CHAPTER

33

Engineering Out the Cost

ABSTRACT

This chapter considers products old and new, existing and nonexistent, from a view-point of a strategy to minimise the possibility of costs being engineered into a product. It describes such a strategy, gives some details of structure and method, and illustrates the basic instability and hence uncertainties of an unstructured approach to artefact production. The chapter concludes by considering that a single strategy or structure is equally applicable to both the aforementioned product types in order to bring about product cost efficiency.

The publishers wish to thank the following for permission to republish this chapter. From *Proceedings of the Sixth International Annual Conference on Design Engineering* (Birmingham, 1983), 121–128.

INTRODUCTION

The question of engineering out the cost will be considered with a new design, without product precedents and, alternatively, with an old or existing design, which by implication means that it is, or has been, in production. The remedy is the same in each case—soundness and thoroughness of approach allied to a range of appropriate techniques and methods which allow cost to become visible but, more important, the avoidance of the creation of cost cankers. Since the title of this chapter means, by definition, that cost has been engineered into a product in the first place, it is the avoidance of the latter which will form the basis of this discussion, since many people propagate techniques which enable one to attempt to cure the disease once it has been diagnosed. Cost prevention is therefore my theme.

SETTING THE BUSINESS SCENE

It should first of all be recognised that businesses are about the creation of products for markets to satisfy needs—and therefore for people—and are not about engineering design, production, and sales as individual activities operated in isolation. If they are run in the aforementioned manner then to talk of cost removal becomes irrelevant, since any attempt at cost improvement will be minimal and probably illusory.

The product or range of products is thus the unifying theme of the business, and unless the business is structured on a modern design basis, adaptable, and flexible to competition, new technology, change, and the like, one can forget the question of cost: there will not be a business left in which to incur costs!

So it is a question of products for markets, and there is an urgent need for industry to adopt a structured approach to business on a comprehensive design base, in order to bring about improvements in efficiency and become, let alone remain, competitive.

To avoid 'engineering the cost in' in the first place is highly dependent on establishing a structured approach. Granted, one can eliminate or reduce costs later, but is the cost base within the product base sound? Most companies have little recognition of this difficulty. Let us now look at a structure which works and leads to the production of efficient products.

Since every factor of business, from production machinery through to the tea lady, affects costs and is ultimately considered in cost terms, the effects of a systematic approach to design and particularly the lack of it are immediate and electric. I have come to this conclusion after many years of studying the topic and also practising design in the comprehensive manner I am about to describe and expand upon.

ESTABLISH A STRUCTURE

All product-producing businesses have either one or two primary characteristics or, alternatively, a combination of each.

The primary output of a business may be an engineered product such as a washing machine, motor car, or pump; alternatively, textile yarn, peas, and paint utilise a significant amount of process machinery, which has to be designed to a cost in order to achieve business efficiency. This is illustrated in Figure 33.1.

In the first case, the product itself is primary and the engineering design as such is directly applied to the product; in the second case, engineering design as such is applied to the process machinery and is thus secondary. However, in order to achieve efficiency and therefore minimum cost, the remedies are the same.

The business should be structured along the lines of the design activity model shown in Figure 33.1, which is given with the appropriate description.

What has this to do with cost, efficiency, and thoroughness, one might ask? I will explain further. It is my experience in putting new products on the market that unless one adopts a structured systematic approach, one cannot hope to compete with products from overseas, and therefore the question of cost, as previously stated, becomes irrelevant.

APPLICATION OF STRUCTURE

Let us now look in detail at the imposition of such a structure on a business and how this can lead to the desirable attributes just described.

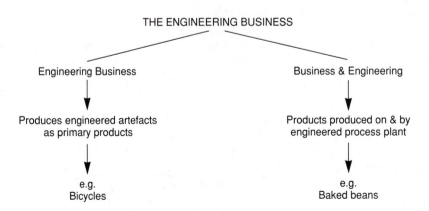

FIGURE 33.1.   The Engineering Business

NEW PRODUCT WITH OUTSIDE COMPETITION

Within the structure just described, the following stages must be followed:

1. a. Market analysis and research using techniques of competition and parametric analysis to fully understand competition in terms of technology and cost (Pugh, 1982),
   b. Parallel investigation of the real need situation in order to ultimately reconcile with (a),
2. The evolution of a comprehensive product design specification upon which to base the future work (Pugh, 1978),
3. The establishment and evolution of concepts from which a choice has to be made (Pugh, 1981).

And so on in sympathy with the structure already described.

It is in these first three stages that the cost scenario is established for the business and its products. Granted one can ruin any business or product by poor detail design and manufacture; currently, in the United Kingdom, however, manufacture is undergoing a revolution on the efficiency and competitiveness front with little regard, if any, for the design of the products to be made within these efficient systems. Therefore, one will succeed in only marginally improving (very necessary) the overall business efficiency, and the results could still spell disaster, aided and abetted by modern technology. I concentrate primarily on the structure and methods necessary to achieve a sound product and business base and not, in this instance, on the panoply of methods to be applied once we have an overcost product. So let me examine more closely stages 1, 2, and 3.

First Stage: Market

Parametric and competition analysis should concentrate not only on establishing the cost patterns for the competition; it should, ideally, try to relate such cost patterns to the basic parameters inherent in any design and different for all designs. The market and the competition must be thoroughly researched.

One must know or deduce why people are doing things in certain ways or, more important, *why they are not*. Detailed analysis to highlight cost cankers in competitor's products should take the following form if possible, based on a dissection of the actual products or, if this is not possible, on estimates based on whatever tangible evidence is available.

- Number of subassemblies: why? why not more? why not fewer?
- Number of major components subdivided according to technology,
- Number of minor components,
- Variety of components,
- Number of junctions,
- Number of joints,

- Overall weight and volume,
- Always a synthesis of parameters to suit the product under consideration,
- Complexity of components—expressed in discontinuity terms (Pugh, 1978),
- Commonality of materials,
- Ease of assembly.

All of the above are directly linked to cost generation in a product.

Second Stage: Product Design Specification

From the foregoing, establish a product design specification which incorporates the primary elements described in the design activity model, including the results of the analysis in the areas just detailed (Pugh, 1982) (Figure 33.1).

Third Stage: Concept Establishment and Selection

In considering the establishment and selection of concepts on a concept or criteria basis via the mechanism of a large matrix, in addition to the criteria derived from the product design specification which forms the basis for such evaluation, the criteria just outlined must be incorporated in order to have any chance whatsoever of minimising cost—by not designing it in.

Generally speaking, to minimise the discontinuities in a system or product leads to minimum cost.

When designing a new product, the opportunity must be taken to do this quite systematically. I have already made reference to several articles and papers which detail the approach I have described. If we consider simply the application of cost-reduction techniques and methods in the accepted fashion, then without the systematic approach just described, the effect will be as shown in Figure 33.2. To achieve a completely sound product base, inclusive of cost, will require working back to the market need situation. If this had not been done properly initially, the attempts at cost control will be based on

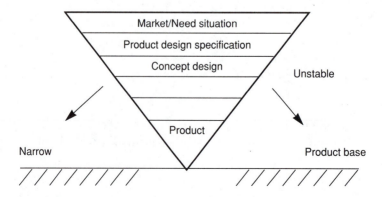

FIGURE 33.2. Business out of Balance with the Market

quicksand. As the figure shows, we can have a situation of unstable equilibrium (a statement designed to warm the hearts of all mechanical engineers). This latter hypothesis leads directly into the alternate situation to the one just considered—old or existing product with outside competition.

OLD OR EXISTING PRODUCT WITH OUTSIDE COMPETITION

Having already made out a case for the systematic evolution of a new product, I have also per se made out the case for the old or existing product. Whilst there is, of course, always a requirement for detailed cost reduction, it is my experience *particularly with existing products* that the product base is unsound for a dynamic market situation, and it is the recognition of this situation that gives rise to my previous statements.

In all cases one has to go back to the beginning and set up the whole situation or structure I have described previously; not to do so increases product or company vulnerability. To not do so may prolong the life of a product and hence a company, but in all probability this will be a short-term gain.

So, one major first step, if the product is overcost, not selling, or whatever, revert to a structured approach, establish a sound product base, and proceed accordingly. In this case one will then have a better chance of engineering out of product cost by revealing just where and why it is engineered in.

Remember the most effective, efficiently produced product that *does not* sell is a disaster in just the same way as one that is overcost.

So as with the previous case, the remedy is the same. If we do this we achieve the effect shown in Figure 33.3. The figure does not, I believe, require any descriptive matter. To achieve a balanced product or business with cost in control and not *engineered in* requires a more professional and radical approach to business structure than has hitherto been in evidence. This requires a switch of attitudes on everyone's part: it is, however, up to management to take the lead. If this does not happen, there will be very few artefact-producing businesses left in existence.

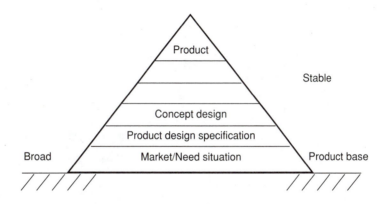

FIGURE 33.3. Business in Balance with the Market

Whilst I have spent a good deal of time over the years establishing approaches to costing and the evaluation of methods to control costs, I have come to realise, perhaps belatedly, that without the establishment of a sound product base most attempts at cost reduction or control are very inefficient operations. It is relatively useless applying ointment and medication to cure the cost cankers if the source of them is not fully understood. It is to be hoped that this chapter throws some light on the problem, enhances understanding, and aids the practitioner with the responsibility. Always start at the beginning, regardless.

REFERENCES

Pugh, S. (1978). 'Quality Assurance and Design: The Problem of Cost Versus Quality.' *Journal of Quality Assurance* 4(1) (March).

Pugh, S. (1981). 'Concept Selection: A Method That Works.' *International Conference on Engineering Design*. Rome: ICED.

Pugh, S. (1982). 'A New Design: The Ability to Compete.' *Proceedings of the Design Policy Conference*. London: Royal College of Art.

CHAPTER

Quality Assurance and Design: The Problem of Cost Versus Quality

ABSTRACT

Two basic factors at the heart of the design-quality-cost trio are considered in detail in this chapter through product design specifications and simplicity of design. The type of information required by a designer in such specifications is formulated, and some tools to enable simplicity to be made practical are presented and discussed. Suggestions for the direction of future work and thinking are given in order to improve the mutual understanding of engineering designers and quality assurance practitioners.

From *Quality Assurance* 4(1) (March 1978), 3–6.

INTRODUCTION

In considering the question of cost versus quality, related to design, I have attempted to examine the question from the point of view of a practising designer, without forgetting the academic aspects, and to avoid the twin pitfalls of zero impact and maximum confusion. The views stated thus stem from design practice and not from quality assurance specialisation; they not only portray some of the increasing problems of the engineering designer but also provide some pointers to the solution of these problems. This chapter represents a distillation of the works of many people in the field of quality assurance, correlated with my own experience in a wide field of engineering design. There is a surprising measure of agreement.

Such comparisons have revealed that industrial quality assurance practitioners are continually saying that, first, adequate product design specifications are the foundation for enhanced quality at minimum cost, and second, simplicity of design is the key to both quality and cost. Papers from such practitioners attempt to grapple with these almost intangible problems and to provide various frameworks for at least dealing with specifications whilst fighting shy of simplicity, which is readily understood by most people but is the more intangible of the two. Academic or semiacademic papers on the subject of quality and reliability, with certain notable exceptions, either deal with the subject almost to mathematical abstraction (which designers often find difficult to put into practise) or refer to specific components when these are known. Most researchers seem to be concerned with components, and, whilst such work is essential in its own right, the questions of specifications and simplicity are possibly more representative of the true nature and magnitude of the problem. This chapter will, I hope, be considered an attempt to unscramble and understand the intangibles.

THE PRODUCT DESIGN SPECIFICATION

General

Many writers on the subject—I am one of them—consider that the lack or inadequacy of the product design specification at the commencement of design gives rise to the problems of cost and quality. Corley (1974) states that 'it is essential that the design engineer be provided with a set of design requirements which are *valid, accurate and complete*. If marketing does not provide a specification for the new product design or major re-design, ask marketing to prepare one or else prepare one yourself for marketing approval. Start your design engineers off on the right foot with a good product design specification.' Langley (1970) confirms this view in a slightly less positive manner, and Bones (1976) refers to a reliability specification. I suggest that the product design specification, if properly formulated, should become a complete statement of the product requirement with its constraints and, by its very nature, should lay the foundations of quality, reliability, and cost and hence represent a statement of targets in these areas. To refer to reliability specifications and quality specifications in their own right without relating them to an overall product design specification may be very misleading and result in undue concentration of effort in these areas without relating them to the total picture. Duggan (1976) refers repeatedly to quality specification and states that 'a more rational approach to the problem is that of establishing a quality specification, and then attempting to design to this specification with the objective of producing the cheapest product with the required quality and reliability.' He makes no reference to an overall product design specification which should encompass all the factors which impinge upon and influence the product design.

Practising designers from industry attending our lectures and seminars during the past seven years have always been asked, 'How many of you commence the design of a product with little more than the statement "Let's design a nuclear power station" and without even the skeleton of a product design specification?' Sadly, even today, the majority of answers are the same: little information is given to designers at the commencement of design other than rather incomplete statements of purely technical requirements. Should we then be surprised at the quality versus cost problems faced by British industry, let alone the problems of wrong products for the wrong markets? One approach to the solutions of these problems must be through educational channels. Engineering teaching and thinking must dwell on this area; otherwise, the situation described above will persist indefinitely.

Within the Engineering Design Centre at Loughborough University of Technology, a good deal of thought and time has been devoted to the question of product design specifications, and one fact is crystal clear: insufficient thought in this area at the commencement of design has usually led to the production of less than adequate designs.

Formulation of Specification

A product design specification should contain statements related to a variety of factors, and, in my opinion, the *same* fundamental factors are applicable in any area of engineering design (Pugh, 1974). This list of fundamental factors has been applied to designs for a variety of machinery, ranging from process plant, construction equipment, and printing presses, to pneumatic and hydraulic valves, pumps and the like. Always the primary factors are the same, but in each specific design the answers to the questions posed are different and thus each specification becomes unique.

Briefly, a product design specification should contain statements on the factors listed in Table 34.1. The factors in the table influence all product design specifications, and the brief list of typical questions addresses issues of quality, reliability, and cost. They cannot, or should not, be added later as they often are because such addition may spoil the design and so result in a product that is not as good as it should be.

TABLE 34.1.   Product Design Specification Factors

Factor	Typical Questions or Comments
1. Performance	How fast, how slow, how often, continuous, discontinuous?
2. Environment	Ambient conditions applicable to product during (1) usage, (2) manufacture, (3) storage, (4) installation, (5) transport Hazards Type of personnel using equipment Degree of abuse.
3. Life expectancy	Life at which performance level measured against what criteria.
4. Maintenance	Is regular maintenance available or desirable? Does the market into which the product will be placed usually operate planned maintenance? Is it aware and appreciative of the benefits of a whole-life costing philosophy?
5. Target product cost	How has it been established and against what criteria? Can it be proved that the target cost is attainable?
6. Packaging	Is packaging required? Effect of volume of product on shipping cost, etc.
7. Shipping	Is product for home market or will it be exported? If exported, by what means of transport? How will it be handled?
8. Quality and manufacture	Mass, batch, or one-off production; effect on tooling policy; manufacturing facility and investment required to support the product.
9. Size and weight	Any restrictions on size and weight? Too heavy for a single lift? Does it need to fit into an aircraft hold? of which aircraft?
10. Aesthetics, appearance	This is always important because, no matter what the product, customers see it first before they switch it on.
11. Materials	Are special materials likely to be needed? Are certain materials forbidden in the market for which the product is intended, such as aluminum and its alloys on exposed surfaces in coalmining equipment?
12. Product life span	Production span of product affects tooling policy and manufacturing methods.
13. Standards and specifications	Is production to be designed to British standards or international standards or both?
14. Ergonomic aspects	All products at some phase of their life interact with people; the extent and nature of likely interactions should be elucidated.
15. Customer	Customer's preferences and views

continued

Factor	Typical Questions or Comments
16. Competition	Nature and extent of existing or likely competition. Does specification being formulated show serious mismatches with competition? If so, why?
17. Quality and reliability	Levels of reliability and quality expected by the market, necessary to ensure product success. Difficult to specify in quantitative terms, particularly in relation to innovative products. DD (Drafts for Development) 10, 11, 12, 13, 14, 15, and 16 help in this area and become particularly useful when the design concept has been established.
18. Shelf life	What is the likely shelf life? Under what storage conditions? Deterioration of packaging and product itself. Perishable components.
19. Processes	Are special processes to be used during manufacture? Do in-company or specific process specifications apply?
20. Testing	Is every product to be tested, or is sample testing likely to be satisfactory? If every product is to be tested, what test equipment is to be used? Who will witness tests? What is the allowable cost of testing each product?
21. Safety	Mandatory and desirable levels of safety related to product market area. Balance between safety, accessibility, and limitation of usage.
22. Company constraints	Is product a departure from current company practice? What effect will this have on personnel within company? Constraints of current production plant.
23. Market constraints	Feedback from market indicating unacceptability of components. Conditions relating to overseas markets. Acceptability or unacceptability of technological level. Is market ready for a technological 'jump'?
24. Patents	Any patents in force in the area of specific product design? Necessity for patent search?

Of all the quality assurance specialists Brewer (1977) comes closest to meeting the above need and recognises that statements of requirement in terms of quality and reliability should 'form a clause in the document which defines the whole project in outline—a document sometimes referred to as the "target specification."'

SIMPLICITY: THE KEY TO BOTH COST AND QUALITY

The second demand from quality assurance practitioners, because of their experience, is for simplicity. Of all the intangible, indefinable words used in relation to engineering design, *simplicity* is probably the most difficult to grasp. The *Concise Oxford Dictionary*'s definition of the word *simple* does, however, throw some light on the use of the word in design: 'Not compound, consisting of one element, all of one kind, involving only one operation or power, not divided into parts, not analysable.' To attempt to persuade engineers in general to think simply is a task in itself, particularly in a world which, in engineering terms, seems to acclaim complexity in its own right whilst denigrating simplicity. Yet it is generally acknowledged by such practitioners that concentration on simplicity leads to enhanced quality at lowest cost.

Carroll and Bellinger (1969) remark that 'Perhaps the major single contributor to operating reliability is simplicity' and 'The superior design is one which encompasses the necessary operating and protective functions with the absolute minimum number of components and connections!' Crouse (1967) says, 'If you simplify something you automatically increase its reliability and cut its cost.' Mihalsky (1975) advises, 'Keep the design simple in order to have good reliability.' Whilst there is overwhelming evidence to support the 'simplicity' argument, how can the designer respond in practise? He needs something much more tangible than pleas to work with. Wood (1966) tells of how his company, having applied a 'zero defects' strategy to production problems, then turned to engineering and to design in particular. The 'simplest is best' philosophy was the basis of the approach, but the company appears to have gone beyond the plea stage. It has defined the most important factor in an approach to simplicity as the exact converse—complexity.

Complexity has been quantified (here we have something the designer can grasp at any stage of the design activity, from the vaguest of concepts to the finality of detail) and defined as being dependent upon

Number of parts, P_p,
Number of different types of parts, P_t,
Number of interconnections and interfaces, P_i,
Number of functions that the product is expected to perform, f,
Where complexity factor $C = K(P_p P_i P_t)^{1/3}$,
K being a constant of convenience.

This approach seems to have met with success for electronic equipment. The result may be summarised as follows: the lower the complexity factor, the greater is the equipment's reliability; the lower is its cost, the higher is its quality. Such an approach must be worthy of consideration in the field of mechanical engineering; it would be particularly useful in the conceptual stage of a design at any level.

Approaches to simplicity of direct use to the designer have been one of my main interests for a number of years, including a 'zero parts' approach to design, based upon the hypothesis that if a function can be attained by designing and manufacturing absolutely nothing, the quality and reliability will be infinite at zero cost. Since nothing costs nothing, it is also inflation-proof! Whilst this is an impossibility in practice, if the designer approaches his task with this attitude, his chances of achieving the simplest design are greatly enhanced.

Let the designer combine this approach with an approach called 'load lines' (Pugh, 1977) and he will soon make inroads into complexity and more rapidly and effectively approach the goal of simplicity. The techniques of 'load line' or 'discontinuity analysis' is based upon the simple doctrine that the shortest distance between two points is a straight line. In engineering terms this means that bent pipes are more costly than straight ones, short members are cheaper than long ones, plain shafts are cheaper than stepped shafts, and so on. The conscious avoidance of discontinuities in systems usually leads to the following benefits:

- A reduction in the total number of parts;
- A reduction in the amount and complexity of machining required on the parts that remain;
- A reduction in material usage and usually an overall reduction of weight;
- *A reduction in the costs of components and hence overall machining costs;*
- A reduction of assembly time because assembly should be easier when components are few;
- A reduction in the number of drawings required;
- An improvement in the overall appearance of the product;
- *An improvement in overall machine quality and reliability.*

Whilst these are the main attributes of such an approach, I am concerned, being undeniably biased, that the approach seems to have so little against it. Perhaps readers of *Quality Assurance* can constructively point out its disadvantages.

Deduction of a complexity factor, from simple examples using the 'load line' technique, shows that the lowest value of complexity factor is attributable to the simplest design in 'load line' terms. We thus have two bases from which to establish a simplicity criterion. One, load lines, is qualitative (at present); the other, complexity factor, is quantitative.

THE FUTURE

Here are a number of suggestions for possible future work on the understanding of design. They necessarily have implications for levels of quality.

Approaches to Reliability

Up to the present day, approaches to reliability and the quality stemming from it seem to fall into two categories:

- A general systems approach to reliability as a whole, as evidenced by the 'drafts for development' which ultimately converge from the total systems to the specific components;
- Approaches based upon specific mathematical techniques for analysis and understanding of data obtained from samples and trials.

The above two categories (and there may be more) both rest on the same premise: they assume that a design exists or, if it does not, that a new design will be based on previous precedents in the field. Detailed knowledge of the design is therefore implied. Admittedly, many 'new' designs are based on developments of existing designs, and only in that sense are they new. For them data on failure, field problems, and the like will assist, and any 'new' product design specification can contain statements on quality and reliability based upon such data.

However, in a limited number of cases, a design specification for a new product will give little or no indication of how the design is to be achieved or, in fact, of which technologies will ultimately emerge in the design. How then can reliability and quality be specified in other than very general terms? Carter (1974) recognised this fact and stated, 'A big departure (from previous practice) creates an untimely new situation in which mathematical processes cannot operate and in which mathematical processes are inadequate.' In fact, to allude to the reliability of, say, electronic equipment in a design specification for a machine which ultimately is devoid of electronics may be particularly misleading to the designer. The initial electronic implication may ensure indeed that electronics are part and parcel of the final design and it could well be assumed that electronics are a necessity—which they may or may not be. If such an approach leads to the best solution or design at the lowest cost and to maximum reliability, all will be well, but if a better solution exists which does not involve electronics, the designer may have been constrained from even looking for it and then all may not be well. We need therefore to give even more thought and effort to consider the effect of Q&R specifications in terms of 'leading' a design. Perhaps all that is needed is correct and unambiguous wording of the particular section of the specification, but the problem is usually not quite as simple as this.

Standards for Quality and Reliability

Various industries and market sectors have British and international standards related to them for reasons which are necessary and obvious, standards which recognise the differing requirements and expectations in those industries. There ought to be also similar specifications of quality and reliability. Standards of quality and reliability acceptable as the norm in some industries would also be acceptable in others with the proviso that the market as such also found the same standards acceptable. Now clearly there is no uniformity of Q&R standards for all industries, and I suggest that, just as technical standards related to specific industries are appropriate, so Q&R specifications similarly related are also necessary and appropriate. To this end research and thought should be given to categorising markets, industries, and products in some manner appropriate to their needs and rate of evolution and hence to producing suitable standards for Q&R. This would lead to a better understanding of the subject by designers at large and make such standards more useful to them. Just as we differentiate between allowable stresses for pressure vessels and mobile cranes, because they are of necessity different, so we must start to put aside the general case for Q&R and become more specific. Lindley

(1976) seems to support this view. It is not a simple task, but one which must be undertaken, if only to enable the designer to respond to the ever-more numerous pleas made to him.

CONCLUSIONS

I have attempted to show how the quality of design might be improved at minimum cost, and I have stressed the relation of design to quality and reliability. I hope I have encouraged my readers to think constructively on the subject.

REFERENCES

Bones, R.A. (1976). 'Designing for Reliability.' *Engineering*, 798–801.

Brewer, R. (1977). 'The Statement of Reliability at the Design Stage.' *Proceedings of the Conference on Information Systems for Designers* (pp. 77–81). Southampton: University of Southampton.

Carroll, J.T., and Bellinger, T.F. (1969). 'Designing Reliability into Rubber and Plastic AC Motor Control Equipment.' *IEEE Transportation Industry and General Applications* (IGA-5, no. 4, pp. 455–464).

Carter, A.D.S. (1974). 'Achieving Quality and Reliability.' *Proceedings of the Conference of the Institution of Mechanical Engineers* 188, 201–207.

Corley, G.W. (1974). 'A Product Safety Checklist for Design Engineering Management.' ASME Paper, December 13.

Crouse, R.L. (1967). 'Graphic Trees Help Study of Reliability vs. Cost.' *Product Engineering*, 48–49.

Duggan, T.V. (1976). 'Quality and Reliability in Design.' *Engineering Designer*, 15–19.

Langley M. (1970). 'Design Badly Now and Pay Later Is a Bad Maxim.' *Engineer, Lond.* 10, 49–50.

Lindley, B.C. 'Design for Reliability in Service.' *Engineering*, 475.

Mihalsky, J. (1975). 'Design-to-Cost Versus Design-to-Customer Requirements Versus Design-for-Safe-Operation: Is There a Conflict?' ASME 75-SAF-2.

Pugh, S. (1974). 'Engineering Design: Towards a Common Understanding.' *Proceedings of the Conference on Information Systems for Designers* (pp. D4–D6). Southampton: University of Southampton.

Pugh, S. (1977). 'Load Lines: An Approach to Detail Design.' *Production Engineer* 56, 15–18.

Wood, W.M. (1966). 'Designers Zero in on Product Reliability.' *Product Engineering*, 100–101.

CHAPTER

Give the Designer a Chance: Can He or She Contribute to Hazard Reduction?

ABSTRACT

This chapter continues with the theme of 'design for X' done through a consideration of elements omitted from the PDS. An example is provided to illustrate the ramifications of this omission, which occurred during the formulation of a PDS, and to introduce the idea of defect classification—such as defect of design, defect of manufacture, defect of method. In order to effectively deal with defects, some form of classification is helpful as part and parcel of systematic design.

From *Product Liability International* 1(9) (1979), 223–225.

Designers these days are in danger of being overloaded and engulfed with more and more responsibilities for the avoidance of practises and procedures likely to lead, ultimately, to defects in the products they are designing. Admittedly, the degree of overload, bewilderment, and glassiness of eye varies according to the level of design at which a particular person is operating and also the type of industry with which he is concerned. Nevertheless, talk to designers used to dealing in facts, figures, definitions, and the like about a contribution from them towards a reduction of potential hazards, reduction of the chances of defects arising, call it what you will, and they will agree with you that they should be able to make a contribution, but how? It is a salutary lesson to anyone concerned with this area to try to explain just what it is that you want done (no need to justify the necessity for the task) in words, pictures, or any form of communication, in a language that is interpretable by designers and which is translatable into positive action, and thereby effect improvements to subsequent designs.

Currently in the United Kingdom we are in danger of committing two cardinal sins. First, we are in danger of emulating the American practise in pursuit of better products—with little evidence that better products are emerging upon the U.S. scene in spite of more than a decade of product liability legislation. Second, we are tending to alight upon detailed arguments in order to be able to 'do something' rather than trying to define the problem in the first place—and from such definitions evolve solutions to the problem.

This chapter will leave the first statement to rest in anticipation that further informed comment will either confirm or deny it, preferably supported by reasoned argument, but will take the second statement and examine it in the context of the designer.

It is now characteristic of U.K. engineering practise that we are more concerned with problem solution rather than problem definition: the elegant solution to the nonproblem is still highly regarded in many circles. What is, in fact, needed is a professional approach to all product design, which commences with problem identification and specification and leads to better ultimate solutions. Do we not have a tendency to start in the middle of a problem and see where it leads? In terms of the designer's contribution to hazard reduction, should not a start be made by examining and defining the problem of product liability in factual terms? Break it down into its constituent parts, evolve a specification of the problem or subproblems, and then, *and only then*, evolve concepts and ultimately detail the chosen concept to produce the whole. In other words, do we not have all the ingredients of a classic design situation? Why is it that we always tend to start things in the middle and then, metaphorically speaking, wander about in a daze looking for direction, which creates confusion and costs money and markets? Dare one say that it is primarily because of our amateurish approach to professionalism?

In the end analysis it is the designer and other doers who have to come up with detailed solutions to problems, resulting in new and, hopefully, better products. The increasing complexity of products, the advance of technology, and increasing competition from abroad are sufficiently difficult to digest, even with a 'start at the beginning' project. Thus, with our tendency to start in the middle, it is surprising that we ever produce satisfactory products for the marketplace.

We, therefore, have two basic propositions:

- Commence product designs from the beginning—that is, start with a product design specification which formulates the problem in breadth and depth together with its constraints. To do so, it is suggested, will lead to better and safer products and hence reduce the risk of hazards.
- In terms of product liability, we must again commence an examination of this problem from the beginning, which will lead to greater understanding and, hopefully, from this understanding, measures or tools that designers and others can utilise in their day-to-day operation—the outcome also being better and safer products.

It is indeed questionable whether there are in fact two basic propositions since the projected outcome of each is identical. Will not, therefore, a thorough investigation and understanding of the first lead to greater understanding, and hence effective progress, in the second? Is not the definition and approach to the solution of the product *liability* problem one and the same thing as the definition and approach to product *design*, since surely the former must emanate from the latter, since without a product, the question of product liability does not arise? Might it be suggested that if this is considered to be a reasonable hypothesis, we run the risk of killing two birds with one stone? So, to give the designer the best chance of responding to increasing demands, it is recommended that attention be paid to the formulation of the initial product design specification (Pugh, 1978). Give him, make him seek, information in the areas appropriate to his product and thus minimise 'defects of the specification' which, ultimately, may give rise to defects of a more specific nature. This smacks of professionalism.

Before expanding further on this question, a brief comment upon the American scene may be appropriate, particularly in respect of information, method, or technique which the designer can effectively use to reduce hazard. A recent report succinctly covers the past decade of U.S. product liability experience (Interagency Task Force, no date available), a report which incidentally, although written by and for lawyers, is surprisingly readable. It contains 686 pages, and twenty are devoted to product liability prevention techniques, primarily describing *what* to do but with no hint as to *how*. This is the magnitude of the problem. Needless to say, no mention is made of product specification.

It is suggested that 'defects of specification' may account for many 'defects of design' and even for subsequent 'defects of manufacture' and that to consider defects under these three headings may lead to a better understanding of the whole question and, ultimately, lead to some tools for the designer and of course others. We, as engineers, have to come to grips with this problem as there is little evidence that lawyers will do so, except at our expense (the published U.S. experience supports this view).

To initiate the topic of 'defect of specification,' an incident relating to a product currently on the market should prove of interest. During the formulation of the original product specification for the 'Giraffe' site placement vehicle, it was decided to make it the safest vehicle on the market from the viewpoint of stability. Since any handling device using forks relies upon gravity to retain its load, the risk to life and limb increases with height and reach, both in terms of the operator and people in the vicinity of the vehicle. Being the first in the field of this design, it was decided to provide a system of operation such that any increase in the extension of the boom at a fixed elevation or, alternatively, depression of the boom at a fixed extension beyond predetermined limits, would automatically override the operator, leaving him with two options: (1) let the load remain at a position in space but still in safety or (2) shorten the boom and withdraw the load into the safety envelope. A system was designed to perform in this manner and is fitted to every vehicle: it works! The essential feature of the system is that the manual lever-operated valves which control the two motions described, automatically centralise when the danger zone is approached, overriding the operator's input.

Two considerations of the above concerning the designer and potential hazards now spring to mind. The first (and this is a true story) relates to such a vehicle operating with an Irish company. An official complaint was lodged with the manufacturer that the levers fitted to the control valves were not strong enough and indeed some had broken. Subsequent investigation revealed that when the safety system had come into operation, with valves centralising, lights flashing, and a buzzer sounding, the operator had attempted to extend the load range by an application of pipes to the levers. Naturally, it required long pieces of pipe and a tremendous effort to even partially overcome the system; the net result, however, was the bending and ultimate fracture of the levers, and hence the request for stronger levers! If the accident had resulted from this practise, under which defect classification would the incident be considered?

- *Defect of specification*   Should it have been foreseen at the outset that operators would attempt to override the safety system by the application of pipes to the valve levers?
- *Defect of design*   If the control levers had snapped, injuring the operator, should they have been made strong enough to accept pipes of infinite length?
- *Defect of manufacture*   Was the lever material not to specification?

It gives interesting food for thought.

Second, a vehicle of similar design has since appeared on the market with a safety system which sounds a klaxon and flashes a light in the event of intrusion of the load outside the safe envelope of operation. If these are ignored, the danger zone can be entered and instability could result. Should an accident occur with this vehicle, it is interesting to speculate whether it would be considered a defect of design or specification. As the Giraffe preempted this latter system, would it give rise to questions in the area of 'reasonable user expectation'?

Yes, further consideration of the aforementioned defects in specification, design, and manufacture is warranted, particularly if in the end the designer's lot and his chances are to be improved.

ACKNOWLEDGEMENTS

Thanks are due to the Liner Company Limited for permission to describe the Giraffe incident.

REFERENCES

Interagency Task Force on Product Liability (no date available). *Final Report.*
Pugh, S. (1978). 'Quality Assurance and Design: The Problem of Cost Versus Quality.' *Quality Assurance* 4, 3–6.

PART VII

Design Research

COMMENTS ON PART VII

The previous parts described total design—the outcome of the type of design research that Stuart Pugh advocated and practiced. Now in this part Stuart turns to the issue of design research itself. This is of timeless interest. There is today much controversy and uncertainty about the productive approach to design research. Many academics, almost always lacking industrial experience, wish to make design research as close to research in the physical sciences as possible. However, often this only results in the aping of the language of research in the physical sciences and a concentration on minutia. Stuart understood the problem very well, and in these three chapters sets forth his views.

Stuart points out that in considering design research we must make the following three distinctions:

- Total design and partial design,
- Static products and dynamic products, and
- Technology-specific methods and generic methods.

Much research into design processes and methods is primarily (perhaps only) applicable to the partial design of static products in some specific technology set. Thus it has only very limited applicability: attempts to apply it elsewhere are usually dangerous. Such methods can be useful in their particular domain. However, they are best viewed as subsets of total design, providing the right details in the context of the more important decisions that have been made by applying the generic methods that Stuart emphasized.

Stuart recognized the vast difference between design and design research, on the one hand, and scientific research in the physical sciences, on the other hand. It was a battle that he fought from the time that he started teaching in Loughborough in 1970, on through his joining the University of Strathclyde in 1986, and until his death in 1993. Traditional academics almost always tried to blur the distinction, with the effect of making design education and research off of the mark—much to Stuart's consternation. Design is inherently complex, while in research in the physical sciences we isolate some relatively simple question that then can be concentrated on to the exclusion of everything else. 'Is the low toughness of high-strength steel the result of plane-strain embrittlement?' It might take a while to answer, but this is a simple question. Design is too complex and interrelated for design research to be decomposed into such simple questions. Attempts to do so throw out the baby with the bath water.

Many of our problems with design and design research and education stem from the attitudes that Stuart addresses in this part. In reading these chapters look for the distinction between universal design methods and specific, limited methods, between design research and physical-science research, and between design and research. Note also the quotations from Dr. G. Taguchi and Stuart's observation that the two of them were converging. Their works, although on rather different aspects of product development, stand out as among the top few in importance during the last half of the twentieth century.

—Don Clausing

INTRODUCTION TO PART VII

Design research is an emotive topic which has been prominent and in the ascendant during the past twenty years. It really falls into two categories: (1) technology specific and dependent methods and (2) methods independent of product or technology. Much design research activitiy in many instances does not differentiate between the two categories: it remains confused and by and large makes little contribution to design practise.

There are now emerging a few examples of the systematic application of the total design process shown in the Figure 37.4. In the main, these usually relate to the partial design of a product (see Chapter 37, Engineering Design: Unscrambling the Research Issues).

The picture is further clouded when we look at the longer term and we introduce the questions of static and dynamic product concepts (see Chapter 38, Long-term R&D Outcomes: Will They Miss the Market?

The subdivision of design into static and dynamic concepts, partial and total design, aids understanding of design per se and hence contributes to the outcome of design research being applicable. This subdivision is particularly relevant and applicable in terms of technology independent design methods. The applicability and outcomes are nonspecific and are not as obviously applicable as technology dependent methods. This part hopefully throws light on the whole issue.

36

Research and Development: The Missing Link—Design

ABSTRACT

This chapter discusses research and design as mutually interactive and interdependent activities with development equally applicable to both; the topics are examined from industrial and academic viewpoints and a hypothesis is proffered to logically relate research to the activity of design. It concludes with an example of research and design interaction.

From *Proceedings of the International Conference on Engineering Design* (Copenhagen: Heurista, 1983), 500–507.

INTRODUCTION

Even today most companies concerned with artefact production have departments or divisions titled research and development (R&D), which seems to mirror to some extent the applied research function of the universities. The fact that the word *design* is rarely used is, in my opinion, illustrative of the confusion surrounding the whole question of research, design, and development, both within industry and academia.

In many companies and institutions the sweeping expression *R&D* is used to embrace all technical functions from fundamental research to product design. It suits many to refrain from seeking a formal structure and approach to what, at this point, I will call the 'product evolution activity'; the lack of structure whilst in some respects being fun to work within and operate, leads to less than successful competitive products.

This chapter considers the question of product evolution based upon structured design activity and into which research and development both fit logically as necessary components to achieve success. The necessary delineation of research is discussed together with the necessity for achieving cross-fertilised research; it is demonstrated that development activity can be called upon, carried out, set in motion, and utilised at any stage of design or research activity and not just as is sometimes depicted at the beginning of a linear progression from research through development to design.

THE RESEARCH QUESTION

In the majority of the U.K. universities and latterly the polytechnics, research activity is keenly sought after since the mechanism and means for the measurement of the same exist and it is a laudable, traditionally respectable activity. That this almost exclusive concentration on research, even applied research is against the national interest has been recognised in the remit of the Engineering Council (1982) which has been established to oversee the totality of the technological institutional structure, both learned societies and universities and the higher educational sector generally. Technology and product design figure large in this remit. I see as one of its main objectives undertaking the almost megalithic task of balancing research activity by some measure of product design activity. The view across the higher education system reveals on the one extreme 'research is design' or 'research isn't design, but something others do' to 'research is part and parcel of design activity'—a minority view. Continuing education of young people against even an applied research background will only continue to add to the confusion since, on arrival in industry, the now qualified engineers will attempt to emulate their mentors and perpetuate the R&D function and so the cycle continues.

Speaking as one who teaches and practises design and also nurtures design and research students, I can state quite unequivocally that research is research, design is design, and development is development: whilst each activity area is essential for product success, they should never become mutually exclusive and separate. They are not interchangeable; they are all different. It is the almost complete lack of recognition of the distinction between and the interactions of these functions that in reality leads to the design of noncompetitive products.

Walker (1978) states, 'Research is often confused with engineering, but it often has a more limited purpose. Research is the collection and analysis of data which may or may not be pertinent to the solution of an immediate engineering problem. Research by itself is not engineering; it is not even a necessary part of engineering. There are times when sufficient facts are not available for the designer/inventor to provide a solution to an engineering problem, and the research must be done to provide the missing facts.'

There is a growing awareness of the need not only for the necessary distinctions to be made, but also that having made such distinctions, action is taken both by industry and education to redress and change the situation. Cooper (1980), in my opinion begs the question of R&D: 'In spite of its strategic importance, product development is plagued by an unacceptably high failure rate. For every six hours spent on industrial R&D five of these are spent on products that fail or are cancelled.' Who is teaching or training people to do this sort of thing? I consider that it stems quite firmly from the engineering educational system and its obsession with research, leading to the confusion of design and development when for the first time, from a research activity one has to evolve a successful product. Research activity based on scientific method is just not geared to do this. Commander (1982), in publicising the award of a university research grant to aid the design and manufacture of new knitting cams and machine mechanisms, says, 'Past research has not provided all the expected results, possibly because each sub-system, for example the cams, have been treated in isolation with invalid assumptions made about input and output forces. Hopefully it all adds up to improved industrial knitting machines for the future.' I doubt it. To design by research alone is almost an impossibility. The *Oxford English Dictionary*'s definition of research is as follows:

1. To search into a matter or subject or to investigate or study closely,
2. To make researches or to pursue a course of research,
3. To seek a woman (in love or marriage) (sixteenth- and seventeenth-century definition),
4. To search again or repeatedly.

If one, therefore, carries this out in a specific narrow area of, say, knitting technology, how can one possibly aspire at one and the same time to design a competitive product? Research activity causes information to be revealed. Work to be done to provide further information related to the area of study, the product design activity, needs much more than this; it needs a strategy, structure, and need in order to attempt to achieve success. Such activity utilises research; it is not in itself research. The random engineering of a well-researched topic or idea will not lead to product success except possibly by chance. The product will be vulnerable to competition.

Muster (1981), again in discussion of the U.S. scene, states: 'Support of technology and research has been heavily skewed towards science rather than engineering and that for engineering has similarly skewed towards research rather than towards solving problems closer to improving manufacturing productivity.' Whither goest the United States also goest the United Kingdom.

Schott (1981) shows a remarkable similarity between U.K. and U.S. performance and considers that development is the application of research and embraces product design.

Bennet and Cooper (1982), in discussing the vulnerability of U.S. products to competitive attacks and speaking solely in terms of management education, states: 'Academic research must be re-orientated to deal with the problems and issues facing industry. Relevance as well as academic vigour should be [addressed]. . . . The audience for academic research should be far more frequently the practitioner and not solely other academics.'

In articulating the problems, possibly the United States is changing its view of R&D and is starting to act; the United Kingdom will no doubt start to follow suit after a suitable time lag.

How do we therefore progress towards a more balanced, logical, and structured approach and in so doing rectify the situation? There is currently only one solution in my view and that is to move towards a product design core activity which utilises research, as previously stated. We should educate and train engineers for competitive product design and development *as well as* training specialists for research.

Setting aside for the moment the question of development, let me further dissect the research scene, not necessarily in terms of what it is but rather what it should be, in order to attempt to unravel a series of complex interactions. Commencing with the hypothesis that there are two types of research:

1. Conventional
   a. Pure research, in search of fundamental knowledge and understanding;
   b. Applied research, where artefacts resulting from (a) are researched further to aid understanding; techniques specific to products and technologies;
2. Nonconventional
   a. Techniques to aid, abet, assist any design activity irrespective of technology, such as costing, material selection, and so on;
   b. Techniques specific to design core activity to enable design activity to be better carried out, such as competition analysis, concept selection, evaluation, and optimisation.

The traditional R&D function sweeps in 'conventional research' in a scrambled manner both in industry and academia. The two items under 'nonconventional' do not yet carry much weight in relation to the former, nor should it since it is very much in its infancy. Because of the nature of successful design activity—multidisciplinary, the nature of the corresponding research activity should also be multidisciplinary. Thus, by definition (and in practise, I might add), one has conflict with type 1 in that such research is usually carried out by the narrower specialist, an essential ingredient for success. To attempt to carry out work in the type 2 area requires a new breed of multidisciplinarians, hence a major dichotomy: where do we obtain such people, since currently they come from academia steeped in type 1 research activity? There is no doubt in my mind that we have to accommodate design on at least the same footing as research and hence breed some multidisciplinarians who can either carry out the design function in industry or the type 2 research activity in the universities in conjunction with industry.

The design activity I refer to in this chapter is as depicted in Figure 36.1. Figure 36.2 shows research activity of the two types superimposed in a logical manner. The research activity feeds both technique and technology and hence the design core, which enables new and better products to emerge as a result of the activity.

Within the Engineering Design Centre we are concentrating our efforts in the type 2 areas, although occasionally we have ventured into the type 1 area, sometimes almost by default. I see no problem with development since as an activity it is equally applicable to research or design. By the *Oxford English Dictionary* definition, development is

1. Gradual unfolding or fuller disclosure or working out the details of something,
2. Evolution or bringing out from a latent or elementary condition,
3. Growth or unfolding of what is in the germ (etymological),
4. Gradual advancement through progressive stages.

So development work can be carried out at any stage of research or design activity. I see this as a very simple premise in the context as defined. Thus ideas stemming from research might be developed further. I see no confusion in this delineation. Effective and efficient operation of the design activity can cause research aims and objectives to be changed and vice versa: they are mutually interactive. Development is all-embracing. Let me now consider an example illustrative of the previously stated R&D situation unscrambled by design activity.

PIGMENT EXTRUSION PLANT

Many industries utilise screw extruders as part of their process: for example, the screw extruder is central to many plastic moulding processes. The screw extruder has been subject to world-wide research, design, and development activity; of all forms of extruder it is the most thoroughly researched.

Some years ago pigment extrusion was examined where control of the parameters of pressure, flow, and temperature and their interaction is essential. Since many of the pigments are thixotropic with non-Newtonian characteristics, the whole question of extruder design to cope with these pigments was considered. As a result of design activity it was concluded that a form of ram extruder would be more likely to have the desired characteristics than a screw type, yet a comprehensive literature search revealed the almost complete absence of research data with which to design. Consequently, a research programme was instituted and the necessary information and laws deduced as a pre-

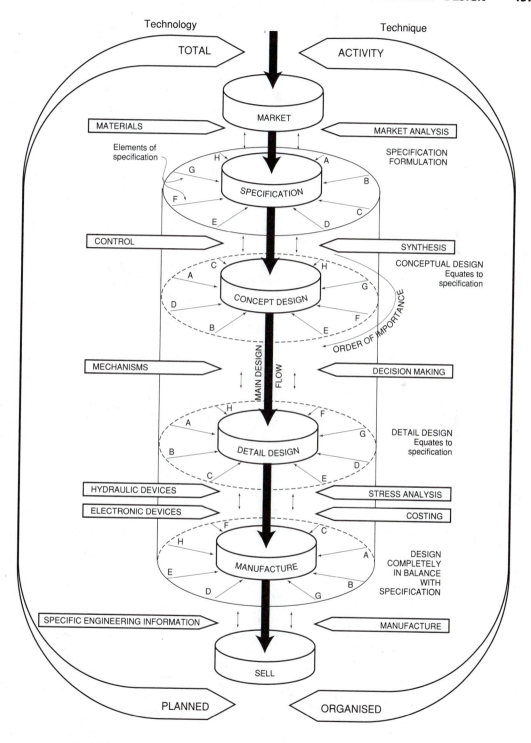

FIGURE 36.1. Design Core Bounded by Product Design Specification

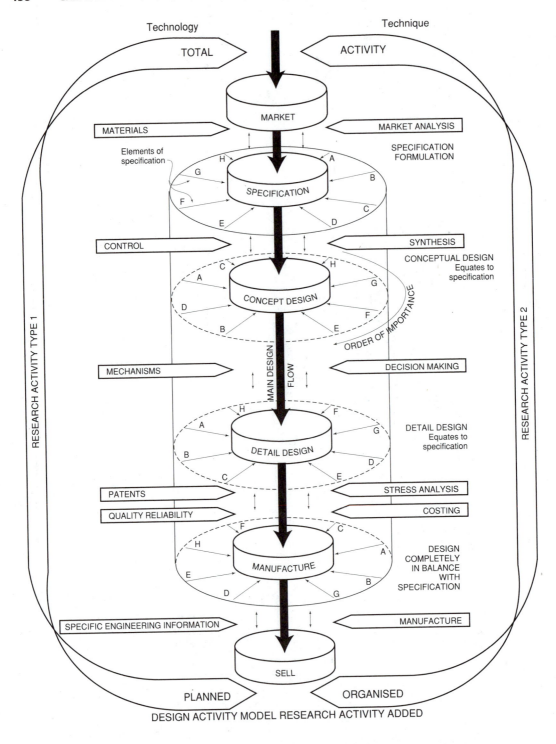

FIGURE 36.2. Design Core Bounded by Product Specification

lude to design. An extruder was designed on the basis of the research activity and has been working successfully in the industry extruding pigments.

This is one of many instances where design activity triggered research activity, which in turn led to the design of the new machine which is now under development. In this case, without the design activity it is unlikely that the necessary research along the lines discussed would have been undertaken.

DISCUSSION

It is hoped that this chapter has contributed to the resolution of the research, design, and development argument and that the interactions are somewhat clearer. Certainly, the clarity we have gained from such understanding has improved student performance and also their product designs. Both design students and industrialists appear to welcome the distinction made between design and research. The problem still remains to attract the multidisciplinary student into the type 2 research, which to have the best chance of success means in reality that the person concerned should have some years experience in competitive design in a comprehensive manner. This, in effect, means attracting engineers from industry back into the academia for a limited period of time, certainly no easy task in the United Kingdom.

REFERENCES

Bennet, R.C., and Cooper, R.G. (1982). 'The Misuse of Marketing.' *McKinsey Quarterly* (Autumn), 52–69.

Commander, M. (1982). 'SERC Cash for Industrial Knitting Research.' *Financial Times*, April 4.

Cooper, R.G. (1980). 'Project New Production: Factors in New Product Success.' *European Journal of Marketing* 14(5/6), 277–292.

Engineering Council (1982). 'Policy Statement.' September.

Muster, R. (1981). 'Relevance of Engineering and Management Education.' *C.M.E.* 28(8), 82.

Schott, K. (1981). 'Industrial Innovation in the United Kingdom, Canada, and the United States.' Pub. British North America Committee.

Walker, E.A. (1978). 'Teaching Research with Teaching Engineering.' *Engineering Education* (January), 303–307.

Engineering Design: Unscrambling the Research Issues

ABSTRACT

This chapter takes as its starting point the premise that applied research outcomes, whatever their labelling, manifest themselves through products. It considers the product as a totality comprising many elements requiring for success technological and non-technological inputs, the appropriate direction and balance coming from research of different kinds—that is, engineering research, engineering design, design process research, market design research, and market research. The interactions and differences between these are discussed and a case made for design process research as being central to them all.

From *Journal of Engineering Design* 1(1) (1990), 65–72. Courtesy of Carfax Publishing Company, Cambridge, MA.

INTRODUCTION

The choice of this chapter's title is quite deliberate since it is considered that one of the major problems facing the academic and industrial communities both now and in the future is the harnessing of scarce resources to bring about better products through the understanding and development of new technologies and products which manifest themselves through sound engineering design. It is, in my view, primarily at the research stage of the modern design process that we lack clarity of understanding, and this chapter will attempt to throw some fresh light on the issue.

THE PRODUCT

If it is accepted that products are made up of many factors, based on many technologies including many nontechnological things, then industry is concerned with the total design of such products; this might be represented as Figure 37.1, where it should be

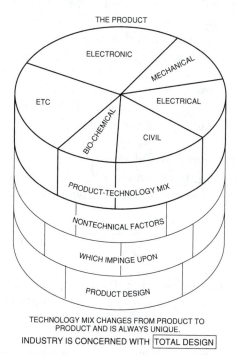

FIGURE 37.1. Factors Making Up the Total Design of a Product

noted that the technological and nontechnological factors always change from product to product and are always unique (Pugh, 1988).

It is a reasonable assumption that research in the technological zones in the upper band is mainly applied, flowing from the fundamental researches which of themselves are not applied, such as particle physics and molecular interaction in chip design.

How do we arrive at the best areas for this research, the ones most likely to bear fruit in a competitive product market? Judging by the number of new product failures, the answer must be 'not very well.'

INITIATION OF RESEARCH

Research is initiated from two primary triggers (Pugh, 1989):

- *Market pull*   Occurring as a natural outcome of customer or user aspirations or where the technology tends to lag the market, as with CAD system design;
- *Technology push*   The outcome creating markets and where the market lags, such as digital audio tape.

From the industrial revolution onwards to the middle of the twentieth century market pull and technology push were indistinguishable as the rate of technological evolution, and hence product evolution was slow relative to today. Thus market feedback was ponderous (if at all), and technology push dominated. An artefact was 'designed' to incorporate the technology and elements that had been conceived so far, which, supported by the appropriate calculations and drawings, were then fed into manufacture, which then figured out a way to make it. There was little, if any, market pull.

The onset of competition, and the massive increase in the numbers and types of products available, have all led to the voice of the customer (VOC) becoming increasingly vociferous and, as always, primarily important. The issues of technology push the market pull, and indeed the nature of the mix is neither well understood nor well controlled. So the technological and nontechnological mixes required of products became increasingly scrambled, and any research in engineering or the market must also address these user aspirations quite systematically, otherwise the outcome of that engineering or market research is likely to mismatch the user expectations.

ENGINEERING RESEARCH

Research in engineering and its associated artefacts is a valid activity in its own right with the proviso that orientation is given to the elements of the artefacts being researched: they should be modern product, technology, and market oriented, but how do we achieve this?

Consider, for example, the current moves to establish special programmes of research in high-speed machinery or, in 'mechatronics.' For the outcomes to be successful they must address the market, the voice of the customer, wherever or whoever he or she may be, and they must do this *before* the research commences and continue to interface and communicate with the VOC during the evolution of the research.

But how do we do this? How is it to be done? Do market researchers know how to analyse markets and increasingly sophisticated VOCs in order to influence such research? One suspects not, as the continuing erosion of the manufacturing base in the United Kingdom would seem to indicate. It is, however, not solely a U.K. problem; it is a European problem: Germany lost the camera market, and the United Kingdom lost the motorcycle market.

If we do not understand the problem of the market/technology mix and design new methods to handle these issues, then ultimately we will become, at best, second best in all markets.

Our engineering research must therefore become more focussed—easy to say, difficult to do.

In terms of the total product model of Figure 37.2, if we do not achieve this bridging between these factors, then we will continue to manufacture nonviable products with the schism of Figure 37.2. Not only will we have a split between the technology and the customer, but our technology mixes will become disarranged and different technologies will start to compete for dominant positions with little influence from the customer. A sophisticated recipe for disaster?

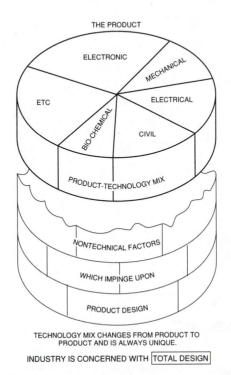

FIGURE 37.2. Effect of Shift of Technology Mix on Product Viability

How are we to avoid the schism, if at all? The current Western response to Eastern competition seems to be focussing upon 'engineering design research' the outcomes of which will 'save the West.' This is highly unlikely unless we understand the ramifications of this research and do not confuse it with genuine engineering research which is *discipline* and *technology dependent* and which may initially only have distinct market orientation.

Is there an area of research which in fact will bridge the gap between the two halves of our idyllic product, Figure 37.1, where everything is in balance? Very clearly the answer is yes—the area of design process or product delivery process research.

DESIGN PROCESS RESEARCH

Studies of the design process or product delivery process have been endemic for at least the past twenty-five years and at a lower level for many decades. It is my firm view that it is the outcome of design process research which should lead first to understanding and then to development of new methods which if applied *systematically* will bridge this gap (see Figure 37.3).

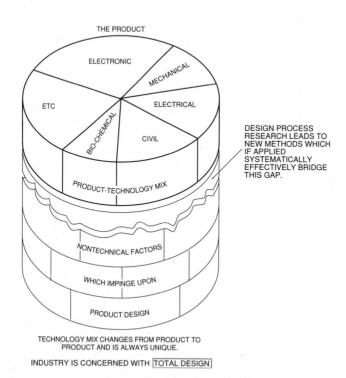

FIGURE 37.3. The Bridging Effect of Research on
Product Outcome

Examination of the outcome of such work spanning the whole of the design core which incidentally is universally axiomatic, at least amongst practitioners, reveals the following truths:

- The methods are independent of technology or discipline;
- The methods that are used in the design core, flow from one to the other: one feeds the other;
- The methods are reversible and can be used generatively for new products or diagnostically to analyse competition;
- There are few mismatches between Western and Eastern methods;
- Degree of implementation of methods depends on visibility and understanding of the methods: if it is not clear and people do not relate to it, it will not be implemented;
- Overlapping, coherent methods lead to reduced product cycle times, utilise fewer people, and lead to better products of robust design;
- Feedback from the above lends focus to the applied research directions to be undertaken.

A sequence of methods is shown in Figure 37.4, but not exhaustively.

It is therefore considered essential that the activity of applied research must interact with and utilise the methods of design process research not only to make that outcome successful but also to maintain the direction of the applied research activity per se. As yet this does not happen, nor in the light of current events does it appear likely to happen.

From whatever the genesis of the applied engineering research—whether industrial or academic—as research moves towards the viable product, it must interact with design process methodology. This vital, necessary, and legitimate interaction may then truly become 'engineering design research.'

ENGINEERING DESIGN RESEARCH

The interaction between design process research and its methods and applied engineering research needs further investigation and may also have a parallel in market research terms.

To return to the market pull/technology push argument. If with market pull, technology lags, and if with technology push, the market lags, then if applied engineering research emanates in the main from technology push, it is vital and essential for product success that the market acceptance of that product is anticipated. It is suggested that this anticipation with increasing certainty can occur only through interaction with the methods and outcomes of design process research in order to give it that market orientation. Engineering design research per se should be the custodian of these interactions.

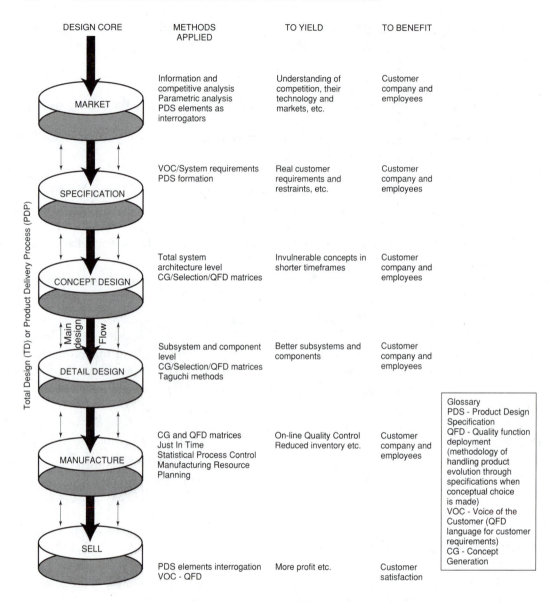

FIGURE 37.4. Methods in Common Use Linked to the
Design Core Strategy

Likewise for market pull products where the technology lags. There must be *the* corresponding interaction between design process research and market research for just the same reasons as for engineering—continuing product success.

In Figure 37.5, it is suggested that for technology push products the engineering research is strong, whilst the market research is correspondingly weak and vice versa for

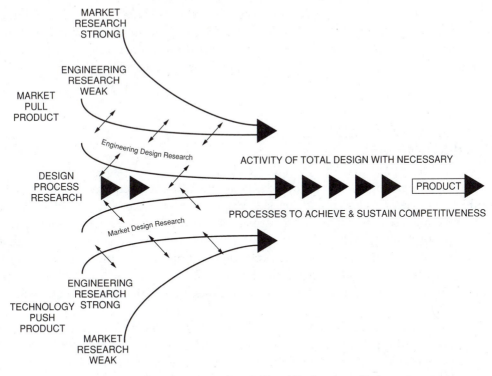

FIGURE 37.5. Market Pull and Technology Push
Interactions

market pull products. We will achieve balance only by recognising these facts and putting in hand the means to rectify the situation.

Applied engineering research must therefore interact with and utilise the methods of design process research, and by so doing this will give rise to a diffusion of understanding and evolution across boundaries. Boundaries between technological disciplines have deliberately not been brought into this discussion, primarily to avoid confusion in the mind of the writer, but an attempt has been made to provide a catalyst for such diffusion to take place—design process research outcomes.

MARKET DESIGN RESEARCH

Similarly, market research fits into the pattern. Whilst the absolute necessity for market research is accepted without question, do the market researchers have adequate tools and methods for dealing with the competitive situations both now and in the future? As with engineering, it is suggested not.

It is suggested that a new area of 'market design research' is necessary which will also interact with the outcomes and evolution of design process research, and it too will provide an adequate vehicle for communication and diffusion amongst disciplines and into market research per se. There is some evidence that this may be starting to happen as market research academics and others address product and process design issues, and in some instances such as my own, useful diffusion is starting to take place. Perhaps new methods will evolve via genuine design process research.

RESEARCH EMPHASIS

To achieve and maintain success in total design in the future will require the cross-coupling of both the engineering design and market design research areas, the former interacting with engineering research and the latter with market research, and the whole coming under the umbrella of design process research. Each area will require the appropriate emphasis, and it is recognised that engineering research will consume most of the necessary research resource *but* relative to the other areas, the gearing is low. If the outcomes of the engineering research appear in products which mismatch the market, then the resource will have been wasted.

The outcomes of design process research requiring a smaller research resource than engineering has the potential to minimise this happening, and thus the gearing is higher: for a much lower investment, the marketability of products will be enhanced.

Do we have the political will and resolve to tackle these issues instead of trying to maintain the status quo under the guise of research in engineering design? I suspect not.

To conclude, I have tried portraying in an inadequate fashion an attempt at unscrambling research issues pertaining to design, by proffering a scenario which seems at least on the face of it to be logical and sensible. If, in so doing, I have made some small contribution to clarifying what to me is currently the very confused state of engineering design, then I am satisfied. The assumption that the engineering design scene per se is not confused is not borne out by the evidence of decline of manufacturing industries both in the United Kingdom and in the United States.

Design process research is central to the success of the whole and can be effectively addressed only by multidisciplinary teams grounded in a common coherent view and understanding of the meaning and ramifications of the total design or product delivery process (Pugh and Morley, 1988).

To quote from the management page of the *Financial Times* discussing new technology making or breaking small businesses, and in particular a 'young French company producing automated inspection machinery,' a market survey reveals some uncomfortable facts. The 'Technical design of the equipment was wrong for the German market and even in France would only cover some unattractive market niches, in addition not enough standard hard and software components were used' (Batchelor, 1989).

I think the above at least gives some credence to the case I have attempted to make, since it would appear that the market research referred to above took place *after* the equipment was in production! The difficulties were not anticipated.

REFERENCES

Batchelor, C. (1989). 'A Route of Realising Innovative Potential.' *Financial Times.* November 7.

Pugh, S. (1988). 'Total Design, Partial Design: A Reconciliation.' *International Journal of Applied Engineering* 4, 203–206.

Pugh, S. (1989). 'Research in Engineering, Research in Design, Research in Engineering Design: They're Not One and the Same Thing.' *Colloquium Research in Engineering Design* (February), I.E.E. London, Digest No. 1989/3/1/1-1/3.

Pugh, S., and Morley, I.E. (1988). *Towards a Theory of Total Design.* University of Strathclyde, Design Division.

38

Long-Term R&D Outcomes: Will They Miss the Market?

ABSTRACT

This chapter considers the processes involved in the assessment and control of research and development projects to ensure that products emanating from such programmes have the best market opportunity.

The systematic application of the technology independent methods developed for use in total design/new product development is shown to be equally applicable to research projects through the vehicle of a current research project. The necessity for such an approach is highlighted by reference to some recent industrial research programmes.

From a paper presented at the Conference on Time-Based Competition: Speeding the New Product Design and Development, Vanderbilt University, Nashville, TN, May 1991.

INTRODUCTION

Over the past twenty-five to thirty years there has emerged a coherent overlapping set of design methods which if applied rigorously and systematically in the product creation process leads to better, more robust products achieved in timescales shorter than with traditional conventional approaches. These methods are completely independent of technology, product, or indeed situation. Let us look now not so much at the methods, which I will briefly review, but at their application to long-term research and development projects.

The Methods Briefly Reviewed

Rather than just list the available methods in possibly a sterile meaningless form, particularly to those unfamiliar with them, it is preferable to relate them to what I will refer to continually as the 'design core' within the process of 'total design' (TD). Total design is the systematic activity necessary, from the identification of the market and user need, to the selling of the successful product to satisfy that need—an activity encompassing product, process, people, and organisation. The product development process (PDP) is another name for the same in the product sense, and hence Figure 38.1 indicates that TD for the product and the PDP are one and the same (Pugh, 1990b).

The design core shows the iterative and sequential stages to which a product is subject on its way to (hopefully) success in the market. All products go through these stages either randomly and uncontrolled, or systematically and controlled; unfortunately the former is still prevalent in the Western world. Even where companies have imported methods and used them, they tend to do so in staccato fashion, not coherently and systematically.

Levels

One attribute of all the methods is that they can be used at the levels of one's own choosing. For the sake of clarity we tend to define three levels: total system architecture (TSA)—the whole system; subsystems (SS)—the subsystems that go to make up the whole; and piece-parts (PP)—the piece-parts that comprise the subsystems.

It is essential in using any method or procedure that the level chosen is fully defined and that *mixed* levels are avoided. If levels are mixed, chaos and confusion will reign and not only will the methods appear not to work, but the resulting confusion will extend the cycle time for the whole process.

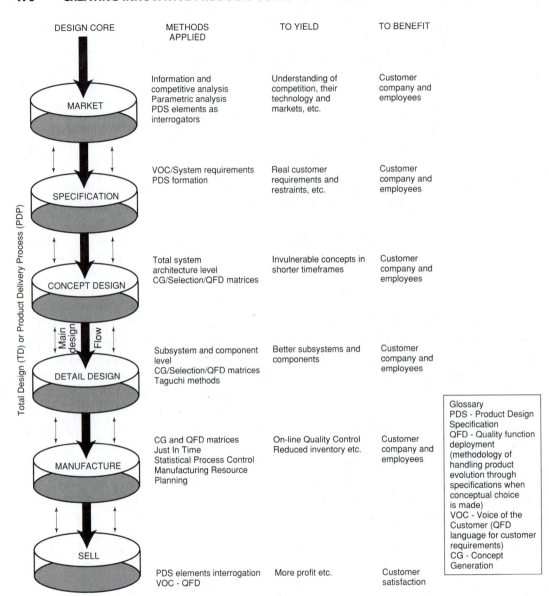

FIGURE 38.1.   The Sequence of Methods for
Successful Product Design

We can prepare a PDS for a total system, a subsystem, or piece-part, and we can carry out concept selection on a piece part, subsystem, or total system.

We can use Taguchi methods or indeed the whole panoply of methods on the production process itself. In fact, the use of coherent methods systematically is equally applicable to the 'design of production systems'—the means of manufacture of the product.

Putting It All Together

Quality function deployment (QFD) is indicative of systematisation where matrices cascade from the voice of the customer through design (in the conventional sense) and production process engineering to production operations planning. This we refer to as *basic QFD*.

Superimposing the use of coherent supportive methods at the appropriate levels gives rise to enhanced quality function deployment *EQFD*, this is depicted in Figure 38.2 (Clausing and Pugh, 1991).

The enhancements to basic QFD may be summarised as follows:

- Parametric analysis,
- Matrix analysis (features),
- Reverse concept selection (competition),
- Static and dynamic status established (Hollins and Pugh, 1990), and
- Rigorous interrogation of data using PDS elements as primary triggers

followed by

- Concept selection at the TSA, SS, and PP levels.

This brings us to the current state of the art in the use of technology-independent methods which must be used systematically to give the opportunity for product success to minimum timescales.

Thus the scene has been set by establishing a coherent, systematic design process, the outcome of which may come to nothing if our developing technologies, novel and innovative though these may be, result in products which miss the market.

Can we, should we, apply the same procedures to R&D? Let us look at this further.

Research and Development Issues

With the plethora of engineering design research work currently starting up or ongoing around the world, I recently published a paper titled 'Engineering Design: Unscrambling the Research Issues' (Pugh, 1990a). In it I attempted to differentiate between market pull products where the market research is strong and in many instances the voice of the customer is well established; and technology push products where the engineering research is strong and the market research is weak since, almost by definition, it must be lagging the market. Figure 38.3 shows this in pictorial form related to the design core, with particular reference to research emphases. To quote from this paper:

> To achieve and maintain success in total design in the future will require the cross coupling of both the engineering design and market design research areas, the former interacting with engineering research and the latter with market research, the whole coming under the umbrella of design process research. Each area will require the appropriate emphasis, and it is recognised that engineering research will consume most of the necessary research resource but, relative to the other areas, the gearing is low. If the outcomes of the engineering research appear in products which mismatch the market then the resource (engineering) will have been wasted.

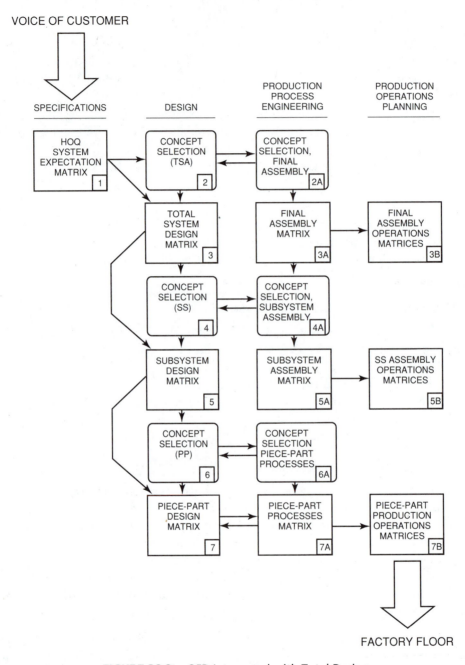

FIGURE 38.2.   QFD Integrated with Total Design

The outcomes of design process research requiring a smaller research resource than engineering, has the potential to minimise this happening, and thus the gearing higher, i.e. for a much lower investment, the marketability of products will be enhanced.

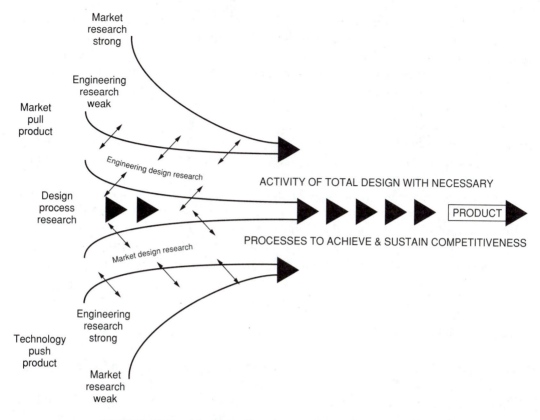

FIGURE 38.3.   Market Pull and Technology Push Interactions

So the message now becomes quite clear: if the outcomes of design process research results in methods which are coherent, are overlapping, and if applied systematically result in better products in shorter timescales, then their application to engineering research, or as it is conventionally known R&D, could result in that R&D being better directed and more focussed. This is what is meant by the synergy and interaction between engineering research and design process research through the mechanism of market design research. It is through this interaction that we will have the opportunity to achieve and sustain market competitiveness.

Tim Costello (1989), in discussing the design processes of General Motors associated with the design of a new pressure regulator, states that 'Our next technology would come from something we were already working on internally.' In many instances this may prove to be a big mistake, but worldwide we do carry out our R&D with a great deal of confidence.

Dr. Genichi Taguchi (1991), in a recent paper entitled 'Technology Development and Robust Design,' discusses the differences between R&D activities in the Western industrialised countries and Japan: 'In general, new products in Japan are usually created by a product development process. Therefore, to develop products quickly and efficiently we (Japanese engineers) need to do research on how to develop new products to meet the targets specified by the product planning effort. I also believe that, if our concepts in R&D activities do not change from the concepts of product development to those of technology development, the efficiency of our product development will be low compared with that of some western industrialised companies.'

Dr. Taguchi appears to be addressing the R&D issue; his conclusions are even more enlightening: 'I believe that the only way for companies (in Japan) to survive in this generation of high technology is to change from product-oriented research to technology-oriented research. To do so, technology research, instead of product design should be done before product planning.'

He then relates the tenets of Taguchi methods to R&D processes at least in tentative form. Perhaps we are on a convergent course? Of course, we should really continue both types of research and not abandon one in favour of the other. It is my view, and the primary objective of this chapter, to establish the relationship with R&D in a complete coherent form, which would naturally embrace Taguchi methods as part of the design core continuum.

RESEARCH PROJECTS

The only logical and just outcome from technology research is to enhance well being through the application of this research to products for the market. For the outcomes of technology research to be properly focussed, we need to subject the processes and outcomes of such research to systematic design procedures, particularly when they are being considered as the nucleus of new products. We need to do this in house, in order to ensure that insofar as is possible at a given point in time, we launch a robust, invulnerable product on to the market.

To demonstrate this I am fortunate to be able to use the mechanism of an ongoing research project at my own university, where colleagues are successfully researching and developing high temperature plasmas having properties useful to the efficient destruction of hazardous waste materials (Campbell, 1991): 'The purpose of research being carried out at Strathclyde University is to develop a small low-cost, low-energy plasma pyrolysis system which could be used by hospitals, small electronic firms and research laboratories to safely destroy biological matter and carcinogenic compounds at their point of production.'

Figure 38.4 outlines in diagramatic form a proposed system configuration.

Now let us assume that such a system has already been designed, manufactured, and marketed, to the configuration shown. What will (or should) happen in the market-place?

I would suggest the following:

1. Companies already in the waste processing field will consider the new system as a competitive threat to their business.

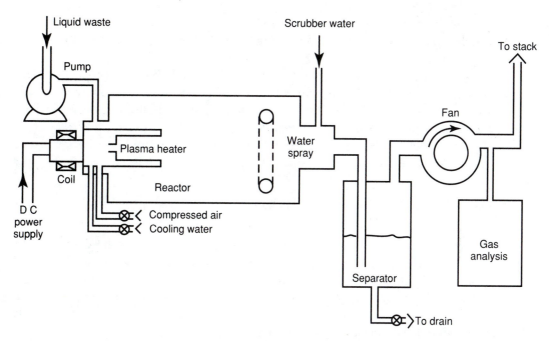

FIGURE 38.4. System Configuration for Liquid Waste Processing

2. If the competition or potential competition is what I call lively or vigilant (Pugh and Morley, 1988), they will obtain information on the new process and its performance, attributes, and omissions. In effect, they will, through the gathering of such information, have the opportunity to systematically assess this new competitive threat against existing systems, through, say, parametric analysis, matrix analysis, and reverse concept selection.

3. Having done this, they can deduce the PDS for the new equipment and fairly readily highlight its deficiencies. In effect, they prepare the PDS backwards from what to them (the competitors) is a prototype. The design process is at this stage working analytically—in reverse.

4. Having prepared this reverse PDS, they are then in a position to operate the design process generatively, to consider their own and other competitive alternatives, including the new one, and thus start a cycle which will have robustness and invulnerability as its major objective.

5. History tells us that invariably this will result in changes to the overall design, at either the TSA, SS, or PP levels.

6. The new design will be found to be vulnerable at either the TSA, SS, or PP levels, but particularly at the SS and PP levels, since to attack at these levels is usually less costly than at the TSA level.

What implications does this have for our R&D team dedicated to liquid waste processing by 'plasma pyrolysis'? Primarily it means that prior to putting a product onto the market, unless the team has subjected their 'prototype' to 'competitive stress' *internally* through the application of coherent systematic design procedures, then the product is likely to be conceptually vulnerable as described above.

The message seems to be quite clear: the team must discipline itself to act like surrogate competition to test out the invulnerability of the product at least in the short term; otherwise, the developers of alternative systems may have the edge over their system, not necessarily in terms of the core technology but at the TSA and SS levels.

All R&D projects as they approach the market must in my opinion be subject to such design processes *internally*, otherwise they will arrive on a market which

- Has passed,
- Has been overtaken by events and alternative technology, or
- Was never there.

Internalisation of initially *reverse* design processes at least gives the company the opportunity to achieve robustness and invulnerability to their own timescales, and under their own control, rather than having to perform reactively to competitive threats, where the timescales and control are out of their hands. Research programmes should thus be formulated using a design process which is systematic, particularly when a core technology is seeking applications.

Support for the Thesis

Sir Robin Nicholson (1988), in his Royal Society lecture discussing 'Success and Failure in Industrial Research Projects' and citing developments in Pilkington Bros., orginators of the float glass process, highlights the following as being the key determinants of success in industrial research:

- Correct timing of projects,
- Correct resourcing of projects,
- Anticipation of scientific needs of projects,
- Early recognition of seminal and relevant scientific advances,
- Constant monitoring of market changes during project, and
- Constant monitoring of product and process cost changes during project.

He particularly highlighted

- The influence of government legislation,
- Underestimation of competitor reaction,
- Underdevelopment of product or process,
- Not equating major improvements with breakthrough, and
- That everything can always be proved.

Schonwald (1990), of Fried Krupp GmbH, relating their experience of simultaneous engineering applied to research and development activities said, 'It is becoming clear that concentration on research alone, although research in its true sense is scarcely ever practised in an industrial institute, is feasible to a limited extent only. Rather, research and development units, and particularly central research institutes, must be structured with the prime aim of translating innovative ideas into economic success. This involves a much greater market bias than hitherto and the integration even of operative and entrepreneurial expertise. Researchers must learn that their ideas can create value only if they can be turned into viable products as quickly as possible . . . technical expertise on its own is no longer enough.'

Digital Equipment clearly recognises such problems and have adopted the design methods described earlier (Nicholls, 1990): 'In hindsight, it was clear that the Company relied too long on its manufacturing strengths and didn't adjust quickly enough to the emerging competitive determinants of the market. It was assumed that:

- You could go on competing with high technology products
- Product life cycles were longer than they really were
- Product options were a direct trade off against lower product costs.'

The efficient conversion of technology push into customer pull (or market pull) is aptly illustrated in Figure 38.3.

CONCLUSIONS

It has been argued that the processes of total design can and must be applied to R&D projects in order to achieve and sustain competitive advantage. This is best achieved by the use of coherent technology independent methods applied systematically. The key determinants identified by Pilkington, Krupp, Digital, and no doubt others support this thesis. The only alternative is to carry on as before and in the end go out of business. Bear in mind there are no short cuts.

REFERENCES

Campbell, L. (1991). *Plasma Pyrolysis of Liquid Waste.*

Clausing, D., and Pugh, S. (1991). 'Enhanced Quality Function Deployment.' *Proceedings of the International Conference on Design Productivity.* Honolulu, HI: ICDP.

Costello, T. (1989). 'A Process for Concept Development.' *Proceedings of ICED 89,* C377/314 (vol. 1, pp. 35–37).

Hollins, W., and Pugh, S. (1990). *Successful Product Design.* London: Butterworth.

Nichols, K. (1990). 'Competing Through Design: Today's Challenge.' *Proceedings of the First International Conference on Simultaneous Engineering* (pp. 85–94). London: Status.

Nicholson, R. (1988). 'Success and Failure in Industrial Research and Development.' Royal Society Lecture.

Pugh, S. (1990a). 'Engineering Design: Unscrambling the Research Issues.' *Journal of Engineering Design* 1(1).

Pugh, S. (1990b). *Total Design*. Reading, MA: Addison-Wesley.

Pugh, S., and Morley, I.E. (1988). 'Towards a Theory of Total Design.' University of Strathclyde, Design Division.

Schonwald, B. (1990). 'The Impact of Simultaneous Engineering on R&D and Manufacturing.' *Proceedings of the First International Conference on Simultaneous Engineering* (pp. 95–105). London: Status.

Taguchi, G. (1991). 'Technology Development and Robust Design.' *Proceedings of the International Conference on Design Productivity*. Honolulu, HI: ICDP.

PART VIII

Total Design:
Summary of the Whole

Here is the grand finale! This part is one long and marvelous chapter that Stuart Pugh coauthored with the social psychologist Ian Morley. It is the result of their joint research activities into design teams. This chapter represents the integration of Stuart's grand edifice of total design with social psychology—so important to the successful practice and management of design teams.

In a subsequent paper by Stuart and myself on EQFD (Chapter 16), we integrated Stuart's work with QFD. These two papers provided important integrations of Stuart's work—with social psychology and QFD. Therefore, after reading this chapter the reader might benefit by then reading Chapter 16. This chapter presents the summary of Stuart's work as it stood in 1988, and the EQFD paper (1991) integrates another important element of total design with Stuart's work.

Chapter 39, Towards a Theory of Total Design, is an outstanding integration of leading viewpoints from total design and social psychology. To the best of my knowledge there are no others like it. It is absolutely essential reading for anyone who practices total design or studies it. All work in total design should start with this chapter.

A careful reading of this chapter will be greatly rewarding. Look for

- Triple synthesis of personal design boundary, product design boundary, and organization design boundary,
- Design as a series of collective decision-making tasks,[1]

- Total design as a quasi-prescriptive theory (rather than normative),
- Creative work, rather than mere creativity,
- The principles of the organization for total design,
- The need for open disagreement in the early stages of collective decision making,
- Distinction between vigilant and nonvigilant information processing, and
- Extensive references.

—Don Clausing

INTRODUCTION TO PART VIII

This part attempts to encapsulate much of design theory and practice into a single theory which rationalises innovation, creativity, decision making, and system theory and shows the convergence towards a systematic approach worldwide with the major differences being semantic and the adoption of different start points for the design activity.

The proposals of this part—Chapter 39, Towards a Theory of Total Design—are now being widely applied in the United Kingdom, although the greatest danger is to 'cut and run'—starting at the wrong point and thereby omitting some important stages, with or without the recognition that this is happening.

To be successful you have to *know* what is happening in the activity at all times.

NOTE

1. QFD is best viewed as a systematic team decision-making process that deploys textual representations of the voice of the customer throughout the development process. For success the emphasis must be on the team decision making, not on filling out forms.

39

Towards a Theory of Total Design

ABSTRACT

This chapter represents the cumulative thinking and work of the past twenty years, made current by researching the performance of some U.K. companies against a developing total design strategy. It was recognised many years ago that researching design per se without a strategy was somewhat less than useful. The greatest drawback being the avoidance of what one means by *design*. The reader may not agree entirely, but at least I have not avoided the issue.

In working with Dr. Ian Morley, we scoured much of the world's literature in the field and find a degree of convergence apparently linked to practise. The models of design formally prescribed here gain ready acceptance amongst industrialists with a handful of semantic variations. For example, Professor Beitz of Berlin may be said to be in close agreement.

This literature is reviewed to put into context and perspective to develop what Morley and I call theory of total design.

In using the models given, we have over the years made many attempts to destroy their logic: we have failed in this endeavour.

In designing new models of the design activity many academics strive to be different and to fill in all the detail.

Since this is a design task, as with any design it is easy to be different, difficult to be complete, and impossible to fill in the detail at the beginning to give a good design.

With Ian E. Morley, Design Division, University of Strathclyde (1988).

In the real world of hardware design one can only approach these facts with well-established products (referred to in Chapter 19 as static concepts).

So to fix all the detail ahead of task definition is pointless and increasingly unnecessary. The essentials for understanding and progress are to establish the correct framework for operation and then to complete the semantics to suit your operation—a model with which you are comfortable and which you can use to convince others.

In fact from Chapter 16 it can be seen that EQFD obtains its enhancement from total design and becomes more complete and successful because of this.

INTRODUCTION

Engineering Design Is Interdisciplinary

In his recent text dealing with *The Leading Edge* in aviation, Payne (1987, p. 34) made the telling observation that 'The Wright Brothers functioned as a unit, two brains acting as one, two personalities bringing out the best in each other. Yet the attainment of the next clear cut leading edge—the Bleriot monoplane—required a whole team of dedicated men.'

The point is that engineering design is an interdisciplinary activity, which usually[1] requires the collective effort of specialists with different kinds of expertise (Morley and Palmer, 1984).

As an academic subject engineering design is also interdisciplinary, with inputs from the various branches of engineering, business studies, organisational behaviour, organisational theory, sociology, and psychology, to name only the most obvious.[2] We shall not attempt to review this literature in its entirety. To do so would require far more scholarship than we possess. And it would mean writing at least one very substantial book. Our aims are rather more modest. We shall attempt to identify some of the more important themes in this literature. And we shall attempt to show how these themes may be integrated into a general theory of total design.

Total Design

By *total design* we mean that design is seen as a broadly based business activity in which specialists collaborate in the investigation of market, the selection of a project, the conception and manufacture of a product, and the provision of various kinds of user support.

Design as Collective Decision Making

We view total design as a special form of collective decision making which is organised or managed to a greater or lesser degree. Consequently, we shall

- Analyse the process of design as a series of collective decision making tasks.
- Analyse design management in terms of the clarity of the tasks, the ordering of the tasks, the choice of methods, and the assignment of resources.
- Argue that the process and management of design is primarily concerned with products and their associated manufacturing processes, and that people and or-

ganisations are critical to success each being constrained (or bounded) by inter-action with the others. To anticipate a little, a general theory of design requires a triple synthesis of the personal design boundary, the product design boundary, and the organisation design boundary (Pugh, 1987).

- Focus on *effective* decision making There are three aspects to this. First, we are attempting to provide a model which will summarise what we know about the best kinds of practise. We shall call such a model *quasi-prescriptive*. Second, we are concerned with the nature of a collective process and what makes it effective. Third, we are concerned with the effective design of manufactured products rather than with creativity in some abstract sense.

The Failure of Existing Design Science

The statement that we are trying to produce a theory which is quasi-prescriptive is very important. We are trying to produce a model which is clearly grounded in empirical re-search. We may have to make the most of a limited amount of research (Morley and Pugh, 1987; Morley, 1987; Morley and Palmer, 1984; Akin, 1986; Wallace and Hales, 1987). But if we can provide the right kinds of descriptions of successful practise, it is very likely that the model will be used by practitioners. We want this to happen. But it has not happened so far. Design science, as Andreasen (1987) says, consists of norma-tive models which have had very little impact on European industry.

There may be several reasons for this. It is clear, for example, that industry has not al-ways appreciated the contribution of good design to business success (*Managing Design*, 1984). It has also been suggested that much of design science is too abstract, so that prac-titioners are unable to link theory and practise (Gregory, 1987). A third possibility is that much of design science describes design in the wrong kind of way. This would have the same kind of effect. A fourth possibility is that design science seems inimical to creativity and innovation. This sort of criticism is so important that we shall devote the next section to a consideration of what it means to be creative and innovative in design.

CREATIVITY, INNOVATION, AND DESIGN

Art, Science, and Technology

Mayall (1979) and Stewart (1987) have charted the history of attempts to reconcile art, science, and technology in design. Historically, there has been a tendency to contrast the constraints of engineering design with the freedom of those engaged in arts and crafts. To quote Mayall (1979, pp. 26–27): '"Design" for the mechanical engineer in par-ticular, meant almost entirely the creation of new devices coupled with their specifica-tion, mainly by "working drawings," so that they could be made. "Design" for those who pronounced upon architectural proprieties and upon the forms of products such as fur-niture, ceramics and textiles . . . was a matter of observing what they regarded as proper combinations of forms and colours.'

There have also been tendencies to contrast what is practical with what is academic and what is inspired with what is formal. Stewart (1987, pp. 33–34) shows the additive effect of these contrasts when he notes that 'The first generation of mechanics such as Richard Trevithick and George Stephenson were empiricists, mistrusting theory, even in basic science and mathematics. . . . They relied upon mechanical aptitude and inventive genius rather than on formal academic training. . . . Stephenson's famous *Rocket* was a triumph of inspiration by which all the best practices of the time were heuristically combined and then improved upon.'

Stewart went on to point out that, even in Stephenson's time, what was needed was some sort of contingency approach to design. Stephenson's empiricist attitude to mechanical engineering was entirely realistic in Stewart's (1987, p. 34) view, because 'in the early days of transport there was no body of mechanical engineering theory to which engineers could turn.' However, his lack of training in the 'exacting science' of civil engineering was a considerable handicap which he 'ignored to his peril' (Stewart, 1987, p. 34).

One contemporary resolution of the tensions between art, science, and technology has come from the recognition that all designers have 'to conceive forms that can be turned into realities with the available resources' (Rawson, 1987, p. 13). Furthermore, the need to integrate industrial and engineering design into a coherent whole has finally been recognised. This was a central recommendation of the Conway Report (1977) and was one reason for the formation of the Design Council in April 1972.

Nevertheless, there is still a tendency to see creative design as an attribute of an individual designer and to see an opposition between creativity and method (Guetzkow, 1965; Jacques, 1981; Broadbent, 1981; Holt, Radcliffe, and Schoorl, 1985; Foster, 1986). Worse still, some of the (alleged) attributes of the creative designer are extrapolated naively from early psychometric studies of cognitive style (Cross and Nathenson, 1981) or from differences in the functions performed by the left and right hemispheres of the brain (Kerley, 1987). The result is an emphasis upon divergent rather than convergent thinking, which Weisberg (1986) has described (correctly, in our view) as one of the myths of creativity. What is required instead is a view of creativity which is grounded in the context of engineering or product design.

Creativity in Context

The first step is to note that creativity is concerned with the design of products which are *better* than (or add value to) those which have gone before (Steiner, 1965, Introduction; Bensinger, 1965). As Bensinger[3] (1965, p. 142) says: '"Innovation" is confused with "creativity." Too frequently that which is new is only different, not better. We in Brunswick do not believe in constantly making change for the sake of change, but we strive to better every aspect of our business.'

The second step is to study different kinds of creativity as they occur in business. And the key point is that creativity takes different forms in different contexts. To quote an ex-president of Bell & Howell, P.G. Peterson (1965, p. 183): 'I cannot emphasise enough the kind of creativity that most of us in business deal with. It is most often not fundamental innovation; it is most often relatively minor rearrangements of things that have already existed. In the language of science, we are most often dealing within the state of the art, rather than with the basic kind of research that Dr. Shockley refers to when he talks of inventing transistors.'

Thus, there may be people who work creatively within the limits of normal science, and there may be people who work creatively because they attempt to get round those limits (Foster, 1986). Tushman and Nadler (1986) go further when they distinguish incremental, synthetic, and discontinuous products and processes. Rothwell and Gardiner (1988) have identified twelve patterns of redesign. But whatever the detail, notice that the emphasis has very clearly shifted from one where we speak of *having creative ideas* to one where we speak of *doing creative work*. When we ground the study of creativity in design work, such as John Roebling's bridges (Petroski, 1987), or in research and development (Shockley, 1976; Atherton, 1988), or in scientific discovery (Watson, 1968; Gruber, 1981), what we see is not divergent thinking but creative failures and perseverance (Atherton, 1988), determination and resourcefulness (Bensinger, 1965), extended revisions and incremental gains (Weisberg, 1986).

The third step is to take seriously the idea that creative design and effective innovation are social rather than solitary activities (Foster, 1986). The image we need to reject is that of the lone designer: 'venturing alone into unknown territory, relying on nothing more than scientific intuition' (Weisberg, 1986, p. 89). It is an image which persists, regrettably, in many influential, contemporary accounts of creative thinking and conceptual design. Consider, for example, Adams's (1979) description of the 'conceptual block-busting' used by Watson and Crick in their work on DNA. Adams's (1979, pp. 61–62) view is that they 'relied heavily on inspiration, iteration and visualization. Even though they were superb biochemists, they had no precedent from which they could logically derive their structure and therefore relied heavily on left-handed[4] thinking.'

In contrast to this Weisberg (1986) has emphasised a social image in which a certain social milieu naturally leads to an interest in certain kinds of problems, in certain kinds of methods, and may provide specific starting points for further research. Furthermore, when research seems to be getting nowhere, information, suggestions, and criticisms from other people may be crucial if further progress is to be made (Weisberg, 1986; Atherton, 1988).

Design Management and the Social Aspects of Design

Once we start thinking of creativity and innovation as social activities, we are beginning to get to grips with what Archer (1981) has identified as design *praxiology*. One central concern is whether social contexts may be organised so that the right kinds of social activities are likely to be performed at the right times.[5] This is one fundamental aspect of design management.

To develop this idea we need to note that design, considered as total design, is not just social it is interdisciplinary (Peyronnin, 1987; Richter, 1987; Morley and Pugh, 1987; Pugh, 1986; Morley and Palmer, 1984; Heany and Vinson, 1984; *Managing Design*,

1984; Bensinger, 1965). This means that different people contribute different kinds of expertise. But it also means that design 'seeks to achieve a variety of interrelated objectives, none of which can be tackled in isolation from any other' (Mayall, 1983, p. 3). Thus, design management is the management of a system of constraints[6] (M'Pherson, 1980, 1981).

And it may be supposed that tradeoffs between constraints need to be performed explicitly[7] rather than intuitively (Peterson, 1965; Rudwick, 1983; Kahneman, Slovic, and Tversky, 1981; M'Pherson, 1981). This is the real point of rational actor models of decision making (Steinbruner, 1974; Morley and Hosking, 1988a).

One objective of design management may very well be to produce a set of organisational routines which lead to imaginative and effective work. We are very much in agreement with Ehrlenspiel and John (1987, p. 31) when they say 'design methodology introduces order into the design process; it creates the framework or organization in which separate intuitive processes take place. Working in a framework of design methodology means working intuitively but at the same time with methodic support. By analogy with CAD this could be termed MAD, that is "method aided design."' Consequently, we disagree with writers such as Guetzkow (1965) who contrast innovation with what is standard, or routine. The point is to design a system which makes effective innovative design routine.

The Design of Organisations: Open Systems Theory

When we think of the design of a system of design management, it may be useful to consider some of the implications of viewing organisations as open systems (Killman, Pondy, and Slevin, 1976; Katz and Kahn, 1978; Cummings, 1980; Gerwin, 1981; Trist, Higgins, Murray, and Pollack, 1983; Barko and Pasmore, 1986; Morley and Hosking, 1988b). We shall have more to say about this later. For the moment it will be sufficient to indicate the kind of attitude we think important. For further detail the reader is referred to Morley and Hosking (1988b).

The first attitudinal component derives from what systems theorists call the principle of equifinality. This means that 'a system can reach the same final state from differing initial conditions and by a variety of paths' (Katz and Kahn, 1978, p. 2). In other words, there may be (in principle) any number of means to achieve the same end. We should therefore be very suspicious about talk that there is any one ideal way of doing things.[8]

The second attitudinal component derives from the fact that social structures are artificial (Simon, 1981) or contrived (Katz and Kahn, 1978). Morley and Hosking (1988b) take the view that such structures evolve from (indeed, are constituted by) the actions of people carrying out projects as members of groups. The point is that as the projects change so the structures will change. Indeed, one major implication of social systems theory is that social structures are dynamic rather than static. Thus, the design of an organisation 'is recognized as an ongoing process rather than a static set of structural decisions' (Kolodny and Dresner, 1986, p. 33). More than this, open-endedness[9] has been recognised as a key characteristic of 'the culture of productivity' (Akin and Hopelain, 1986).

The third component defines a way of thinking about social control. We see projects as regulated by rules (roles), by norms, and by values. When we speak of social systems, we are speaking, therefore, of open systems of rules (roles), norms, and values. The nature of the rules, the norms, and the values determines the relative strength of tendencies for systems to split into clearly differentiated subsystems or for them to remain as integrated wholes. In design terms, when differentiation occurs, there are problems of transfer. To quote Oakley (1984, p. 9), 'Once an effective design has been achieved in the design department, there remains the problem of successfully transferring it to the production system and from there to the customer.'

Transfer problems of this kind may be extremely severe. Not surprisingly, therefore, some writers pay special attention to interactions between departments (Berridge, 1977; Hein, 1987; Hein and Andreasen, 1983; Andreasen and Hein, 1987; Andreasen, 1987; Ehrlenspiel, 1987). Others describe the design of integrated systems based on interdisciplinary teams (Richter, 1987; Gunz and Pearson, 1977a, 1977b); Pearson and Gunz, 1981; Barth and Steck, 1979). To quote Clausing (1984), 'A process that has evolved in which information about customer needs is handed from customer to market researcher to product planner and then to the design engineer has been tried and failed. The design engineer must accept the responsibility for being a leader in the planning of new products.'

The fourth component of our attitude to organisational design is that organisational designs will always be incomplete. This is one of the reasons why there is always the potential for conflict between subsystems within an organisation. It is also one of the reasons why leadership is central to the dynamics of behaviour within and between subsystems or groups. Other reasons are spelled out in Hosking and Morley (1988). Design, we might say, 'is not a socially neutral process. Rather it is political, having to overcome sectional views and perceived interests, and meeting resistance to change. This means that the successful follow through of design innovations may require powerful backing from senior management' (Child, 1984).[10]

The Effects of Environment

It is a characteristic of systems thinking that it gives priority to the study of organisations and their environments. The general approach is to suggest that the relationship between an organisation and its environment affects the relative development of subsystems concerned with operations, maintenance, regulation,[11] and the like.

However, systems thinking also shows that there is no general method of characterising the environment of an organisation (Morley and Hosking, 1988b). Indeed, what counts as the system and what counts as the environment are very much a matter of the purposes of the analysis (Carter, Martin, Mayblin, and Munday, 1984).

Organic Versus Mechanistic Forms of Organisation

The most obvious concerns have been to identify sets of environmental constraints (which may be classified as cultural, political, economic, technological, and ecological[12]) and to locate them along dimensions of turbulence-stability, diversity-homogeneity, clustering-randomness, and scarcity-munificence (Emery and Trist, 1965; March and Simon, 1958; Katz and Kahn, 1978). This produces what might be called a profile of the environment.

It may then be asked whether certain kinds of organisational arrangements are more appropriate to certain kinds of profiles. Morgan (1987) has provided an excellent summary of the state of the art. Clearly, there may be varieties of excellent organisations.

However, most[13] writers believe that shifts from stability to turbulence, from uniformity to diversity, and from clustering to randomness require a shift in organisation from some sort of *mechanistic* form to some sort of *organic* form. Very broadly speaking, what is involved is a shift from the hierarchical control of standardised jobs to an open form of organisation in which 'jobs [are] allowed to shape themselves, people being appointed to the organisation for their general ability and expertise and allowed and encouraged to find their own place and to define the contribution they could make' (Morgan, 1987, p. 51). This sort of emphasis is also found in the culture of productivity identified by Akin and Hopelain (1986).

However, the data needs to be evaluated with considerable care. It is, for example, sometimes suggested that the adoption of innovations is more likely in organic than mechanistic organisations (Hage and Aiken, 1967). But recent research suggests that this may only be true for innovations which are revolutionary rather than evolutionary (Cohn and Turyn, 1984). The reader should also note that organisational structures may have a variety of effects, depending on the performance measures which are used (Might and Fischer, 1985; Hull, Hage and Azumi, 1985).

The findings most germane to our present concerns come from Might and Fischer's (1985) questionnaire, given to over 100 managers in charge of development projects in thirty American firms. This is an excellent study, and because of its importance, we shall quote its conclusions at some length (p. 76):

> Organisational design was one of the structural variables found . . . to be associated with project management outcomes. Although not strongly significant, it was clear, at a statistically significant level, that some form of decentralised management structure (specifically a matrix format) was positively related to project management success, as measured by overall impressions of project performance or by cost performance. There did, however, appear to be some evidence that the choice of decentralised management structure was a particularly sensitive one, where the selection of an 'intermediate' project management form, rather than the matrix structure, could lead to disadvantageous rather than advantageous results. This finding, coming on the heels of recent discussions of the difficulties inherent in establishing matrix organisations, suggests that much more attention needs to be paid to the question of project organisation. In addition, . . . the findings of this study on the relationship between organisational design and project success indicate that technical success was not related in any way to organisational design.'

Life Cycles and Dominant Designs: The Maturity of the Sector

A second set of concerns may be identified in the work of Whipple and Clark (1986). They believe that earlier approaches to innovation and design have failed adequately to examine the relationships between product design, production processes and the organisation of work. If they are correct, too little attention has been given to 'multiple structures' within the organisation (p. 14). Furthermore, studies of design and innovation have not been located 'within the appropriate historical perspective and comparative contexts' (p. 14). Part of the reason, according to Whipple and Clark, is that the concept of a *sector* has rarely been used in studies of the organisation of work (p. 26). In contrast, they make the sector[14] a 'key analytic construct' (p. 25).

They are surely correct to do so. For example, there are now a number of papers emphasising the role of the end user in successful innovation and design. Some of the most prominent are those by Rothwell, Freeman, Jervis, Horsely, Robertson, and Townsend (1974), Rothwell and Gardiner (1983, 1988), Gardiner and Rothwell (1985), and von Hippel (1976, 1978, 1979). The general message is that 'tough customers + robust designs = good designs' (Gardiner and Rothwell, 1985).

The emphasis on robust design is probably of general importance (Pugh, 1984b). However, the evidence suggests that the way in which the end user is involved varies from sector to sector and is more important in some cases than in others. Von Hippel (1978), for example, notes a number of contrasts between sectors manufacturing process equipment, scientific instruments, polymers, and additives.

Whipple and Clark's analysis of design and innovation within a sector contains several elements which are important here. First, they suggest that, for various reasons, sectors develop their own distinctive languages, frameworks, and strategies, with their own distinctive weaknesses and strengths. Second, they distinguish three modes of change: innovatory production, incremental innovation, and styling innovation. Third, they take up the concepts of a design *hierarchy*[15] and a *dominant* design. A key question is whether the design hierarchy leads to a dominant design. Fourth, they use these concepts to discuss the relationship between product, process, and work organisation, building on the work of Abernathy and his colleagues (Abernathy and Utterback, 1975; Abernathy, 1978; Abernathy, Clark, and Kantrow, 1981, 1983; Hayes and Abernathy, 1980). The central idea is that dominant designs, within a sector, or within a market segment of that sector, lead to *mature* industries. Their basic characteristic is that of 'stability, of technology and market needs, and the resulting ease of copying the technology or responding to market needs' (Dowdy and Nikolchev, 1986, p. 38). Dowdy and Nikolchev (1986) argue that companies in mature industries are more likely to display signs of premature senility, such as management inflexibility, overoptimistic projections, complacency, low morale, mismanaging the experience curve, using inappropriate measures of performance, overstaffing, poor communications, and financial mismanagement.

Thus they are particularly unlikely to recognise ways in which new technology can change the nature of the design hierarchy leading to new, fast-growing, dynamic segments of the industry. Dowdy and Nikolchev describe this change as the 'de-maturing' of an industry.

Differentiation and Integration: Organising for Product Innovation

Our purpose here is to introduce the work of Lawrence and Lorsch (Lawrence and Lorsch, 1967a, 1967b; Lorsch, 1970; Lorsch and Lawrence, 1970). Their general strategy may be described as follows. First, consider an organisation as a set of subsystems, each dealing only with part of the environment. Second, for each subsystem measure the extent of uncertainty[16] about what needs to be done. Third, explore the relationships between performance, uncertainty and organisational structure, for each subsystem and for each sector studied.

Lawrence and Lorsch's first major finding was that the more uncertainty faced by each subsystem the more that subsystem built a distinctive identity for itself: that is, the more that subsystem was differentiated[17] from others.

Their second major finding followed from the (very sensible) recognition that 'the environment of an organization imposes requirements other than differentiation upon the organization. One of these is the dominant competitive issue' (Lorsch, 1970, p. 8).

They studied three sectors empirically: the plastics industry, the food industry, and the container industry. In the first two sectors the dominant competitive issue was innovation: that is, finding new products and processes. In the third sector the dominant competitive issue was scheduling and allocating production facilities to meet market demands.

Apparently, different competitive issues require different patterns of integration.[18] Because the conclusion is important, we quote Lorsch's (1970, pp. 8–9) summary of what has been found at some length:

> In all three environments the tightness of integration required was found to be identical. However, there was an important difference in the pattern around which this integration was occurring. In plastics and food, where innovative issues are dominant, the tight integration was required between sales and research and production and research. In the container industry, the tight integration was required between production and sales and between production and research. [Lawrence and Lorsch] report that in each industry the high-performing organizations achieved more effective integration around these critical interdependencies than their less effective competitor. Thus, the effective organization more satisfactorily met the demands of its environment for both differentiation and integration than did the less effective organization(s) in the same environment.

These results have not always been replicated when different measures have been used (Katz and Kahn, 1978). Also, it is not always easy to say what the critical interdependencies are. Nevertheless, the basic ideas deserve further study and merit detailed consideration of how they might be extended to the design field.[19]

DESIGN AS COLLECTIVE DECISION MAKING

Morley and Hosking (1988a) have recently considered some of the ways in which theories of decision making may contribute to a more general theory of *organising*. They have argued that the ideas of decision making need to be grounded in particular tasks (such as engineering product design) and particular contexts (that of the wider business activity of the firm). They have identified four steps in this kind of analysis, which we shall apply to the special case of engineering design. It is not suggested that any one theorist or set of theorists has proceeded exactly in this way, although they might have done. The framework is simpy a convenient way of arranging what we have to say:

- *Step 1*   The first step is to produce an appropriate model of stages in the process of decision making. Essentially, this means producing a model which translates stages of *identification*, *development*, *selection*, and *implementation* into the language of design.
- *Step 2*   The second step is to consider whether different translations are necessary for different contexts. The process may have different stages, or the same stages may occur in a different order.
- *Step 3*   The third step is to consider what properties are imparted to the process of design because design is managed as one part of a wider management task. This means that the core activities of product design are located within, and regulated by, a business design boundary (Pugh, 1986).
- *Step 4*   The final step is to relate what has been said to a more formal model of collective decision making, such as that of Friedlander and Schott (1981) or Janis (1982).

We consider each of these steps in turn. When we have done so, we shall have moved a long way towards setting out a theory of total design.

STEP 1: STAGES IN DESIGN

Stages in Decision Making

Writers on group decision making (Hare, 1985; Burnstein and Berbaum, 1983), organisational behaviour (Mintzberg, Raisinghani, and Theoret, 1976; Hickson, Butler, Cray, Mallory, and Wilson, 1986), and design management (Pelz, 1983) have examined some of the ways in which decision processes extend over time. What they say brings out three main themes.

The first is that decision making is organised, or managed, to a greater or lesser degree. This leads to simple block structures of the kind outlined by Hill (1984) and to more complex variations on the same kind of theme (Gill, 1986). At the simplest level Hill (1984) has identified stages which define the problem: select members, brief the group, structure the group, establish deadlines, and lead to a decision. At the most complex level Gill (1986) has given a fairly detailed description of the methodology built into a Lucas's project to develop a stop control antilock braking system (the SCS system).

The second theme is that, however the process is managed, we may identify certain core elements or processes. At the most general level these may be described as *identification* (of problems), *development* (of solutions), and *selection* (of policies) (Mintzberg, Raisinghani, and Theoret, 1976; Burnstein and Berbaum, 1983).[20] Alternatively, we may distinguish stages of *intelligence*, *design*, and *choice* (Simon, 1965; Minkes, 1987). From

our point of view it is important to complete the analysis by adding a stage of implementation of policies, carried out by other people (Witte, 1972; Pelz,[21] 1983; Morley,
1986).

The third theme is that progress through the stages, perhaps even the nature of the
stages, will be affected by the politics of the organisation (Dill and Pearson, 1984;
Hickson et al., 1986; Morley, 1986). Following Lawrence and Lorsch (1967a, 1967b),
much of this work has focussed on the ways in which different structures enhance or
impede the creation of shared meanings (Dill and Pearson, 1984; Kanter, 1984; Kolodny
and Dresner, 1986; Tushman and Nadler, 1986; Ebert, Slusher, and Ragsdell, 1987;
Bouwen, Visch, and Steyaert, 1987).

It is important to note that these provide only general guidance about how the
process of decision making is to be analysed (Morley and Hosking, 1988a). The crucial
first step is to ground these themes in the appropriate context, moving from the most
general categories to those which may apply in particular kinds of cases. Of course, we
shall still be operating at a fairly high level of generality. That is to say, we shall be operating with stages that contain at least one round (Humphreys, 1984).[22]

Stage models exist for a number of areas of design (e.g., Rudwick, 1983; Pelz, 1983;
Pugh, 1986). It may be illuminating to compare and contrast these models at a later
date, extending Pugh's (1986) analysis. For the moment, however, we shall focus on the
nature of activity models of engineering (product) design.

Activity Models of Engineering Design

There are three main kinds of activity models of engineering product design, which we
consider in turn.

Models Based on VDI Guidelines

Taken as a whole the *VDI Guidelines* emphasise systems approaches to problem solving
which show the process of design as a series of steps, hierarchically arranged, so that
superordinate problems are tackled first and then decomposed 'into parallel paths . . .
at as early a stage as possible' (*VDI Guideline 2221*, p. 4). If the analysis is sufficiently
thorough, superordinate problems will then necessarily be solved before subordinate
ones (Grotloh and Rothlin, 1983).

VDI Guideline 2221 is very much associated with Pahl and Beitz and based upon
their work (Pahl and Beitz, 1984; Beitz, 1987). Two aspects of their work have been extremely influential. To begin with they see the *design* function as receiving inputs from
planning and sending outputs so that the product can be manufactured, sold, and maintained before the cycle of activity starts again. Design is explicitly recognised as part of
the process of product creation. Thus conceived, the process of design is then broken
down into seven general working stages, each with a defined output:

1. To clarify the requirements of the design until a comprehensive product design
   specification is produced;

2. To establish the most important functions required in the design and to group these into structures;
3. To establish general solution principles and to document these in a principle solution;
4. To divide the design into modules and group them according to type;
5. To develop preliminary layouts for each of the modules, meaning scale drawings, circuit diagrams, and the like;
6. To add further information until there is a definitive layout which would allow the product to be made; and
7. To convert this information into standard forms for others to use.

Similar stages have been identified by Hubka (1982, 1983).

For certain purposes it is convenient to group the stages into slightly overlapping phases. In mechanical engineering these would be called something like clarification (stage 1), conceptual design (stages 2–3), embodiment design (stages 4–6), and detail design (stages 6–7) (*VDI 2221*, pp. 11–13). This simpler version of the original has been used in a number of influential publications (such as *Managing Design for Competitive Advantage*, 1986; Wallace and Hales, 1987) and in the draft of a British Standard guide to managing product design (*British Standards Institution Document 87/67900*). Hubka's model also distinguishes between phases of planning, conceptual design, layout design, and detail design (Hubka, 1982).

A rather different classification scheme is used by Grotloh and Rothlin (1983), who describe stages of planning, conception, draft design, and implementation based (they say) upon *VDI Guideline 2222* and *VDI Guideline 2225*. Their model makes problems of embodiment a much more central feature of conceptual design than the model of Pahl and Beitz. It describes commercial practice at Sulzer Brothers, a Swiss firm.

Models of Integrated Product Development

Andreasen and his colleagues (Andreasen, 1987; Hein, 1987; Eekels, 1987; Hein and Andreasen 1983; Andreasen and Hein, 1987) have argued that 'The condition for the establishment of product development is the creation of company goals and strategies. The strategy creation is in itself a design activity' (Andreasen, 1987, p. 174). Recently, their work has crystallised into a model in which there are six phases involving recognition of need, investigation of need, determination of product principles, the design of the product, preparation for production, and execution. Rather general terms are used because the object of the exercise is to explore what these terms mean in the market domain, the design domain, and the production domain. When this has been done, the resulting descriptions may be used to improve the integration of market activity, design activity, and production activity.

Similarly, the New Product studies of Cooper and his associates have attempted to identify those factors implicated in the successful development of new products (Cooper, 1979a, 1979b, 1980, 1986, 1988; Cooper and Kleinschmidt, 1986, 1987a, 1987b, 1987c). A complete new product process is defined in six stages akin to

Andreasen—assessment, definition, development, testing, trial and commercialisation—described as the stage-gate new product process. Before proceeding from one stage to the next a review is carried out, hence the expression *gate*, analogous to stages of design review Pugh (1979). The process described also strives for the integration of the production, technical, and marketing activities into the whole.

The research showed that successful products come from firms that systematically paid attention to more of these key activities. To quote Cooper (1988, pp. 242–243): 'There are serious gaps—omission of steps, and poor quality of execution in the new product process. These serious gaps are the rule rather than the exception. And they are strongly tied to product failures. . . . These research findings are the strongest evidence [that] . . . a formal and systematic approach is required to guide and facilitate a new product as it moves from the idea stage to launch.'

Design Boundary Models

Pugh (1986) has recently reviewed models of design activity intended to apply to different disciplines and to different products. He has concluded that the key ideas may be summarised and integrated by his own model of product design activity (Pugh, 1985a; Kimber, Coulthurst, Hamilton, Sharp, and Smith, 1985), represented graphically in Figure 39.1 (Chalk, 1981; Turner, 1977; Holt, 1980). There are nominally six stages in the model (Chalk, 1981; Turner, 1977; Holt, 1980).

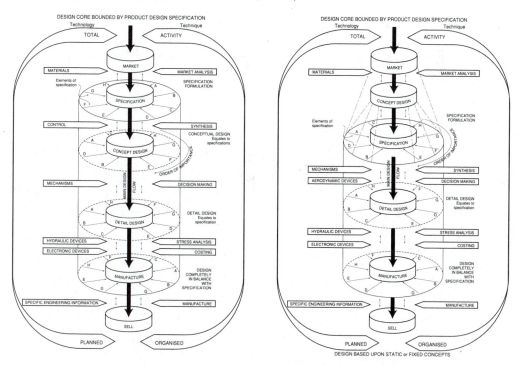

FIGURE 39.1. Pugh's Product Design Activity Model:
Dynamic Concepts and Static Concepts

1. Investigation of the market; leading to
2. The development of a product design specification, which constrains and controls stages of
3. Concept design,
4. Detail design, and
5. Manufacture.
6. The final stage is selling the product, which is the reason the product was made.

Pugh calls this kind of model a design boundary model because it shows a central core of design activity, disciplined (Rawson, 1976) or restrained (Atkinson, 1972) by the nature of the agreed product design specification[23] (Figure 39.2). The general idea is that the product design specification should be consulted regularly and used to guide, regulate, and control the design activity which follows. It is supposed that without the constraints of an adequate product design specification an enterprise is likely to produce a product nobody wants, either because it leaves too much out or because it puts too much in (Oakley, 1984; Zarecor, 1975). It is implicit in Pugh's model, and explicit in his writing, that development of a comprehensive product design specification should follow from, and be linked to, a comprehensive review of the users' needs[24] (Taguchi, 1978). Without such a review, a product that sells will be more a matter of good luck than good management.

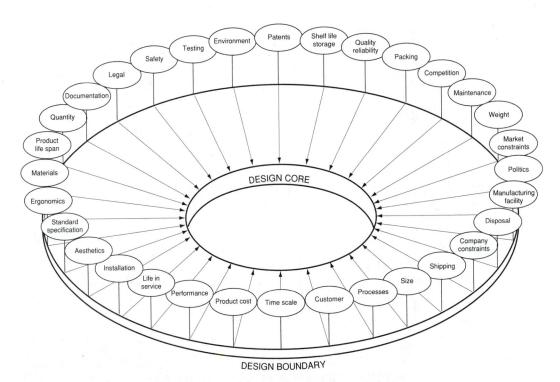

FIGURE 39.2. Elements of Product Design Specification

Many problems which emerge in the final stages of design are really built in at an early stage (Turner, 1985; Lindley, 1983; Cabinet Office Advisory Council for Applied Research and Development, 1982; Maidique and Zirger, 1984; Rudwick, 1983; McDonough and Kinnunen, 1984; Oakley, 1984; Craig, 1986). McDonough and Kinnunen's (1984, p. 19) study of new product development projects showed, for example, that 'the looser the product specifications were to begin with, the later in the project they became more tightly defined, the more problems developed with respect to product price and reliability, and the greater the likelihood of the product failing the marketplace.'

Similarly the New Prod series of research studies suggest that 'the outcome of new product projects are largely decided in the early stages of the new product process' (Cooper 1988, p. 249). Pugh assumes that the core stages are universal: that is, common to all kinds of design, albeit with some flexibility. In his view, it is other areas of design activity which differentiate engineering product design from other kinds of design. More precisely, different kinds of design may require different kinds of *information*, *techniques*, and *management*. Thus, much of the material included in Figure 39.2 is intended for purposes of illustration only. Strictly speaking, the particular inputs to the design core need to be reconsidered for each new case, as in Gill (1986), for example.

It is assumed that all other models may be readily assimilated to one or the other of the above cases. For example, Oakley's (1984) four main stages of formulation, evolution, transfer, and reaction may be readily translated into Pugh's design boundary model. Models such as those of van den Kroonenberg[25] (1987) provide points of comparison with the work of Hubka and Pugh. And so on.

STEP 2: STAGES IN CONTEXTS OF INNOVATION

The second step in our analysis is to consider whether different arrangements of the stages occur in different engineering contexts. One distinction which has very general support is that between designs which require different degrees of innovation or novelty. Thus, *VDI 2221* distinguishes tasks according to whether they are completely new, further developments, or adaptive; Pelz (1983) distinguishes between origination, adaptation, and borrowing; Whipple and Clark (1986) distinguish between innovatory production; incremental innovation, and styling innovation; and so on.

We shall follow Pugh (1983a, 1984b) and argue that we can consider two kinds of case as the end points of a continuum. One represents the case in which design starts, essentially, with a clean sheet of paper; the other represents the case in which many of the details are already known. Pugh has described the first case as involving dynamic concepts and the second as involving static concepts.[26] The design core model appropriate for dynamic concepts has already been set out in Figure 39.1 (left). The version of the model which treats static concepts is set out in Figure 39.1 (right). The essential difference is that with a static concept a detailed product design specification may be drawn up in the light of the known and accepted concept and written in terms of this known and understood context. With products tending towards the dynamic, this becomes less and less possible and in the ultimate is impossible.[27] The distinction between static and dynamic contexts, or something like it, is of considerable importance when we consider how to make design studies interdisciplinary.

STEP 3: THE BUSINESS DESIGN BOUNDARY

It is now necessary to consider how the process of design is located within the wider context of the business activity of the enterprise or firm. Recently, Pugh (1986) has described two business design boundary models which attempt to do just this job: one for static concepts and one for dynamic concepts. The version of the model which treats dynamic concepts is shown in Figure 39.3. This kind of model is very much an idealisation, drawn up to show the general kinds of moves which have to be made. A detailed example has been provided by Gill (1986), in effect, although he does not refer to Pugh's work and does not use Pugh's terminology. As we shall report, we have proceeded, ourselves, to ask members of an enterprise to draw up their own business design activity models, using blank templates as shown in Figure 39.4.

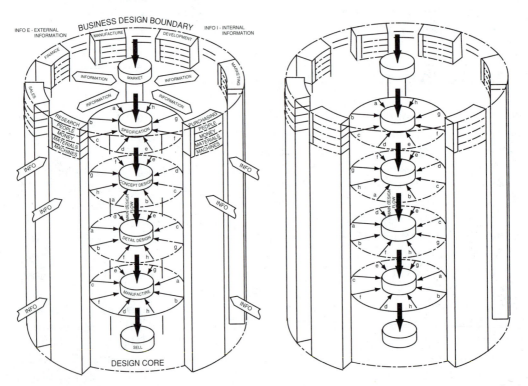

FIGURE 39.3. Business Design Activity Model **FIGURE 39.4.** Business Design Activity Blank

When we speak of a business design boundary, we are noting that explicit attempts are made to manage the activities in the design core, so that they are given a structure and provided with appropriate information, resources, and support. Following Hubka, we may say that the task of management is to give the design core an anatomical structure that will be the carrier of certain desired properties (Hubka, 1982, p. 34).

The Anatomical Structure of the Design Core

Our next task is to consider what kinds of properties are to be desired, on the basis of existing theories and research. One way of looking at the task is to say that it is to examine how it coheres because coherence in the 'workscape' is one key element in the culture of productivity examined by Akin and Hopelain (1986). Coherence is also a key concept in the evolutionary design methodology described by Rzevski (1981).

Structure in Process: Stages in Design

Design activity models are quasi-prescriptive models. They do not necessarily describe how actual designs get from start to finish. Quite possibly (Hickson et al., 1986, p. 98),

> One phase does not lead to another in a logical order, so that processes as a whole do not move steadily onwards phase by phase in an inexorable procession. They jump about. They hop to and fro. They turn back. Fresh information forces a rethink, something unforeseen happens which opens up a new alternative, powerful voices close off an otherwise attractive course of action. So there are rediagnoses, reconsiderations, reassessments, and people find themselves going through the same thing all over again. At the extreme, processes can seem to veer uncontrollably, and solutions come before problems, in a veritable 'garbage can' of problems, solutions, and participants, in which decisions are arrived at that appear to have little to do with the process of arriving at them, the choices being uncoupled from the processes.

However, Hickson et al. were speaking of decision making in general, rather than design in particular. Whilst a certain amount of iteration is inevitable, one main point of design activity models is to suggest that orderly processes will be more likely to lead to effective designs, other things being equal of course. In particular, coupling choices to the correct sequence of stages is more likely than uncoupling to lead to designs which are effective, at least in the sense that process losses are minimised.[28] To the best of our knowledge this hypothesis has not been tested formally, although it is consistent with some case research (Oakley, 1984).

We would expect order in the stages to lead to designs which were effective in other ways, although caution is clearly required. One general argument has been provided by Akin and Hopelain (1986, p. 27). As they say, 'One aspect of integration is the extent to which all elements work together to give the same overall message. In images that lack coherence, different elements may signal different things, thus making it difficult to decide what is appropriate.'

This is one reason for having a clear sequence of stages in the design core, perhaps one in which the output from each stage is formally accredited before moving further.[29] It may also be a reason for following systematic procedures, or using various kinds of decision support system, within each stage. Our own view is that within each stage it is important to establish ways of working which mean that disagreements converge productively onto solutions all can understand and all can accept. Some systematic procedures appear to have this effect (e.g., Pugh, 1981; Kuppuraju, Ittimakin, and Mistree, 1985; Morley and Pugh, 1987; Rzevski, 1981). However, existing studies of design or innovation have done little more than document some of the methods which might be used. We do not know to what extent they are used and to what effect.

Perhaps the best evidence available about the effects of stages comes from Pelz's (1983) study of adaptation[30] of urban innovations in energy conservation, noise control, and solid waste. Pelz's data, based on telephone interviews with key actors, showed that stages appeared in a fairly clear order but that clearer separation between the stages did not lead to greater success, measured in fairly broad terms. However, Pelz's report does not reveal to what extent process losses may have been involved, nor to what extent the organisations muddled through despite them. Nor is the adoption of innovation quite the same as engineering design.

The Integration of Different Functions

At several points we have used the phrase *total design*. This reflects several concerns. One is that we regard design as beginning with the investigation of a market, whoever does the investigating, and ending with a product to be marketed and sold, whoever does the marketing and whoever does the selling. By definition, design is thus made central to the business success of an enterprise.[31] A second concern is that all facets of the business interact with the design core. If Pugh (1983b) is correct, this means that to narrow or bias the design core towards any one of these facets is to increase the risk of designing a product that does not have, and will not create, a market niche. Thus, talk of total design implies that it is essential somehow to integrate the work of specialists from different disciplines and from different parts of the enterprise. There is, for example, good evidence that successful product innovation requires that 'The create, make and market functions are well interfaced and coordinated' (Maidique and Zirger, 1984, p. 201) (see also McDonough and Kinnunen, 1984).

One way of achieving this is to have a product development policy which is *market led* in the sense that the analysis of the market involves all key functional specialists. Johne and Snelson (1988) have shown that a number of outstanding product innovators (large firms in four sectors of manufacturing industry) are market led in this sense.

It is possible, however, to go further than this. When we speak of total design, we are speaking of a process which involves all the most important functional specialists from the beginning to the end of the process of product design and development. This is sometimes called simultaneous design (Winstanley and Francis, 1988). It may be that no one kind of organisational form is inherently more suited to design than another, as

Tushman and Nadler's (1986) paper would seem to imply. What is much more important, in their view, is that organisations 'develop formal linking mechanisms, which are important vehicles for creativity and innovation' (p. 83).

Thus, one possibility is to have a separate design department and take explicit steps to identify and manage the problems of transfer which then arise (Oakley, 1984). The general idea is to get high differentiation with high integration (Lawrence and Lorsch, 1967a, 1967b) by means of formal linking mechanisms such as coordinating departments, cross-functional committees, task forces, formal liaison roles, and so on (Lorsch and Lawrence, 1970; Tushman and Nadler, 1986).

However, Lorsch and Lawrence's (1970) data is at least consistent with the view that linking mechanisms of this kind work better when the units to be differentiated are arranged organically rather than mechanistically. Overly mechanistic forms have also been widely canvassed as demotivating, especially for those professional people, such as engineers and research scientists (Badawy, 1975; Zachary and Krone, 1984; Abetti, 1986; Kolodny and Dresner, 1986). What is required is a relaxed working atmosphere, which means that the basic groundwork is completed in a context of open communication and interpersonal trust (Bensinger, 1965; Bower, 1965; Barko and Pasmore, 1986) guided by norms emphasising individual autonomy, participation (Zachary and Krone, 1984) and providing opportunities to exercise one's own expertise (Cherns, 1976; Bass and Barrett, 1981; Katz and Kahn, 1978; Morley, 1979; Zachary and Krone, 1984; Kolodny and Dresner, 1986). This is almost a description of an organic form of organisation. Nor surprisingly, therefore, some writers have simply stated that mechanistic forms are unlikely to provide the right kinds of anatomical structures for the design core (Oakley, 1984).

Might and Fischer (1985) have presented a more balanced case, and provided empirical comparisons of the effectiveness of development projects carried out within four kinds of structure: traditional functional organisations (FO), matrix organizations (MO), and two forms of project structure (P1 and P2). Might and Fischer argue that they have established a quasi-continuum of project structure, running from FO through MO and then P1 to P2. However, they predicted 'a direct relationship between organisational structure and project success, with some form of decentralized format (MO-P2) being most successful and the purely functional and project formats FO and P1 being least successful' (p. 73). Of the various structures, the matrix form (MO) was most successful overall and most successful in controlling costs. However, the other decentralised structure (P1) was the least successful overall and most likely to be behind schedule.

Decentralisation is only one element in an organic structure, however, and the possible advantages of multifunction (multidisciplinary) teams need to be explored more fully. After all, they open up the possibility of minimising the transfer problem by giving specialists from other functions a permanent place within the design team. Might and Fischer's data shows that this is by no means a simple matter (otherwise the project teams would have been more successful): perhaps because interdisciplinary teams bring political problems to the fore (Lorsch and Lawrence, 1970; Dill and Pearson, 1984; Hickson et al., 1986; Peyronnin, 1987), which may require organisations to develop in some quite unconventional ways (Dill and Pearson, 1984).

It is also worth noting that considerable attention has and is being given to the use of computer-aided design as an integrating device (Amkreutz, 1984; Otker, 1984; Pugh, 1984a). The integration issue was discussed in some detail in Pugh (1985a), with implications for organisational structure. Bridges (1981) considers involving the end user in the process of design and providing all participants with consistent documents and information (Daub, 1987).

Project Control Systems

Might (1984) has distinguished four aspects of project control systems concerned with the level of: detail in planning, use of specific techniques, periodic reviews and reports, and management resources required to monitor and evaluate progress. The level of detail in technical planning was the only variable positively related to success on more than one of the six measures of success used. Otherwise, what was very evident was that the success of a project depended upon interactions between the control system, the project manager,[32] the project team,[33] and various channels of communication within the firm. Results were extremely complex.

The interaction between the types of control and patterns of communication is of some general significance, perhaps. To quote Might (1984, p. 131),

> Unfortunately for the project manager, the implementation of certain control systems can interact with communication patterns to produce a negative impact on success. . . . Since many control systems place an added level of communication, often a 'bureaucratic' type of communication, on an organization, it is possible that this imposition reduces the value of an existing effective informal communication network. This produces a paradoxical situation for the project manager who needs some form of a formal control system, but already has an effective communication pattern within the firm . . . If project team members know what has to be done, how difficult it is, and that they will be pressed to get it completed in a satisfactory manner, the added burden of supporting a reporting system on accomplishments may be too much.

The point is well taken. However, from our perspective what counts as an added burden will depend upon whether existing control procedures are integrated into, and provide a clear logic for, the activities of the design core. To date studies of project control have not been integrated with studies of design activity. What is needed is a much clearer coupling between the study of management and the study of what is being managed.

STEP 4: COLLECTIVE DECISION MAKING: MODELS OF SMALL GROUP PERFORMANCE

The final stage in this part of our analysis is to show how consideration of design as a process of collective decision making may be informed by models of small group performance (Kirkpatrick, Davis, and Robertson, 1976; Janis and Mann, 1977; Friedlander and Schott, 1981; Golembiewski and Miller, 1981; Janis, 1982; Steiner, 1972; Morley and Hosking, 1984). These attempt to show how the performance of groups is affected by the nature of their primary task, by personal constraints, and by environmental constraints. A similar model of the process of engineering design would require the triple synthesis of a product design activity model, a personal design boundary model, and a business design boundary model (Morley and Pugh, 1987; Pugh, 1987).

The Personal Design Boundary

Morley and Pugh (1987), and more recently Pugh (1987), have taken the view that just as there is a product design boundary and a business design boundary there is also a personal design boundary. It represents the constraints placed upon the process of design by the personal resources of the designers: that is, whatever they bring with them in the way of dispositions, outlooks, knowledge, skills, and power.

It is important to note that the various stages of activity within the design core may be performed by different groups of people. However, for any particular group the personal design boundary refers to the *distribution* of resources among the members of the group. This means that the personal design boundary defines various aspects of the *structure* of the group (Golembiewski, 1962). That is, the personal design boundary is an *interpersonal* design boundary.

In terms of total design it is the *diversity of resources* within a group or team which is of central importance. Essentially, the diversity within the group should reflect diversity within the organisation so that different facets of the business activity are brought together within the design core (Friedlander and Schott, 1981).

It is diversity which promotes the creative expression of different points of view, and is the *sine qua non* of successful performance (Hoffman, 1982; Belbin, 1981; Friedlander and Schott, 1981). If design is decision making at work the first rule of decision making is that *not to disagree means not to solve the problem* (Drucker, 1970; Morley, 1981, 1986). Without disagreement a group may commit itself prematurely to a decision (design) which has not been sufficiently well appraised (Janis, 1982; Morley and Pugh, 1987). Without *organised* disagreement group members may find that what is intellectually clear to them is not at all clear to others (Morley, 1984). They may also find that disagreement turns into antagonism, which degrades the process of decision or design (Burnstein and Berbaum, 1983; Morley and Pugh, 1987). In other words, unless disagreement is properly organised, the advantages of working together will be lost because of faulty processes which degrade the cognitive and political[34] activity of the group. This is one major message of social psychological research (Steiner, 1972; George, 1974; Janis, 1982; Morley, 1986; Morley and Pugh, 1987).

The Personal Design Boundary and the Political Activity of the Group

Diversity implies that people have different kinds of substantive knowledge. It also implies that they will have different values, motivations, commitments, and power. This means that design is a process of political decision making.[35] Four themes are evident in the literature:

- Attempts to integrate different specialists into a process of total design may be welcomed, tolerated, or actively resisted, depending on circumstances (Whipple and Clark, 1986; Peyronnin, 1987; Dill and Pearson, 1984). However, individuals should be selected for a team because of their skills rather than because they happen to be available (Heany and Vinson, 1984).
- Design managers, project managers, and product managers need to be skilful negotiators (Hodgetts, 1968; Sayles, 1976; Dill and Pearson, 1984; Morley and Pugh, 1987). For example, a product manager may have to negotiate with market research and manufacturing. To quote Sayles (1976, pp. 11–12), 'This means assessing how important their technical criteria are, which ones are modifiable and, overall, what is best for the new product's introduction.' A useful treatment of the skills of negotiation is given in Morley (1986), although it has not been applied directly to negotiation within the context of design.
- Design needs to be represented at board level and consistently supported by top management (Oakley, 1984).
- In many circumstances the model of design as a process of rational decision making needs to be replaced by a model of design as organisational politics (Dill and Pearson, 1984).

Decision Making Models: Stages and Strategies in Context

Dill and Pearson's (1984) analysis of the effectiveness of project managers contains two components. The first locates the manager's primary task on a continuum ranging from coordination to leadership. However, the continuum is complex. The term *coordination* is used when it is clear who has to do what, why, and when and when conflict for resources does not adversely affect its progress. That is to say, the term *coordination* refers to the management style which is appropriate when a project is well formed. The term *leadership* refers to the management style which is appropriate when a project is not well formed: that is, when tasks are frequently changed, when there is crippling competition for scarce resources, and when commitment to a task is the outcome of a negotiation within or between groups. The second component identifies two strategies for making decisions: one involving rational analysis (as in the *rational actor* paradigm), the other involving bargaining and negotiation (as in the *organisational politics* paradigm).

The two components are different ways of making the point that decisions made in organisations are political decisions: first, because they are made in a political context,[36] and second, because some decision making strategies make explicit use of bargaining routines.[37]

The major claim of Dill and Pearson's paper comes from matching a more political strategy (bargaining rather than analysis) to a more political context (requiring leadership rather than coordination). If they are correct 'changing environmental conditions will inevitably shift the emphasis from coordination towards leadership and make the organizational politics model a closer description of reality than the rational actor model' (p. 138). This paper has much to commend it. It represents one way of thinking about decision making strategies, most fully developed, perhaps, in Pfeffer (1981). However, the distinction between the different strategies is not entirely clear, if only because bargaining is often treated from the perspective of the rational actor model (Bacharach and Lawler, 1981). Consequently, we prefer to follow Janis and Mann (1977), and distinguish different decision strategies in terms of different *attitudes* to processing information. We shall assume that each stage in the process of total design may be carried out using *more or less vigilant strategies for processing information.*

One advantage of this kind of conceptualisation is that it provides a direct link with models of decision making under stress (Janis and Mann, 1977; Wohl, 1981). This seems particularly important in the context of engineering design because attempts to cope with stress often mean that 'individuals under stress display reduced information search, consider fewer alternatives, overreact to isolated pieces of information and generally engage in what would otherwise be suboptimal choice generation and selection' (Wohl, 1981, p. 631). The effect of stress is thus to make information processing[38] less vigilant. Furthermore, this tendency becomes more pronounced in certain groups and in certain contexts (Janis, 1982; George, 1974). From our point of view, the choice of vigilant or nonvigilant strategies for processing information is a function of the interpersonal design boundary and the business design boundary.

Vigilant information processing strives towards the ideal of rational choice explicated in the rational actor model. It is assumed that each individual[39]

1. Collects information before, rather than after, a choice is made. Thus, the decisions follow from the analysis (the product design specification controls the design) rather than the analysis from the decisions[40] (the design controls the product design specification);

2. Documents the existence of a problem and the need to solve it (the product design specification follows from a thorough investigation of the market);

3. Seriously considers more than one policy, option, or possible course of action (design strategy, product strategy, conceptual design);

4. Carefully considers the full range of goals and constraints, and the values implicated by each (thoroughly canvasses the elements in the product design specification);

5. Carefully considers the costs and benefits of each alternative course of action (with many concepts, many criteria, in a concept selection matrix);

6. Strive for a comprehensive, coherent, and detailed analysis of the elements of the problem, and of systemic relationships between them[41] (as in *VDI 2221*, for example);

7. Intensively searches for new information and is sensitive to judgements of outside experts, even when that information or judgement is unpalatable, or not what the decision maker hoped to find (evolution of product design specification, concept selection, changes in concepts as process iterates through the design core);

8. Reexamines the consequences of all known alternatives, including any which had been discounted initially, before making a final choice (reestablish and rerun concept selection matrix);

9. Attempts to make the best possible choice within the constraints of the task (using a vector regret criterion to establish system cost effectiveness);[42] however, the particular method used is less important than the fact that there is some explicit attempt to work out how best to trade off benefits against costs;

10. Considers in detail how the policy chosen is to be put into practice (designing for manufacture), paying special attention to the contingency planning required to take account of known uncertainties and risks (dual sources of supply).

This implies that the decision maker is seeking the best choice possible within the constraints of the task, recognising that to achieve some benefits others may have to be sacrificed.

However, Simon (1957) has given an account of *bounded rationality* which suggests that decisions made according to the rational actor model are likely to be relatively rare because of personal limitations in mental capacity,[43] and because of external constraints on choice (Janis and Mann, 1977; Pfeffer, 1981). These combine to produce decision makers who are not even trying to make decisions which are optimal. They are concerned only with what is acceptable. More precisely, when operating in a nonvigilant mode a decision maker (designer)

1. Collects information to justify choices which have already been made, without explicit analysis (the design controls the product design specification);

2. Fails adequately to document the existence of a problem and the need to solve it (leading to designs nobody wants);

3. Considers a restricted number of policies, frequently only one (design based on one concept, one idea);

4. Considers a restricted number of goals and constraints, paying insufficient attention to the values implicated by each (fails thoroughly to canvass the elements in the product design specification);

5. Considers a restricted number of costs and benefits, sometimes even focussing on one cost or one benefit (ignores a formal product design specification and its implied constraints);

6. Does not strive for a comprehensive, coherent, and detailed analysis of the elements of the problem, and of systemic relationships between them but relies on intuition to take short cuts (rapidly discounts one concept without serious analysis);

7. Fails intensively to search for new information and is insensitive to judgements of outside experts, especially when that information or judgement is unpalatable, and not what the decision maker hoped to find (ignores early warning signs that a design is going wrong);

8. Evaluates each alternative only once. Options are evaluated sequentially as they arise (single ideas evaluated serially, staccato design);

9. Regards a policy as acceptable if it satisfies a few major constraints, so that it is easy to justify in ordinary language, without recourse to systematic methods or sophisticated analyses. There is no attempt to work out a comparative balance sheet of the pros and cons associated with the main alternatives (lack of systematic design methods, staccato design, first idea is best);

10. Fails adequately to solve the problem of interfacing designs to manufacturing, marketing, sales, etc. because the problems of manufacturing, marketing, sales, etc. are not integrated into the activities of the design core (major modifications in design required at the production stage).

Morley and Pugh (1987) have suggested that the process of total design requires vigilant information processing. It is more likely to occur in dynamic contexts but is required even when the context is static, and the designs which emerge are variations on old themes. This also has significant implications for CAD application (Pugh, 1985a). To quote Pugh (1983a, p. 6), 'One should always be operating . . . the principle of controlled dissatisfaction—the designs of today can always be bettered tomorrow. The assumption is made that until proved otherwise, existing concepts are dynamic.'

Fortunately, a great deal is known about the kinds of interpersonal design boundary and business design boundary which make vigilant information processing more or less likely to occur (Janis, 1972, 1982; Janis and Mann, 1977; George, 1974; Wohl, 1981). This is one reason that we suppose that a general theory of total design will contain within it an application of a general theory of the performance of small groups. Further, it is possible to spell out the kind of analysis which may be involved.

To begin with, let us suppose that, other things being equal, effective performance is more likely when we have groups composed of people who have sentiments of mutual liking and respect for each other's expertise. Conventional wisdom suggests that project teams are chosen to be *cohesive* in just this way. To begin with, therefore, the interpersonal design boundary is designed to minimise the likelihood of *antagonism* within the group. The clear message from the literature in the social sciences is that groups of this kind will sometimes perform extremely well and sometimes extremely badly. The other things that have to be equal are that

- There is sufficient expertise within the group and sufficient diversity of expertise (Friedlander and Schott, 1981);
- Group members have clear ideas about the nature of the task and their place within it (Pearson and Gunz, 1981);
- The group has norms and standard operating procedures which facilitate critical enquiry (Janis, 1972; Morley and Hosking, 1984). This means that assumptions which are false or incomplete will be rooted out, and there is controlled convergence upon solutions all can understand and all can accept (Morley and Pugh, 1987);
- The group has an appropriate form of leadership (Abetti, 1986; Nicholls, 1986; Hosking and Morley, 1988). This is especially important in teams composed of highly creative individuals (Belbin, 1981);

- The group contains members who between them carry out a variety of psychological roles (Clutterbuck, 1979; Belbin, 1981).[43] Which roles are most important will depend, presumably, upon which operating procedures are in place;
- The group is kept closely in touch with other parts of the organisation rather than insulated from it (Janis, 1984; George, 1974).

CONCLUSION

We have attempted to set out the general form of total design. We believe that what is required is a triple synthesis between the design of the product, the organisation and the people concerned, which we consider will be best achieved by a consideration of the boundaries in each area and the interactions resulting therefrom, ie the product design boundary, the business design boundary and the (inter-) personal design boundary. We have also attempted to show how such a synthesis might be achieved and based on the results of existing research. The major weakness of the synthesis is that there are very few empirical studies of engineering design which allow us to build a quasi-prescriptive model with any degree of confidence (Morley and Palmer, 1984).

It is clear that any investigation of structure in the processes which lead to successful rather than unsuccessful designs must use a variety of criteria for success. However, if we are to make the most of the available data it is important that we are able to give theoretical explanations of any results obtained. This is easier for some criteria than others. Perhaps the major contribution we have made is to identify, and synthesise, a body of research and theory which brings losses from faulty process centre stage.

ENDNOTES

1. However, Oakley (1984) has claimed that "Until a few years ago, it was not uncommon to find that the design of a product and its subsequent manufacture was the responsibility of a single individual" (p. 93).
2. At a recent meeting of the Design Research Society, Hales (1987) attempted to map inputs from over twenty disciplines.
3. Bensinger was chief executive officer of the Brunswick Corporation, then regarded as a paradigm case of organisational success. He was one of sixteen eminent academics, practising scientists, and business executives who attended a seminar on 'The Creative Organisation' at the Graduate School of Business, the University of Chicago. The proceedings of the seminar are reported in Steiner (1965).
4. Left-handed thinking means thinking which is guided by the *right* hemisphere of the brain, primarily. An excellent summary of the literature on hemispheric specialisation is given in Wood (1983). Broadly speaking, the left hemisphere is *more* involved in the sequential, analytic, processing of linguistic information. The right hemisphere is *more* involved in the wholistic, synthetic, processing of nonlinguistic information.
5. It may also be very important that creative *failures* are allowed to happen without penalty to those involved (Goldsmith and Clutterbuck, 1985).

6. Goldsmith and Clutterbuck (1985) quote the view of the CEO of PA Technology that '[technological creativity] flourishes under the constrained regime set by market and financial factors—so much so that the establishment of such constraints becomes the foundation stone of innovation rather than its boundary' (p. 109).

7. This is not to deny what is sometimes called type 3 error: solving the wrong problems precisely. Nor is it to deny that formal procedures can sometimes have deleterious effects (Rapoport, 1968; Dreyfus, 1982; Silverman, 1985). Methods need to be managed and managed with care (Pugh and Smith, 1976).

8. This may be what Wallace and Hales (1987) mean when they say that we need 'contingency models with multiple levels of resolution' (p. 95).

9. Meaning 'opportunities for new events and for meanings to be overlaid on familiar events' (Akin and Hopelain, 1986, p. 28).

10. In March 1984 the Economic and Social Research Council of Great Britain sponsored a workshop on 'The Use of Social Science to Improve the Process of Design in U.K Industry.' Participants were active social scientists or engineers or designers or fairly senior industrialists. Before the workshop each participant was asked to write one or two sides of A4 stating how they thought design activity should be organised within industry, and how this activity might be improved through the application of the social sciences. A complete set of the replies is included in Appendix 1 of Morley and Palmer (1984). The above comments are taken from Professor Child's reply.

11. This particular classification is taken from Miller and Rice (1967). Many other classifications are possible (Morley and Hosking, 1988b).

12. This classification is taken from Katz and Kahn (1978).

13. It is still possible, however, to argue that 'No single organisational form is inherently more conducive to innovation than the next' (Tushman and Nadler, 1986).

14. A sector is a collection of enterprises which often but not always compete to provide similar goods and services (see Whipple and Clark, 1986, p. 26).

15. The concept came from Alexander's (1964) work on architectural design. It has been modified subsequently by Abernathy (1978) and others.

16. Uncertainty was measured in terms of the certainty of information, the rate of change in the environment, and the time range of the task.

17. Lawrence and Lorsch use the term *differentiation* rather than the term *identity*. What they mean is that it is relatively easy to distinguish departments in terms of degree of departmental structure (low versus high), members' orientation towards time (long versus short), members' orientations towards others (permissive versus directive), and members' views of the nature of their primary task (Lawrence and Lorsch refer to this as members' orientation towards environment). One important shift in emphasis from other work, therefore, is that organisational structure is defined partly in terms of the perceptions of the members of the organisation.

18. Lawrence and Lorsch define integration as 'the quality of the state of collaboration that exists among departments that are required to achieve unity of effort by the environment' (Lorsch, 1970, p. 5).

19. This conclusion may be generalized, perhaps. For example, Whipple and Clark

(1986, p. 39) take the view that 'Many of the studies from business policy and from business history are useful, yet they require location and extensions to satisfy the requirements of our framework of inquiry. In particular they do not normally address the problems of design and innovation, and their relation to structural repertoires, corporate cultures and capital/labour dynamics. For these reasons, additional features are needed.' We agree, although we are not attempting the same kind of analysis as Whipple and Clark.

20. These are global terms. It is also possible to argue that some sort of decision sequence occurs within each of these general stages. This leads to models which represent the activity of design as spiralling through a series of decision making stages, which lead to progressive elaboration of the design (e.g., M'Pherson, 1981).

21. Some models go even further. For example, Pelz's (1983) study of stages in urban innovation added stages of incorporation and diffusion.

22. Humphreys takes the term *round* from Kunreuther (1983). It is regarded as a convenient analytic device to indicate that there has been a change in the focus of discussion because a decision has been made or because new evidence or new parties have been introduced.

23. Hubka (1982) has also developed a design boundary model using similar stages to those in *VDI 2221*.

24. This sort of emphasis is very much a feature of what M'Pherson (1980, 1981) calls *systems engineering design*.

25. Van den Kroonenberg has also given a very interesting and useful discussion of processes within each stage in terms of a divergence, ordering, convergence sequence.

26. Distinctions of this kind may apply to the system of constraints considered as a whole or to various subsets of those constraints. *VDI 2221* makes a similar point.

27. *VDI 2221* also implies that the sequence of stages may change in different contexts, but does not show how, or why, this would occur (see pp. 5–6).

28. This is similar to Might's sixth criterion for success in project management—namely, that a successful project is one that meets cost and schedule goals but with little or no crisis management (Might, 1984, p. 128; Might and Fischer, 1985, p. 71).

29. If design is treated as a separate function, it will also be necessary to have control systems which say what information is to be passed across the design boundaries and in what form.

30. Rather than origination or borrowing.

31. A very great deal of literature shows empirically that design, defined more narrowly, is central to the business success of an enterprise (see, e.g., Rothwell and Gardiner, 1983; *Managing Design*, 1984).

32. Also see McDonough and Kinnunen (1984).

33. Specifically, whether the project manager was given an increase in responsibilty, whether he or she demonstrated technical expertise and administrative competence, and whether he or she was able to change their leadership style to suit the situation.

34. Specifically, whether the members find their work challenging and whether the team is enthusiastic and supportive, considered as a whole.

35. The cognitive activity functions to help people organize their intellectual activity and think clearly about the problems they face; the political activity functions to manage differences in outlook generated within and between groups (Pruitt, 1981; Morley, 1986).

36. Some contexts are more obviously political than others. This is the point of the co-ordination-leadership continuum. Other writers make the same kind of point in different ways (e.g., Morley and Hosking, 1984). But the central point is that collective decision making is always political decision making (Steinbruner, 1974). That is to say, whichever decision strategy is followed the decision maker is always concerned to influence others and to justify to them the moves he or she has made.

37. In the rational actor model decisions are justified by explicit reference to a process of analysis. In the organisational politics model decisions are justified by explicit reference to a process of bargaining. The outcome of the bargaining is a set of jointly constructed rules.

38. Akin (1986) has set out a rather different approach to producing an information processing theory of (architectural) design. This sort of model is very useful for identifying micro skills, involving various search heuristics to partition the problem space. For an interesting example of protocol analysis in engineering design see Stauffer, Ullman, and Dietterich (1987). Our concerns are pitched at a more macro level.

39. The analysis that follows is based on the work of Steinbruner (1974), Janis and Mann (1977), Pfeffer (1981), Morley and Pugh (1987), and Morley and Hosking (1988a).

40. See Weick (1968).

41. Steinbruner (1974) calls this constructing a blueprint.

42. This method is described in M'Pherson (1981).

43. This is another reason for following systematic methods in design.

44. Clutterbuck has argued that R&D groups collectively need to perform four roles: generating, integrating, developing, and perfecting. Belbin's research on management teams has suggested a more complex scheme. Without going into too much detail, perhaps the most useful thing to say is that Belbin adds two leadership roles (chairman and shaper), an administrative role (company worker), and a liaison role (resource investigator).

REFERENCES

Abernathy, W.J. (1978). *The Productivity Dilemma*. Baltimore: Johns Hopkins University Press.

Abernathy, W.J., Clark, K.B., and Kantrow, A.M. (1981). 'The New Industrial Competition.' *Harvard Business Review* (September/October), 68–81.

Abernathy, W.J., Clark, K.B., and Kantrow, A.M. (1983). *Industrial Renaissance*. New York: Basic Books.

Abernathy, W.J., and Utterback, J.M. (1975). 'A Dynamic Model of Process and Product Innovation.' *Omega Management Science* 3, 6.

Abetti, P.A. (1986). 'Fostering a Climate for Creativity and Innovation in Business Oriented R&D Organisations: An Historical Project.' *Creativity and Innovation Network* (January-March), 4–16.

Adams, J.L. (1979). *Conceptual Blockbusting* (2nd ed.). New York: Norton.

Akin, G, and Hopelain, D. (1986). 'Finding the Culture of Productivity.' *Organizational Dynamics* 14(3), 19–32.

Akin, O. (1986). *Psychology of Architectural Design.* London: Pion.

Alexander, C. (1964). *Notes on the Synthesis of Form.* Cambridge, MA: Harvard University Press.

Amkreutz, J.H.A.E. (1984). 'CAD and the Future of Design Practices.' *Proceedings of the Second European Conference on Developments in CAD CAM.* Amersfoort.

Andreasen, M.M. (1987). 'Keynote Paper: Design Strategy.' In Eder, W. (ed.), *Proceedings of the 1987 International Conference on Engineering Design (vol. 1).* New York: American Society of Mechanical Engineers.

Andreasen, M.M., and Hein, L. (1987). *Integrated Product Development.* London: IFS Publications.

Archer, B. (1981). 'A View of the Nature of Design Research.' In Jaques, R., and Powell, J.A. (eds.), *Design: Science: Method.* Guildford: Westbury House.

Atherton, W.A. (1988). 'Pioneers 15: William Shockley, John Bardeen and Walter Brittain: Inventors of the Transistor.' *Electronics and Wireless World* 94(1625), 273–275.

Atkinson, J.R. (1972). 'An Integrated Approach to Design and Production.' *Philosophical Transactions of the Royal Society of London* A273, 99–118.

Bacharach, S.B., and Lawler, E.J. (1981). *Bargaining: Power, Tactics, and Outcomes.* San Francisco: Jossey-Bass.

Badawy, M.K. (1975). 'Organizational Designs for Scientists and Engineers: Some Research Findings and Their Implications for Managers.' *IEEE Transactions on Engineering Management*, EM-22, 4, 134–138.

Barko, W., and Pasmore, W. (1986). 'Introductory Statement to the Special Issue on Sociotechnical Systems: Innovations in Designing High Performance Systems.' *Journal of Applied Behavioral Science* 22(3), 195–199.

Barth, R.T., and Steck, R. (1979). *Interdisciplinary Research Groups: Their Management and Organization.* International Research Group on Interdisciplinary Programs.

Bass, B.M., and Barrett, G.V. (1978). *People, Work, and Organizations: An Introduction to Industrial and Organizational Psychology.* Boston: Allyn & Bacon.

Bass, B.M., and Barrett, G.V. (1981). *People, Work, and Organizations: An Introduction to Indusrial and Organizational Psychology.* Boston: Allyn and Bacon.

Beitz, W. (1987). 'General Approach of Systematic Design: Application of *VDI Guideline 2221.*' In Eder, W. (ed.), *Proceedings of the 1987 International Conference on Engineering Design (vol. 1).* New York: American Society of Mechanical Engineers.

Belbin, R.M. (1981). *Management Teams: Why They Succeed or Fail.* London: Heinemann.

Bensinger, B.E. (1965). 'A Creative Organization.' In Steiner, G.A. (ed.), *The Creative Organization.* Chicago: University of Chicago Press.

Berridge, T. (1977). *Product Innovation and Development.* London: Business Books.

Bouwen, R., Visch, J. De, and Steyaert, C. (1987). 'Intrapreneuring and Entrapreneuring: Managing Involvement Through the Creation of Shared Meaning.' Paper presented at Third West European Congress on the Psychology of Work and Organization, Antwerp, 14–16 April.

Bower, M. (1965). 'Nurturing Creativity in an Organization.' In Steiner, G.A. (ed.), *The Creative Organization.* Chicago: University of Chicago Press.

Bridges, A.H. (1981). 'The Application of Computer Based Models to Building Design.' In Jaques, R., and Powell, J.A. (eds.), *Design: Science: Method.* Guildford: Westbury House.

Broadbent, G. (1981). 'The Morality of Designing.' In Jaques, R., and Powell, J.A. (eds.), *Design: Science: Method.* Guildford: Westbury House.

Burnstein, E., and Berbaum, M.L. (1983). 'Stages in Group Decision Making: The Decomposition of Historical Narratives.' *Political Psychology* 4, 531–561.

Cabinet Office Advisory Council for Applied Research and Development (1982). *Facing International Competition: The Impact on Design of Standards, Regulations, Certification and Approvals.* London: HMSO.

Carter, R., Martin, J., Mayblin, B., and Munday, M. (1984). *Systems, Management and Change.* London: Harper & Row/Open University.

Chalk, W.S. (1981). 'Design Methodology and Teaching Engineering Design.' In Hubka, V. (ed.), *Proceedings of the International Conference on Teaching Engineering Design.* Rome: Workshop Design & Construction.

Cherns, A. (1976). 'The Principles of Sociotechnical Design.' *Human Relations* 29, 783–792.

Clausing, D.P. (1984). Product Development Process: Overview and Needs. *Proceedings of the ASME International Computers in Engineering Conference.* Las Vegas, Nevada.

Clutterbuck, D. (1979). "R&D Under Management's Microscope." *International Management,* February.

Cohn, S.F., and Turyn, R.M. (1984). 'Organizational Structure, Decision-making Procedures, and the Adoption of Innovations.' *IEEE Transactions on Engineering Management,* EM-31, 4, 154–161.

Cooper, R.G. (1979a). 'The Dimensions of Indusrial New Product Success and Failure.' *Journal of Marketing* 43(3), 93–103.

Cooper, R.G. (1979b). 'Identifying Industrial New Product Success: Project New Prod.' *Industrial Marketing Management* 8, 124–135.

Cooper, R.G. (1980). 'Project New Prod: Factors in New Product Success.' *European Journal of Marketing* 14(5/6), 277–292.

Cooper, R.G. (1986). *Winning at New Products.* Reading, MA: Addison-Wesley.

Cooper, R.G. (1988). 'The New Product Process: A Decision Guide for Management.' *Journal of Marketing Management* 3(3), 238–255.

Cooper, R.G., and Kleinschmidt, E.J. (1986). 'An Investigation into the New Product Process: Steps, Deficiencies and Impact.' *Journal of Production Innovation Management* 3(2), 71–85.

Cooper, R.G., and Kleinschmidt, E.J. (1987a). 'What Makes a New Product a Winner: Success Factors at the Project Level.' *R.&D. Management* 17(3), 175–189.

Cooper, R.G., and Kleinschmidt, E.J. (1987b). 'Success Factors in Product Innovation.' *Industrial Marketing Management* 16(3), 215–223.

Cooper, R.G., and Kleinschmidt, E.J. (1987c). 'New Products: What Separates Winners from Losers.' *Journal of Product Innovation Management* 4(3), 169–184.

Council of Engineering Institutions (1968). *A National Design Council. Conway Report.*

Craig, S.R. (1986). 'Seeking Strategic Advantage with Technology? Focus on Customer Value.' *Long Range Planning*, 19(2), 50–56.

Cross, N., and Nathenson, M. (1981). 'Design Methods and Learning Methods.' In Jaques, R., and Powell, J.A. (eds.), *Design: Science: Method.* Guildford: Westbury House.

Cummings, T.G. (ed.) (1980). *Systems Theory for Organization Development.* Chichester: Wiley.

Daub, H.R. (1987). 'Development of an Integrated 3D-CAD-System for Structural and Mechanical Engineering.' In Eder, W. (ed.), *Proceedings of the 1987 International Conference on Engineering Design (vol. 2).* New York: American Society of Mechanical Engineers.

Dill, D.D., and Pearson, A.W. (1984). 'The Effectiveness of Project Managers: Implications of a Political Model of Influence.' *IEEE Transactions on Engineering Management*, EM-31, 3, 138–145.

Dowdy, W. L., and Nikolchev, J. (1986). *Long Range Planning*, 19(2), 38–49.

Dreyfus, S.E. (1982). 'Formal Models Versus Situational Understanding: Inherent Limitations on the Modeling of Business Expertise.' *Office: Technology and People* 1, 133–165.

Drucker, P.F. (1970). *The Effective Executive.* London: Pan Business Management.

Ebert, R.J., Slusher, E.A., and Ragsdell, K.M. (1987). 'Modeling the Engineering Design Process.' In Eder, W. (ed.), *Proceedings of the 1987 International Conference on Engineering Design (vol. 1).* New York: American Society of Mechanical Engineers.

Eekels, J. (1987). 'On Strategy and Engineering Design.' In Eder, W. (ed.), *Proceedings of the 1987 International Conference on Engineering Design (vol. 1).* New York: American Society of Mechanical Engineers.

Ehrlenspiel, K. (1987) 'Reduction of Product Costs in West Germany.' *Proceedings of the 1987 International Conference on Engineering Design (vol. 2).* New York: American Society of Mechanical Engineers.

Ehrlenspiel, K., and John, T. (1987). 'Inventing by Design Methodology.' *Proceedings of the 1987 International Conference on Engineering Design (vol. 1).* New York: American Society of Mechanical Engineers.

Emery, F.E., and Trist, L. (1965). 'The Causal Texture of Environments.' *Human Relations* 18, 21–32.

Foster, R.N. (1986). *Innovation: The Attacker's Advantage.* London: Macmillan.

Friedlander, R., and Schott, B. (1981). 'The Use of Task Groups and Task Forces in Organizational Change.' In Payne, R., and Cooper, C.L. (eds.), *Groups at Work.* Chichester: Wiley.

Gardiner, P., and Rothwell, R. (1985). 'Tough Customers: Good Designs.' *Design Studies* 6(1), 7–17.

George, A.L. (1974). 'Adaptation to Stress in Political Decision Making: The Individual, Small Group, and Organizational Contexts.' In Coeltho, G.V., Hamburg, D.A., and Adams, J.E. (eds.), *Coping and Adaptation*. New York: Basic Books.

Gerwin, D. (1981). 'Relationships Between Work and Technology.' In Nystrom, P.C., and Starbuck, W.H. (eds.), *Handbook of Organizational Design, vol. 2, Remodeling Organizations and Their Environments*. Oxford: Oxford University Press.

Gill, A.K. (1986). 'Engineers Matter: The Lucas Key to International Success.' *Proceedings of the Institutute of Mechanical Engineers*, 200(B3), 205–213.

Goldsmith, W., and Clutterbuck, D. (1985). *The Winning Streak*. Harmondsworth: Penguin Books.

Golembiewski, R.T. (1962). *Behavior and Organization: O&M and the Small Group*. Chicago: Rand McNally.

Golembiewski, R.T., and Miller, G.J. (1981). 'Small Groups in Political Science.' In Long, S.L. (ed.), *The Handbook of Political Behavior (vol. 2)*. New York: Plenum Press.

Grotloh, K., and Rothlin, E. (1983). 'Techno-economic Product Designing.' *Design Studies*, 4(3), 177–182.

Gruber, H. (1981). *Darwin On Man* (2nd ed.). Chicago: University of Chicago Press.

Guetzkow, H. (1965). 'The Creative Person in Organizations.' In Steiner, G.A. (ed.), *The Creative Organization*. Chicago: University of Chicago Press.

Gunz, H.P., and Pearson, A.W. (1977a). 'Matrix Organization in R&D.' In Knight, K.W. (ed.), *Matrix Management*. London: Saxon House.

Gunz, H.P., and Pearson, A.W. (1977b). 'Introduction of a Matrix Structure into an R&D Establishment.' *R&D Management* 7(3), 173–181.

Hage, J., and Aiken, M. (1967). 'Program Change and Organizational Properties: A Comparative Analysis.' *American Journal of Sociology* 72, 503–519.

Hales, C. (1987). 'Report on the International Congress of Planning and Design Theory.' Paper presented to Design Research Society Workshop, Manchester Polytechnic, 17 October.

Hare, A.P. (1985). *Social Interaction as Drama: Applications from Conflict Resolution*. Beverly Hills: Sage.

Hayes, C., and Abernathy, W.J. (1980). 'Managing Our Way to Economic Decline.' *Harvard Business Review* (July/August), 69–77.

Heany, D.F., and Vinson, W.D. (1984). 'A Fresh Look at New Product Development.' *Journal of Business Strategy* 5(2) (Fall), 22–31.

Hein, L. (1987). 'Boosting Product Development Ability.' In Eder, W. (ed.), *Proceedings of the 1987 International Conference on Engineering Design (vol. 1)*. New York: American Society of Mechanical Engineers.

Hein, L., and Andreasen, M.M. (1983). 'Integrated Product Development: A New Framework for Methodical Design.' *Proceedings of ICED83, Copenhagen*. Zurich: Heurista.

Hickson, D.J., Butler, R.J., Cray, D., Mallory, G.R., and Wilson, D.C. (1986). *Top Decisions: Strategic Decision-making in Organizations.* Oxford: Basil Blackwell.

Hill, P.H. (1984). 'Decisions in the Corporate Environment.' In Swap, W.C., and Associates, *Group Decision Making.* Beverly Hills: Sage.

Hodgetts, R.M. (1968). 'Leadership Techniques in the Project Organization.' *Academy of Management Journal* 11 (June), 211–219.

Hoffman, L.R. (1982). 'Improving the Problem-Solving Process in Managerial Groups.' In Guzzo, R.A. (ed.), *Improving Group Decision Making in Organizations.* New York: Academic Press.

Holt, J.E., Radcliffe, D.F., and Schoorl, D. (1985). 'Design or Problem Solving: A Critical Choice for the Engineering Profession.' *Design Studies* 6(2), 107–110.

Holt, M. (1980). 'Using the Computer to Control Design Costs.' Session 13, Paper 1, presented at the Design Engineering Conference, Birmingham.

Hosking, D.M., and Morley, I.E. (1988). 'The Skills of Leadership.' In Hunt, J.G., Baliga, B.R., Dachler, H.P., and Schriesheim, C.A. (eds.), *Emerging Leadership Vistas.* Lexington, MA: Lexington Books (D.C. Heath).

Hubka, V. (1982). *Principles of Engineering Design.* London: Butterworth Scientific.

Hubka, V. (1983). 'Design Tactics=Methods+Working Principles for Design Engineers.' *Design Studies* 4(3), 188–195.

Hull, F.M., Hage, J., and Azumi, K. (1985). 'R&D Management Strategies: America Versus Japan.' *IEEE Transactions on Engineering Management*, EM-32, 2, 78–83.

Humphreys, P. (1984). 'Contribution to ESRC Joint Committee Workshop on Processes Within Design Teams.' Economic and Social Research Council, London, 6 June.

Janis, I.L. (1972). *Victims of Groupthink.* Boston: Houghton Mifflin.

Janis, I.L. (1982). *Victims of Groupthink* (2nd ed.). Boston: Houghton Mifflin.

Janis, I.L. (1984). 'Counteracting the Adverse Effects of Concurrence Seeking in Policy-Planning Groups: Theory and Research Perspectives.' In Brandstatter, H., Davis, J.H., and Stocker-Kreichgauer, G. (eds.), *Group Decision Making.* New York: Academic Press.

Janis, I.L., and Mann, L. (1977). *Decision Making: A Psychological Analysis of Conflict, Choice and Commitment.* London: Collier Macmillan.

Jaques, R. (1981). 'Introduction.' In Jaques, R., and Powell, J.A. (eds.), *Design: Science: Method.* Guildford: Westbury House.

Johne, A.F., and Snelson, P. (1988). 'Marketing's Role in Successful Product Development.' *Journal of Marketing Management* 3(3), 256–268.

Kahneman, K., Slovic, P., and Tversky, A. (eds.) (1981). *Judgment Under Uncertainty: Heuristics and Biases.* Cambridge: Cambridge University Press.

Kanter, R. (1984). *The Change Masters: Corporate Entrepreneurs at Work.* London: Allen & Unwin.

Katz, D., and Kahn, R.L. (1978). *The Social Psychology of Organizations* (2nd ed.). Chichester: Wiley.

Kerley, J.J. (1987). 'Retroduction: A New Structured Approach to Mechanical Design.' In Eder, W. (ed.), *Proceedings of the 1987 International Conference on Engineering Design* (vol. 1). New York: American Society of Mechanical Engineers.

Killman, R.H., Pondy, L.R., and Slevin, D.P. (eds.). (1976). *The Management of Organization Design, vol. 1, Strategies and Implementation.* New York: North Holland.

Kimber, M.S., Coulthurst, A., Hamilton, P.H., Sharp, J.E., and Smith, D.G. (1985). 'Curriculum for Design: Engineering Undergraduate Courses.' *Proceedings of Working Party, SEED.*

Kirkpatrick, S.A., Davis, D.F., and Robertson, R.D. (1976). 'The Process of Political Decision-Making in Groups.' *American Behavioral Scientist* 20(1), 33–64.

Kolodny, H.F., and Dresner, B. (1986). 'Linking Arrangements and New Work Designs.' *Organizational Dynamics* 14(3), 33–51.

Kuntreuther, H. (1983). 'A Multi-attribute, Multi-party Model of Choice: Descriptive and Prescriptive Considerations.' In Humphreys, P.C., Svenson, O., and Vari, A. (eds.), *Analysing and Aiding Decisions.* Amsterdam: North Holland.

Kuppuraju, N., Ittimakin, P., and Mistree, F. (1985). 'Design Through Selection: A Method That Works.' *Design Studies* 6(2), 91–106.

Lawrence, P.R., and Lorsch, J.W. (1967a). 'Differentiation and Intregration in Complex Organisations.' *Administrative Science Quarterly* 12, 1–47.

Lawrence, P.R., and Lorsch, J.W. (1967b). *Organization and Environment: Managing Differentiation and Integration.* Boston: Division of Research, Harvard Graduate School of Business Administration.

Lindley, B.C. (1983). 'Facing International Competition.' *Design Studies* 4(3), 151–154.

Lorsch, J.W. (1970). 'Introduction to the Structural Design of Organizations.' In Dalton, G.W., and Lawrence, P.R. (eds.), *Organizational Structure and Design.* Homewood, IL: Irwin. Georgetown, Ontario: Dorsey Press.

Lorsch, J.W., and Lawrence, P.R. (1970). 'Organizing for Product Innovation.' In Dalton, G.W., and Lawrence, P.R. (eds.), *Organizational Structure and Design.* Homewood, IL: Irwin. Georgetown, Ontario: Dorsey Press.

Maidique, M.A., and Zirger, B.J. (1984). 'A Study of Success and Failure in Product Innovation: The Case of the U.S. Electronics Industry.' *IEEE Transactions on Engineering Management,* EM-31, 4, 192–203.

Managing Design: An Initiative in Management Education (1984). London: Council for National Academic Awards.

March, J.G., and Simon, H.A. (1958). *Organizations.* New York: Wiley.

Mayall, W. (1979). *Principles in Design.* London: Design Council.

Mayall, W. (1983). 'Problems, Problems.' *Engineering Designer* (November-December), 3.

McDonough, E.F. III, and Kinnunen, R.M. (1984). 'Management Control of New Product Development Products.' *IEEE Transactions on Engineering Management,* EM-31, 1, 18–21.

Might, R. (1984). "An Evaluation of the Effectiveness of Project Control Systems." *IEEE Transactions on Engineering Management,* EM-31, 3, 127–137.

Might, R.J., and Fischer, W.A. (1985). 'The Role of Structural Factors in Determining Project Management Success.' *IEEE Transactions on Engineering Management,* EM-32, 2, 71–77.

Minkes, A.L. (1987). *The Entrepreneurial Manager: Decisions, Goals, and Business Ideas.* Harmondsworth: Penguin Books.

Mintzberg, H., Raisinghani, D., and Theoret, A. (1976). 'The Structure of "Unstructured" Decision Processes.' *Administrative Science Quarterly* 21, 246–275.

Morgan, G. (1987). *Images of Organization.* Beverly Hills: Sage.

Morley, I.E. (1979). 'Job Enrichment, Job Enlargement, and Participation in Work.' In Stephenson, G.M., and Brotherton, C.J. (eds.), *Industrial Relations: A Social Psychological Approach.* Chichester: Wiley.

Morley, I.E. (1981). 'Negotiation and Bargaining.' In Argyle, M. (ed.), *Social Skills and Work.* London: Methuen, 84–115.

Morley, I.E. (1984). 'Bargaining and Negotiation.' In Cooper, C.L., and Makin, P. (eds.), *Psychology for Managers.* London: British Psychological Society/Macmillan.

Morley, I.E. (1986). 'Negotiation and Bargaining.' In Hargie, O. (ed.), *Handbook of Communication Skills.* London: Croom Helm.

Morley, I.E. (1987). 'Psychology and Engineering Design: Some Comments Prompted by the International Congress on Planning and Design Theory.' Paper presented to Design Research Society Workshop, Manchester Polytechnic, 17 October.

Morley, I.E., and Hosking, D.M. (1984). 'Decision Making and Negotiation: Leadership and Social Skills.' In Gruneberg, M., and Wall, T.D. (eds.), *Social Psychology and Organizational Behaviour.* Chichester: Wiley.

Morley, I.E., and Hosking, D.M. (1988a). 'Decision Making Approaches to Social Psychology and Organisational Behaviour.' Working Paper, Department of Psychology, University of Warwick.

Morley, I.E., and Hosking, D.M. (1988b). 'Systems Approaches to Social Psychology and Organizational Behaviour.' Working Paper, Department of Psychology, University of Warwick.

Morley, I.E., and Palmer, H. (1984). *Report of the ESRC Workshop of the Process of Design at Arden House, Warwick University.* London: Economic and Social Research Council.

Morley, I.E., and Pugh, S. (1987). 'The Organization of Design: An Interdisciplinary Approach to the Study of People, Process and Context.' In Eder, W. (ed.), *Proceedings of the 1987 International Conference on Engineering Design* (vol. 1). New York: American Society of Mechanical Engineers.

M'Pherson, P.K. (1980). 'Systems Engineering: An Approach to Whole-System Design.' *Radio and Electronic Engineer* 50(11), 545–558.

M'Pherson, P.K. (1981). 'A Framework for Systems Engineering Design.' *Radio and Electronic Engineer,* 51(2), 59–93.

Nicholls, J. (1986). 'Beyond Situational Leadership: Congruent and Transforming Models for Leadership Training.' *European Management Journal* 4(1), 41–50.

Oakley, M. (1984). *Managing Product Design.* London: Weidenfeld & Nicholson.

Otker, T. (1984). 'The Introduction of Engineering Know How Through CAD/CAM and the Implications to the Organisation.' *Proceedings of the Second European Conference on Developments in CAD CAM.* Amersfoort: ECD-CAD CAM.

Pahl, G., and Beitz, W. (1984). *Engineering Design.* London: Design Council.

Payne, W.J. (1987). *The Leading Edge.* London: Macdonald Orbis.

Pearson, A.W., and Gunz, H.P. (1981). 'Project Groups.' In Payne, R., and Cooper, C. (eds.), *Groups at Work.* Chichester: Wiley.

Pelz, D.C. (1983). 'Quantitative Case Histories of Urban Innovations: Are There Innovating Stages?' *IEEE Transactions on Engineering Management*, EM-30, 2, 60–67.

Peterson, P.G. (1965). 'Some Approaches to Innovation in Industry.' In Steiner, G.A. (ed.), *The Creative Organization.* Chicago: University of Chicago Press.

Petroski, H. (1987). 'Design as Obviating Failure.' In Nadler, G. (ed.), *1987 Congress on Planning and Design Theory: Plenary and Interdisciplinary Lectures.* New York: American Society of Mechanical Engineers.

Peyronnin, C.A. (1987). 'Keeping Contemporary with the Changing Nature of Inter-disciplinary Design.' In Eder, W. (ed.), *Proceedings of the 1987 International Conference on Engineering Design* (vol. 1). New York: American Society of Mechanical Engineers.

Pfeffer, J. (1981). *Power in Organizations.* Boston: Pitman.

Pruitt, D.G. (1981). *Negotiation Behavior.* New York: Academic Press.

Pugh, S. (1979). 'The Design Audit: How to Use It.' *Proceedings of the Design Engineering Conference, NEC.* Birmingham: NEC.

Pugh, S. (1981). 'Concept Selection: A Method That Works.' *Proceedings of the 1981 International Conference on Engineering Design.* Rome: ICED.

Pugh, S. (1983a). 'The Application of CAD in Relation to Dynamic/Static Product Concepts. *Proceedings of the 1983 International Conference on Engineering Design.* Copenhagen: ICED.

Pugh, S. (1983b). 'Design Activity Model.' Engineering Design Centre, Loughborough University of Technology, June 28.

Pugh, S. (1984a). 'CAD-CAM: Hindrance or Help to Design.' *Proceedings of Conception et Fabrication Assisteé par Ordinateur.* Brussels: Université Libre de Bruxelles.

Pugh, S. (1984b). 'Further Development of the Hypothesis of Static/lDynamic Concepts in Product Design.' *Proceedings of the International Symposium on Design and Synthesis.* Tokyo: ISDS.

Pugh, S. (1985a). 'CAD/CAM: Its Effect on Design Understanding and Progress.' *Proceedings of the International Conference CAD/CAM Robotics and Automation.* Tucson, AZ: IC-CAD/CAM.

Pugh, S. (1985b). 'Systematic Design Procedures and Their Application in the Marine Field: An Outsider's View.' *Proceedings of the Second International Marine Systems Design Conference* (pts. 1–10). Copenhagen: IMSDC.

Pugh, S. (1986). 'Design Activity Models: Worldwide Emergence and Convergence.' *Design Studies* 7(3), 167–173.

Pugh, S. (1987). 'Visions of Design Practices for the Future: Design Synthesis.' Paper presented to the joint Workshop Xerox Design Institute/MIT, Boston, MA, October 18–20.

Pugh, S., and Smith, D.G. (1976). 'CAD in the Context of Engineering Design: The Designer's Viewpoint.' *CAD 76 Proceedings* (pp. 193–198).

Rawson, K.J. (1976). 'The Art of Ship Designing.' *Proceedings of Europort '76, International Maritime Conference.* Amsterdam: IMC.

Rawson, P. (1987). *Creative Design: A New Look At Design Principles.* London: Macdonald Orbis.

Rapoport, A. (1968). 'Editor's Introduction.' In Rapoport, A. (ed.), *Clausewitz On War.* Harmondsworth: Penguin Books.

Richter, W. (1987). 'To Design in an Interdisciplinary Team.' In Eder, W. (ed.), *Proceedings of the 1987 International Conference on Engineering Design* (vol. 1). New York: American Society of Mechanical Engineers.

Rothwell, R., Freeman, C., Jervis, P., Horsely, A., Robertson, A.B., and Townsend, J. (1974). 'SAPPHO-Updated: Project SAPPHO Phase II. *Research Policy* 3(3), 258–291.

Rothwell, R., and Gardiner, P. (1983). 'The Role of Design in Product and Process Change.' *Design Studies* 4(3), 161–169

Rothwell, R., and Gardiner, P. (1988). 'Re-Innovation and Robust Design: Producer and User Benefits.' *Journal of Marketing Management* 3(3), 372–387.

Rudwick, B.H. (1983). 'Pitfalls in the Architectural Design Process and a Defensible Method for Overcoming Them.' *IEEE Transactions on Engineering Management*, EM-30, 3, 128–139.

Rzevski, G. (1981). 'On the Design of a Design Methodology.' In Jaques, R., and Powell, J.A. (eds.), *Design: Science: Method.* Guildford: Westbury House.

Sayles, L.R. (1976). 'Matrix Organizations: The Structure with a Future.' *Organizational Dynamics* (Autumn), 2–17.

Shockley, W. (1976). 'The Path to the Conception of the Junction Transistor.' *IEEE Transactions*, ED-23, 597–620.

Silverman, B.G. (1985). 'Expert Intuition and Ill Structured Problem Solving.' *IEEE Transactions on Engineering Management*, EM-32, 1, 29–33.

Simon, H.A. (1957). *Models of Man.* New York: Wiley.

Simon, H.A. (1965). 'Administrative Decision Making.' *Public Administration Review* 25, 1.

Simon, H.A. (1981). *The Sciences of the Artificial, Second Edition.* Cambridge, MA: MIT Press.

Stauffer, L.A., Ullman, D.G., and Dietterich, T.G. (1987). 'Protocol Analysis of Engineering Design.' In Eder, W. (ed.), *Proceedings of the 1987 International Conference on Engineering Design* (vol. 1). New York: American Society of Mechanical Engineers.

Steinbruner, J.D. (1974). *The Cybernetic Theory of Decision.* Princeton, NJ: Princeton University Press.

Steiner, G.A. (1965). *The Creative Organization.* Chicago: University of Chicago Press.

Steiner, I. (1972). *Group Process and Productivity.* New York: Academic Press.

Stewart, R. (1987). *Design and British Industry.* London: John Murray.

Taguchi, G. (1978). 'Off-Line and On-Line Quality Control System.' *Proceedings of the International Conference on Quality Control.* Tokyo: KQC.

Trist, E., Higgins, G., Murray, H., and Pollack, A.B. (1983). *Organizational Choice.* London: Tavistock.

Turner, B.T. (1977). 'Stetting up and Managing a Design Team.' *Proceedings Engineering Design Conference 1977.* London: Fairs & Exhibitions.

Turner, B.T. (1985). 'Managing Design in the New Product Development Process: Methods for Company Executives.' *Design Studies* 6(1), 51–56.

Tushman, M., and Nadler, D. (1986). 'Organization for Innovation.' *California Management Review* 28(3) (Spring), 74–92.

van den Kroonenberg, H.H. (1987). 'The Development of CAD from Product-Orientation to Process Orientation. In Eder, W. (ed.), *Proceedings of the 1987 International Conference on Engineering Design* (vol. 1). New York: American Society of Mechanical Engineers, p. 434–445.

Verein Deutscher Ingenieure (1987). *VDI Guidelines 2221 Systematic Approach to the Design of Technical Systems and Products.* Dusseldorf: VDI-Verlag.

von Hippel, E. (1976). 'The Dominant Role of Users in the Scientific Instrument Innovation Process.' *Research Policy* 5, 3.

von Hippel, E. (1978). 'Users as Innovators.' *Technology Review* 80(3), 30–39.

von Hippel, E. (1979). 'A Customer-Active Paradigm for Industrial Product Idea Generation.' In Baker, M.J. (ed.), *Industrial Innovation.* London: Macmillan.

Wallace, K.M., and Hales, C. (1987). 'Detailed Analysis of an Engineering Design Project.' In Eder, W. (ed.), *Proceedings of the 1987 International Conference on Engineering Design* (vol. 1). New York: American Society of Mechanical Engineers.

Watson, J. (1968). *The Double Helix.* New York: Signet.

Weick, K.E. (1979). *The Social Psychology of Organizing.* Reading, MA: Addison-Wesley.

Weisberg, R.W. (1986). *Creativity: Genius and Other Myths.* New York: Freeman.

Whipple, R., and Clark, P. (1986). *Innovation and the Auto Industry: Product, Process and Work Organization.* London: Francis Pinter (Publishers).

Winstanley and Francis (1988). *Engineering* (March), 133–135.

Witte, E. (1972). 'Field Research on Complex Decision-making Processes: The Phase Theorem.' *International Studies of Management and Organization* 2(2), 156–182.

Wohl, J.G. (1981). 'Force Management Decision Requirements for Air Force Tactical Command and Control.' *IEEE Transactions of Systems, Man, and Cybernetics*, SMC-11, 9, 618–639.

Zachary, W.B., and Krone, R.M. (1984). 'Managing Creative Individuals in High Technology Research Projects.' *IEEE Transactions on Engineering Management*, EM-31, 1, 37–40.

Zarecor, W.D. (1975). 'High Technology Product Planning.' *Harvard Business Review.*

Appendix: Chronology

'Engineering Design: Towards a Common Understanding.' *Proceedings of the Second International Symposium of Information Systems for Designers*, University of Southampton, July 1974, pp. D4–D6. (Chapter 26)

'Marathon 2550: A Successful Joint Venture' (with Douglas G. Smith). Engineering Design Centre, Loughborough University of Technology, 1976. (Chapter 6)

'The Dangers of Design Methodology' (with Douglas G. Smith). Presented at the First European Design Research Conference - Changing Design, Portsmouth Polytechnic, April 1976. (Chapter 12)

'CAD in the Context of Engineering Design: The Designer's Viewpoint' (with Douglas G. Smith). *Proceedings of the Second International Conference on Computers in Engineering and Building Design CAD76*, London, 1976, pp. 193–198. (Chapter 17)

'The Engineering Designer: His Tasks and Information Needs.' *Proceedings of the Third International Symposium on Information Systems for Designers*, University of Southampton, March 1977, pp. 63–66. (Chapter 30)

'CAD and Design Education: Should One Be Taught Without the Other?' Paper presented at the International Conference on Computers and Design Education, London, 1977. (Chapter 18)

'Manufacturing Cost Data for the Designer.' *Proceedings of the Engineering Design Conference*, London, 1977, pp. 17-1–17-16. (Chapter 32)

'Quality Assurance and Design: The Problem of Cost Versus Quality.' *Quality Assurance* 4(1) (March 1978), 3–6. (Chapter 34)

'Engineering Design Teaching Ten Years On' (with Douglas G. Smith). *Engineering Design Education* (Spring 1978), 20–22. (Chapter 2)

'Engineering Design Education with Real-Life Problems.' *European Journal of Engineering Education* 3(2) (Summer 1978), 135–147. (Chapter 4)

'Engineering Design at the Postgraduate Level.' Paper presented at the Tomorrow's Engineering Design Conference, Loughborough University of Technology, 1979. (Chapter 3)

'Design Audit: How to Use It.' *Proceedings of the Design Conference*, Birmingham, (October 1979), Session 4a, paper 3, pp. 4a/3/1–4a/3/6. (Chapter 27)

'Give the Designer a Chance to Contribute to Hazard Reduction.' *Product Liability International* 1(9) (1979), 223–225. (Chapter 35)

'Concept Selection: A Method That Works!' *Proceedings of the International Conference on Engineering Design ICED'81*, Rome, Techniche Nuove, (March 1981), pp. 497–506. (Chapter 14)

'Design Is the Biggest Exposure.' *Proceedings of the Conference on the Improvement of Product Safety*, London, (October 1981), pp. 29–30. (Chapter 31)

'Engineering Design: Time for Action.' In Evans, B., Powel, J.A., and Talbot, R. (eds.), *Changing Design* (London: Wiley, 1982), pp. 85–97. (Chapter 1)

'Design: The Integrative Enveloping Culture, Not a Third Culture.' *Design Studies* 3(2), (April 1982), 93–96. (Chapter 8)

'Projects Alone Don't Integrate: You Have to Teach Integration.' *Engineering Design Education* (Autumn 1982), 14–16. (Chapter 5)

'State of the Art on Optimization in G.B.' *Proceedings of the International Conference on Engineering Design ICED'83*, Heurista, Copenhagen, (August 1983), pp. 389–394. (Chapter 15)

'The Application of CAD in Relation to the Dynamic/Static Product Concept.' *Proceedings of the International Conference on Engineering Design ICED'83*, Heurista, Copenhagen, (August 1983), pp. 564–571. (Chapter 19)

'Engineering Out the Cost.' *Proceedings of the Sixth Annual Conference on Design Engineering*, Birmingham, October 1983, pp. 121–128. (Chapter 33)

'Research and Development: The Missing Link—Design.' *Proceedings of the International Conference on Engineering Design ICED'83*, Heurista, Copenhagen, (August 1983), pp. 500–507. (Chapter 36)

'A New Design: The Ability to Compete.' *Design Policy, Vol. 4, Evaluation.* London: Design Council, 1984. (Chapter 13)

'Further Development of the Hypothesis of Static/Dynamic Concepts in Product Design.' *Proceedings of the International Symposium on Design and Synthesis*, Japan Society of Precision Engineering, Tokyo, 1984, pp. 216–221. (Chapter 23)

'CAD/CAM: Hindrance or Help to Design?' Paper presented at the Conference on Conception et Fabrication Assisteé par Ordinateur, Université Libre de Bruxelles, (October 1984). (Chapter 20)

'CAD/CAM: Its Effect on Design Understanding and Progress.' *Proceedings of the CAD/CAM, Robotics and Automation International Conference*, Tucson, (February 1985), pp. 385–389. (Chapter 21)

'Systematic Design Procedures and Their Application in the Marine Field: An Outsider's View.' *Proceedings of the Second International Marine Systems Design Conference Theory and Practice of Marine Design*, Lyngby, (May 1985), pp. 1–10. (Chapter 9)

'Engineering Design in Practice.' In Wall, R.A. (ed.), *Product Information Finding and Using: From Trade Catalogues to Computer Systems*. Gower, 1986, ch. 12. (Chapter 24)

'Design Activity Models: Worldwide Emergence and Convergence.' *Design Studies,* 7(3), 1986, 167–173. (Chapter 10)

'Integration by Design Is Achievable.' *Engineering Design Education* (Autumn 1986), 30–31. (Chapter 11)

'Total Design, Partial Design: A Reconciliation.' *Proceedings of the International Conference on Engineering Design ICED'87*, ASME, Boston, (August 1987), pp. 1005–1011. (Chapter 7)

'The Organization of Design: An Interdisciplinary Approach to the Study of People, Process and Context' (with Ian E. Morley). *Proceedings of the International Conference on Engineering Design ICED'87*, ASME, Boston, (August 1987), pp. 210–222. (Chapter 25)

'Knowledge-Based Systems in Design Activity.' Paper presented at the International Conference Modern Design Principles in View of Information Technology. Trondheim, Norway, (June 1988). (Chapter 22)

'Towards a Theory of Total Design' (with Ian E. Morley). Design Division, University of Strathclyde, 1988. (Chapter 39)

'Organising for Design in Relation to Dynamic/Static Product Concepts' (with Ian Morley). *Proceedings of the International Conference on Engineering Design ICED'89*, Institution of Mechanical Engineers, Harrogate, (August 1989), pp. 313–334. (Chapter 28)

'Engineering Design: Unscrambling the Research Issues.' *Journal of Engineering Design* 1(1) (1990), 65–72. (Chapter 37)

'Enhanced Quality Function Deployment ' (with Don Clausing). *Proceedings of the Design Productivity International Conference*, Honolulu, 1991, pp. 15–25. (Chapter 16)

'Long-Term R&D Outcomes: Will They Miss the Market?' Paper presented at the Conference on Time-Based Competition: Speeding the New Product Design and Development, (May 1991). (Chapter 38)

'Balancing Discipline and Innovation.' *Manufacturing Breakthrough* 1(1) (1992), 9–13. (Chapter 27)

Index

-A-

Abbott, Howard, 405
Abernathy, W. J., 239, 496
Adams, J. L., 492
Adie, J. F., 180
Advanced development
    generic improvements, role of, 202
    versus product engineering, 200–203
AIDA, 145, 147
Aitchison, T. E., 281
Akin, O., 495, 505
Alcatel Business Systems Ltd., 367
Allerhand, M. E., 363
Allwood, R. J., 279
Amkreutz, J. H. A. E., 249
Analogy, 146, 148
Andreasen, M. M., 490, 500
Archer, Bruce, 89, 92, 492
Art
    design viewed as the combination of
        science and, 92–94
    relationship to technology, 490–92
Ashton, J. N., 181
Atkinson, J. R., 100
Attribute listing, 146, 148

-B-

Barnard, L., 180
Beheshti, M. R., 117, 120, 121
Beitz, W., 487, 499
Belbin, R. M., 336, 337, 362
Bellinger, T. F., 439
Bennet, R. C., 455
Bensinger, B. F., 491
Berbaum, M., 336
Bertoncelj model of innovation, 53
Bäzier, P. E., 249
Bleay, J. A., 180
Bleker, B., 118
Boekholt, J. T., 120
Bones, R. A., 436
Botma, E. F., 120

Boundaries
    design boundary models, 103–5, 123,
        346, 356, 396, 501–3
    personal design boundary, 330–33,
        509–10
Bounded rationality, 512
Brainstorming, 147
Brewer, R., 438
Bridges, A. H., 508
Brooking, A. G., 285
Burnstein, E., 336
Business design activity model, 125–28,
    330–33, 363–67, 504–8

-C-

CAD. *See* Computer-aided design
Caldwell, J. B., 180
CAM. *See* Computer-aided manufacture
Carroll, J. T., 439
Carter, A. D. S., 441
Castellano, E. J., 180
Chaddock, Denis, 19, 24
Checklists, 147
Chryssostomidis, C., 100
Clark, P., 496, 503
Clausing, Don, xxxi, 188, 494
Clutterbuck, D., 336–38
Coates, P. S., 258
Coherence, 505
Cole, B. N., 181
Competitive analysis, 387, 406, 429–30
Competitive benchmarking (CBM), 196,
    197
Competitive Strategy (Porter), 160
Complexity, costing and, 439–40
Computer(s)
    defined, 229
    in dynamic design, 205
    as an integrating tool, 110–11
    marine design industry by, 102–3, 108,
        110–11
    in static design, 205

in total design, 205–6
Computer-aided design (CAD)
    design activity model and, 250
    efficiency in, 263–65
    in existing practice versus in innovative
        projects, 248–49
    iteration based on dynamic concepts,
        266–69
    iteration based on fixed concepts,
        265–66
    knowledge-based systems and, 275–77
    mechanical engineering and, 256–57
    origins of, 249–50, 253
    relationship between design activity and,
        253–55, 269–70
    static and dynamic concepts and, 263–65
    user needs and, 257–58, 269
Computer-aided design (CAD), conceptual
        design and, 211–17, 236
    dynamic (nonplateau) designs, 237,
        240–43
    static (plateau) designs, 237, 238–40
Computer-aided design (CAD), design
        education and
    benefits of, in teaching, 227–29
    guidelines for, 234
    levels of computing, 229–30
    Marathon 2550 project, 230–33
    specialist versus generalist at the under
        graduate level, 225–27
Computer-aided design (CAD), engineering
        design and
    conceptual stage of design and, 211–17
    detail stage of design and, 217–19
    factors to consider when using, 219–20
    Giraffe site placement vehicle, 217
    Marathon 2550 project, 215–17
    parabolic dish aerial example, 213–15
Computer-aided manufacture (CAM), 257
Concept selection, xxix, 353
    conceptual weakness/vulnerability in
        designs, 169–70
    costing and, 430–31
    decision-making stage, 172
    defined, 167
    example of how student groups used,
        173
    importance of, 173, 175–76
    need for numerous evaluations, 172–73
    rules and procedure for minimizing
        vulnerability, 170–72

Concept selection process, Pugh
    advanced development versus product
        engineering, 200–203
    compared to quality function
        deployment, 186
    enhancements introduced, 186–87
    matrix analysis, 196–97
    piece-part expectations, 191–93
    reverse, 187, 197
    static/dynamic conceptual status,
        186–87, 193–94
    subsystem, 191
    total system architecture, 187–91
Conceptual design
    boundaries A and B, 211–12
    computer-aided design and, 211–17,
        236–43
    defined, 211
    as a design core stage, 106, 108
    Giraffe site placement vehicle, 217
    of Marathon 2550 vehicle, 64–68, 215–17
    parabolic dish aerial example, 213–15
    teaching of, 52
Conceptual envelope, 211–12, 237, 275
Conceptually dynamic (nonplateau)
        designs, 237, 240–43
Conceptually static (plateau) designs, 237,
        238–40
Contextual analysis, xxix, 196–97
Controlled convergence, 109, 407
Conventional products
    ball valve example, 296–97
    generic base of, 298–300
    inflatable motor car tire example, 297
    as static or dynamic, 295–300
Conway Report, 491
Cooke, P., 123–24, 269
Cooper, R. G., 454, 455, 500–501
Corfield, K. G., 15–16
Corley, G. W., 435–36
Cost(s)/costing
    of components, 417–21
    concept establishment and selection
        and, 430–31
    curves, 421
    diagram describing procedure for,
        420–21
    formula for estimating, 389–400
    German system of, 422
    Giraffe site placement vehicle as an
        example of, 414–17

lack of knowledge regarding, 413, 436
market analysis and, 429–30
a new product, 429–31
an old/existing product, 431–32
other methods for determining, 421–22
by Pareto distribution, 421
procedure for preparing cost patterns,
    419–20
procedure for target, 416
product design specification and, 430
sources of information on, 422–23
stages for controlling, 429
structured approach to business needed
    to control, 427–28
target, 414
updating, 418–19
Costello, Tim, 479
Costs and quality
approaches to reliability, 440–41
complexity and, 439–40
product design specification and, 435–37
simplicity and, 438–40
standards for quality and reliability,
    441–42
Coyne, R. D., 279
Cranfield Institute of Technology, 72, 231
Creativity
need for disciplined, 351
role of, 491–92
Cross, Nigel, 89
Cross-plotting, examples of, 161, 162–65,
    196
Crouse, R. L., 439

-D-

Dankel, D. D., 281
David, D. F., 335
Day, J. G., 109
Decision making
group, 337, 508–14
models, 510–14
total design as collective, 489–90, 497–514
under stress, 337–38
Decision making and design steps
activity models of engineering design,
    499–503
business design activity model, 504–8
decision models, 510–14
design boundary models, 501–3
different arrangements of stages, impact
    of, 503–4

factors affecting, 498–99
group performance and, 508–14
integrated product development, models
    of, 500–501
integration of different functions, 506–7
personal design boundary, 509–10
project control systems, 508
VDI Guideline 2221, 499–500
Decision matrix, examples of
used in designing energy-saving devices,
    151–53
used in methods of applying heat,
    148–51
Decision trees, 147
Defect(s)
of design, 447
issues and Giraffe site placement
    vehicle, 447
of manufacture, 447
need for definitions before changing, 445
product design specification and, 446–47
relationship between product design and
    product liability, 446
of specification, 447
who is responsible for fixing, 446
de Kluyver, C. A., 238, 243
Delft University of Technology, 49
Design
See also Engineering design; Total design
basic factors when creating, 91
boundaries, 103–5, 123, 346, 356, 396,
    501–3
communication problems regarding,
    345, 395
comparison of, in other fields, 117–28
criteria for successful, 307–8
defined, xxviii
flow, 396
integrative approach to, 91–94, 99,
    506–8
life cycles and dominant, 496
praxiology, 492
process research, 466–67
relationship between computer-aided
    design and, 253–55, 269–70
relationship between knowledge-based
    systems and, 277–81
simultaneous, 506
viewed as a combination of art and
    science, 92–94, 490–91
viewed as a third culture, 89–90

Design activity model
  See also Design models
  business, 125–28, 330–33, 363–66, 504–8
  characteristics of an effective, 102, 119
  computer-aided design and, 250
  description of, 10–12, 111–13, 310–12,
    330–34, 396–97
  design core (anatomical structure), 505
  design core (techniques related to), 12,
    21–22, 312, 330, 353–55, 396, 398
  evolution of, 123
  hierarchical structures, 312–13, 398–401,
    496
  information types and levels needed by
    operators of the design core,
    313–14, 398–401
  integration and, 103–4, 121–23
  interaction between functions, 356–57,
    506–8
  personal design boundary, 330–33
  role of, xxviii, 1–2, 328
  stages, 505–6
  structure versus method, 50
  technology and, 12, 22, 396, 398
Design audit/review
  activity of, 385–86, 409
  operation of, 383–85
  operators of, 386–87
  product design specification and, 383,
    384
  product safety and, 408–10
  purpose of, 383
  stages of, 384–85, 408
  teams, 410
Design core
  anatomical structure, 505
  techniques related to, 12, 21–22, 312,
    330, 353–55, 396, 398
Design core stages
  computer use in, 110–11
  conceptual design, 106, 108
  evaluation and optimization, 109–10
  market/user needs, 105–6
  product design specification, 106
Design engineering, confusion over, xxviii
Design for assembly (DFA), 391
Design methods
  conclusions, 155
  effective engineering design defined,
    143–44
  effectiveness of qualitative, 146–47

effectiveness of quantitative, 147–48
  lack of information on the effectiveness
    of, 143
  scope of, 145–46
Design models
  See also Design activity model
  Bleker's, 118
  Boekholt's, 120
  common features in, 121
  conflicting views over designing, 117–21
  Cooke's, 123–24
  Swinkels' integrated, 117–18
  three-dimensional, 123
Design research. *See* Research; Research
    and development (R&D)
Design science, failure of, 490
Design Spiral (Atkinson), 100
Design Spiral (Eames and Drummond
    simplified), 100
Design teaching
  See also Computer-aided design (CAD),
    design education and
  broad-based courses, 28–30
  departmental educational requirements,
    81–83
  future for, 14–16
  importance of minor projects for
    students, 52–53
  industry needs and, 31, 83
  issue of integration, 47–53
  models used in, 50
  partial and total design used in, 81–83
  postgraduate level, 9–13, 28
  undergraduate level of, 13–14, 50, 52
  who is responsible for, 28
Detail design
  computer-aided design and, 217–19
  of Marathon 2550 vehicle, 68–71
  teaching of, 52
DFMA, 355
Digital Equipment, 483
Dill, D. D., 510–11
Discontinuity analysis, 440
Dowdy, W. L., 496
Dowson, D., 179
Drummond, T. G., 100
Duggan, T. V., 436
Dyer, M. G., 279
Dyer, W. G., 363
Dym, C. L., 278
Dynamic design/concepts

computer-aided design and, 263–65
computers in, 205
conceptually, 237, 240–43
conventional products, 295–300
defined, xxx
embodiment design and, 376–78
example of, xxx
iteration based on, 266–69
in product design, 333
vigilant versus nonvigilant processing and, 337–38

-E-

Eames, M. C., 100
Economic and Social Research Council of 16 Great Britain, 327
Edosomwan, J. D., 279–80
Education
  See also Computer-aided design (CAD), design education and; Design activity model; Engineering Design Centre at Loughborough University of Technology
  confusion over design in education and design in industry, 133–34
  description of the Engineering Design Centre course, 17–24
  distinction between engineering education and design education, 7–8
  future for, 14–16
  postgraduate level course for integrating practice and engineering, 9–13, 25–32
  present situation in engineering, 8–9
  problem with partial design approach to, 134
  role of design in, xxxi–xxxii, 1–2
  undergraduate level of design teaching, 13–14, 225–27
  university degrees as prerequisites, 8
Ehrlenspiel, K., 123, 270, 493
Ellis, J., 181
Embodiment design and static/dynamic 17 concepts, 376–78
Engineering, difference between research and, 9
Engineering Council/The Design Council, 363, 453
Engineering design
  See also Computer-aided design (CAD), engineering design and
  activity models of, 499–503

communication problems regarding, 345, 395
  core phases, 310–12
  defined, 307, 395–96
  design activity model in, 310–12
  information requirements related to operational level, 312–14
  as a interdisciplinary, 489
  research, 467–69
  use of term, 93
Engineering Design Centre (EDC) at Loughborough University of Technology, 143
  background of postgraduate course, 9–13, 19, 57
  broad-based courses, 28–30
  changes in course structure, 35, 94
  description of course, 17–24, 144–45
  description of student project 18 (materials handling system for pigment processing), 37–45
  design activity model, use of, 20–22
  formulation and understanding problems in course content, 19–20
  how projects are decided upon, 57–58
  implementation of, 22–24
  increase in design content in courses, 49
  Marathon 2550 project, 55–75
  student assessment, 24, 30–31
  student performance during first phase, 35–36
  student performance during second phase, 36–37
Engineering design education
  See also Computer-aided design (CAD), design education and; Design activity model; Engineering Design Centre at Loughborough University of Technology
  broad-based courses, 28–30
  description of the Engineering Design Centre course, 17–24
  future for, 14–16, 31–32
  importance of, 27
  industry needs and, 31
  narrow-based courses, 27–28
  need to combine total design and engineering skills, 84–85
  postgraduate level course for integrating practice and, 9–13, 25–32
  present situation in, 8–9

"Engineering Design: Unscrambling the Research Issues" (Pugh), 477–78
"Engineering Philosophy: The Third Culture" (Lewin), 90
Engineering research, 464–66
    See also Research; Research and development (R&D)
Enhanced quality function deployment (EQFD)
    advanced development versus product engineering, 200–203
    basic process of, 187–97
    benefits of, 203
    generic improvements, role of, 201–3
    piece-part expectations, 191–93
    research and development and, 477
    role of, xxx
    subsystem, 191
    total system architecture, 187–91
    total system design matrix, 189–91
    Xerox copier example, 188–94
Environment, effects of the, 494
Equifinality, principle of, 493
Erichsen, S., 99, 103, 105, 108, 123
Ertürk, S., 119
Ertürk, Z., 119
Evaluation(s)
    as a design core stage, 109–10
    matrix, example of, 170
    need for numerous, 172–73

-F-

FDM, 147
Feilden Report (1963), 19
Fells, I., 180
Fenner, R., 250
Fenves, S. J., 278, 284
Fischer, W. A., 495, 507
Fixed concepts, iteration based on, 265–66
Foister, P., 255
Ford Motor Co., 183, 417
Frazer, J. H., 258
Frazer, J. M., 258
Freeman, C., 496
French, M. J., 90
Friedlander, F., 335, 498
Frieling, D. H., 119
Fry, Jeremy, 250
Fuchi, K., 268
Fulmer Research Institute, 422

-G-

Gairola, A., 281
Gallin, C., 100
Galloway, D. F., 15
Gardiner, P., 492, 496
GEC-Marconi Electronics Ltd., 221
General Motors (GM), 183, 479
German system of costing, 422
Gill, A. K., 498, 503, 504
Giraffe site placement vehicle, 162, 165, 186, 217, 256
    defect issues and, 447
    as an example of target costing, 414–17
Golembiewski, R. T., 335, 338
GRASPIN, 280
Grotloh, K., 500
Group(s)
    decision making, 337, 508–14
    performance, how they work, 335
    personal design boundary, 509–10
    research on, 335
    structural versus style panel of variables, 335
    successful versus unsuccessful, 336–37
    vigilant versus nonvigilant processing of information, 337–38, 511
Guetzkow, H., 493
Guide to Design for Production, A, 315
Gulvanessian, H., 180
Gunz, H. P., 336

-H-

Habraken, N. J., 117
Hauser, John R., 188
Hayward, A., 367
Hazard reduction. *See* Defect(s)
Heany, D. F., 336, 339
Heirung, E., 99
Helms, C. P., 363
Henry, T. A., 421
Hermanns, Harry, 36
Hewitt, A. D., 180
Heyderoff, P., 280
Hibino, S., 120
Hickson, D. M., 505
Higher National Certificate, 7, 8
Hill, P. H., 498
Hinrichs, C. L., 119
Holland, M., 181
Honeywell Control Systems Ltd., 367
Hopelain, D., 495, 505

Horsely, A., 496
Hosking, D. M., 335, 367, 493, 494, 497–98
House of quality, 188–89
    product design specification and, 197–200
Howard, H. C., 282
Hubka, V., 49, 500, 505

-I-

Industrial design, use of term, 94
Industry, confusion over design in educa-
    tion and design in, 133–34
Industry needs
    acceptability of students to, 31
    design and engineering teaching and, 31
    total design and, 83
Information
    computer aids and presentation of, 321
    computer aids and type and level of,
        321
    current systems for abstracting, 322
    original equipment manufacturers'
        product data and quality of, 320–21
    requirements related to operational
        level, 398–401
    vigilant versus nonvigilant processing of,
        337–38, 511–13
Institution of Mechanical Engineers, 8
Institution of Production Engineers, 315
Institution of Technician Engineers, 8
Integrated product development, models
    24 of, 500–501
Integration
    design activity model and, 103–5, 121–23
    differentiation and, 497
Integrative approach to design, 91–94, 99
    computers as integrating tools, 110–11
International Conference on Engineering
    Design (1981), 89
International Design Participation
    Conference, 115
Inversion, 146, 148

-J-

Jacobs, R. M., 383
Janis, I. L., 338, 498, 511
Jenkin, Patrick, 127
Jervis, P., 496
John, T., 493
Johne, A. F., 506
Joselin, A. G., 226

-K-

Kimber, 364
Kinnunen, R. M., 503
Kirk, G., 367
Kirkpatrick, S. A., 335
Knowledge-based systems (KBS)
    acquisition of knowledge, 281–86
    application orientation, 279–81
    as a component of computer-aided
        design, 275–77
    components, 282–83
    constraints on design, 281, 283
    difference between diagnostic and
        generative design systems, 283–86
    partial, 281, 282–83
    relationship of design to, 277–81
    "Ten Design Rules for KBES," 279–80
    theoretical orientation, 278–79
Kotter, J. P., 362

-L-

Langenberg, H., 106
Langley, M., 436
Lateral thinking, 147
Lawrence, P. R., 363, 497, 499, 507
Leading Edge, The (Payne), 489
Levitt, T., 238
Lewin, Douglas, 90, 94
Life cycles
    dominant designs and, 496
    product, 238, 243
Lindley, B. C., 441–42
Liner Concrete Company Ltd., 55, 57, 221,
    448
Lippitt, G. L., 363
Livesly, R. K., 72
Load lines approach, 440
Lorenz, C., 159–60, 238
Lorsch, J. W., 363, 497, 499, 507
Loughborough University of Technology.
    See Engineering Design Centre at
    Loughborough University of
    Technology

-M-

MacCallum, K. J., 102–3
McDonough, E. F., III, 503
McKenzie, A. B., 180
Maher, M., 284
Management

design activity and, 330
social aspects of design and design, 492–93
subjects, 12
Managing Design for Competitive
Advantage, 363, 377
Mandell, P., 100
Mann, M., 338, 511
Marathon 2550 project
background of, 55–58
computer-aided design and, 215–17,
230–33
conceptual design of vehicle, 64–68,
215–17
detail design of vehicle, 68–71
features of, 75
formulation of work program, 63–64
market analysis for, 58–59
market introduction of, 75
preparation of product specification,
59–62
production and test of prototype, 74–75
program implementation, 64–74
students grouped into project teams,
62–63
test rig, completion of, 71–74
Marine design industry
by computer, 102–3, 108, 110–11
conceptual design in, 106, 108
design activity model and, 103–5
Design Spiral (Atkinson), 100
Design Spiral (Eames and Drummond
simplified), 100
Gallin design, 100
product design specification in, 106
search for a design model, 100
Marinissen, A. H., 49
Market design research, 469–70
Market pull
engineering design research and, 467–69
research initiation and, 464
Market research
components of, 160
costing and, 429–30
as a design core stage, 105–6
importance of, 157, 159, 506
lack of information given to designers,
157, 160–61
Marples, D. L., 147
Materials handling system for pigment
processing (student project), description of, 37–45

Materials Optimiser (Waterman), 422
Mathews, A., 282
Matrix analysis, 109
concept selection and, 196–97
use of, 355
Maver, T. W., 119
Mayall, W., 490
Mechanical engineering, computer-aided
design and, 256–57
Meek, M., 106
Merry, U., 363
META cards, 147
Might, R. J., 495, 507, 508
Mihalsky, J., 439
Miller, G. J., 335, 338
Mittal, S., 281
Model T Ford, 238, 239
Morgan, G., 495
Morgan, N., 263
Morley, C. T., 180
Morley, Ian E., 325, 327, 335, 362, 485,
487, 493, 494, 497–98, 509
Murray, Peter, 127–28
Muster, R., 455

-N-

Nadler, D., 492, 507
Napier, M. A., 383
Nicholson, Robin, 482
NIH (not invented here) syndrome, xxxi
Nikolchev, J., 496
Nonconventional products, 300–302
Nordenstrom, N., 105

-O-

Oakley, M., 374, 494, 503
Open systems theory, 493–94
Optimization
constraints of mathematical, 180
defined, 179
as a design core stage, 109–10
middle ground between evaluation and,
180–81
who should be responsible for, 181
Organization, organic versus mechanistic
forms of, 495
Organizations, as open systems, 493–94
Original equipment manufacturers' (OEM)
product data
computer aids and presentation of, 321

computer aids and type and level of, 321
current systems for abstracting, 322
description of, 315–19
information quality and, 320
Otker, T., 249

-P-

Pahl, G., 499
Palmer, H., 327
Parabolic dish aerial example, 213–15
Parametric analysis (PA)
    cross-plotting in, 161, 196
    defined, xxix, 161
    examples of cross-plots, 162–65, 196
    rules and guidelines for, 162
    static/dynamic conceptual status and, 187, 194
    use of, 355
Pareto distribution, costing by, 421
Partial design
    defined, 82
    higher education and, 134
    used in teaching design, 82
Payne, W. J., 489
PDS. *See* Product design specification
Pearson, A. W., 336, 510–11
Pelz, D. C., 503, 505
Personal design boundary, 330–33, 509–10
PERT chart, use of, 64
Peterson, P. G., 491–92
Pfeffer, J., 511
Piece-part (PP), xxx
    design matrix, 192–93
    expectations, 191–93
    research and development and, 475
Pilkington Brothers, 482
Pirnie, B., 119
Porter, M. E., 160
Potterton International Ltd., 367
Problem-solving model, 50, 94, 100
Procter & Gamble, 183
Product(s)
    conventional, 295–300
    life cycles, 238, 243
    nonconventional, 300–302
    static and dynamic concepts in design of, 333
    techniques for successful, 353–55
    total design and successful, 351–52, 463–64

Product design specification (PDS)
    boundary, 406, 407
    costing and, 430
    costs and quality and, 435–38
    defects and, 446–47
    defined, 309
    design audit/review and, 383, 384
    as a design core stage, 106, 329
    early model of, 346
    elements of, 108, 197–98, 309, 328, 336–38, 355, 362
    factors, 437–38
    formulation of, 436–38
    how to meet, 407
    importance of, 333
    original equipment manufacturers' product data and, 315–20
    poor, 355
    preparation of, for Marathon 2550 project, 59–62
    product safety and role of, 406–7
    structured formulation, xxix
    target, 438
Product development process (PDP). *See* Total design (TD)
Product engineering
    advanced development versus, 200–203
    problems with, 201–2
Product safety
    competitive analysis, 387, 406
    design audit/review and, 408–10
    problem with applying latest technology and, 405
    product design specification and, 406–7
    thoroughness and, 405
Project control systems, 508
Pugh concept selection. *See* Concept selection process, Pugh

-Q-

Qualcast Garden Products Ltd., 367
Qualitative design methods, 146–47
Quality
    See also Costs and quality
    original equipment manufacturers' product data and, 320
Quality function deployment (QFD)
    See also Enhanced quality function deployment (EQFD)
    basic, 185–86, 353, 477

benefits of, 203
compared to concept selection process
    (Pugh), 186
deployment steps for, 185
development of, 183
Quantitative design methods, 147–48
Questionnaire for design assessment
    constructing the initial interview
        schedule, 362–67
    embodiment design and static/dynamic
        concepts, 376–78
    findings of, 368–75
    general approach, 361–62
    generating questions, 362–64
    losses due faculty process, 374–75
    production of questionnaire, 367
    revising initial schedule, 367
    selecting the questions, 364
    structure of the schedule, 364–67
    testing initial schedule in use, 367

-R-

Randomness
    conventional products and, 295–300
    nonconventional products and, 300–302
Rawson, K. J., 103, 123
Rehak, D. R., 282
Reliability
    approaches to, 440–41
    standards for, 441–42
Research
    See also Market research
    confusion over design research and,
        453–55
    conventional versus nonconventional,
        455
    defined, 454
    design process, 466–67
    difference between engineering and, 9
    done on groups, 335
    engineering, 464–66
    engineering design, 467–69
    importance of, 470
    initiation of, 464
    market design, 469–70
    needed for, in design, 100
    relationship of design to research
        (pigment extrusion plant example),
        456–59
voice of the customer and, 464, 465

Research, interdisciplinary
    design activity models, 328–34
    role of, 327
Research and development (R&D)
    confusion over design research and,
        453–55
    enhanced quality function deployment
        and, 477
    factors affecting the success of, 482–83
    issues, 477–80
    piece-parts and, 475
    project example, 480–82
    quality function deployment and, 477
    subsystems and, 475
    Taguchi methods and, 476, 479–80
    total design and, 475
    total system architecture and, 475
Reverse concept selection, 187, 197,
    355
Rhodes, R. G., 422
Ritchie, G. S., 180
Robertson, A. B., 496
Robertson, R. D., 335
Robust design, 496
Rogers, P. J., 179–80
Rolls-Royce Ltd., 367
Roozanburg, N., 49
Rosenbrock, H. H., 249, 253, 269
Rothlin, E., 500
Rothwell, R., 492, 496
Rowbottom, M. D., 181
Rozenblit, J. W., 280
Russell, C. M. B., 180
Rychener, M. D., 282, 283
Rzevski, G., 505

-S-

Safety. *See* Product safety
Sanville formula, 180
Sayles, L. R., 510
Schonwald, B., 482
Schott, B., 335, 498
Schott, K., 455
Science
    design viewed as the combination of art
        and, 92–94
    relationship to technology, 490–91
Sharing Experiences in Engineering
    Design (SEED), xxviii, xxxii, 78, 133,
    135

Shortliffe, E. H., 283
Simon, H. A., 512
Simmons, M. K., 286
Simplicity, costing and, 438–40
Sinclair C5, 85
Siram, D., 284
Smith, Douglas, xxxii, 133, 367
Snelson, P., 506
Snow, C. P., 78, 89, 95
Social aspects of design and design man-
      agement, 492–93
Social psychologists, groups and, 335–38
Spence, J., 181
Starkey, C. V., 421
Static design/concepts
   computer-aided design and, 263–65
   computers in, 205
   conceptually, 237, 238–40
   conventional products, 295–300
   defined, xxx
   embodiment design and, 376–78
   example of, xxx
   in product design, 333
   products that are, 238–39, 296
   vigilant versus nonvigilant processing
      and, 337–38
Static/dynamic (S/D) conceptual status
   advanced development versus product
      engineering, 200–203
   computer-aided design and, 236–43
   concept selection used to check, 186–87
   multiple levels of a system and, 187
   parametric analysis and, 187, 194
   total system architecture, 187–91
   Xerox copier examples, 193–94
Steele, K. A., 180
Steiner, I. D., 368
Stewart, R., 362, 490, 491
Student project (materials handling system
      for pigment processing), description
      of, 37–45
Sturridge, H., 257
Subsystem (SS), xxx
   concept selection and, 191
   design matrix, 191–92
   research and development and, 475
Sulzer Brothers, 500
Swift, K. G., 282
Swinkels, T., 117–18
Syan, C. S., 282
Synthetics, 147

-T-

Taguchi, Genichi, xxix, xxxi, 353, 355, 476,
      479–80
Target costing
   description of, 414
   Giraffe site placement vehicle as an
      example of, 414–17
   procedure for, 416
Target product design specification, 438
T-charts, 146, 148
TD. *See* Total design
Teams
   See also Group(s)
   design audit/review, 410
   product development, 357
   successful versus unsuccessful, 336–37
Technical Indexes Limited, 322
Technology
   design activity model and, 12, 22, 396,
      398
   generation, 202
   product safety and the problem with
      applying latest, 405
   push and engineering design research,
      467–69
   push and research initiation, 464
   relationship of art and science to,
      490–91
Technology-dependent methods, differ-
      ence between technology-independent
      methods and, xxx–xxxi
"Technology Development and Robust
      Design" (Taguchi), 479–80
Technology-independent methods
   difference between technology-
      dependent methods and, xxx–xxxi
   use of, 477
Thoroughness
   competitive analysis, 387, 406
   design audit/review and, 408–10
   product design specification and, 406–7
   product safety and, 405
Total design (TD)
   See also under Design
   as collective decision making, 489–90,
      497–514
   computers in, 205–6
   defined, 81–82, 351, 475, 489
   efficiency and, 135–36
   how to implement, 357–58
   industry needs and, 83

interaction between functions, 356–57, 506–8
need for, 351–52
need for engineering skills and, 84–85
principles of, 352
problem of higher education and, 134–35
product model, 134
product success and, 351–52, 463–64
research and development and, 475
structured approach to, 134–35
techniques, 353–55
used in teaching design, 82
Total Quality Development (TQD), xxxi
Total Quality Development: World-Class Concurrent Engineering (Clausing), xxxi
Total system architecture (TSA), xxx, 187–91, 194, 475
Total system design matrix, 189–91
Townsend, J., 496
Toyota Autobody, 202–3
Trend analysis, 105
Tushman, M., 492, 507

-U-

UMIST, 419
Undergraduate level of design teaching, 13–14, 50, 52, 225–27
University of Southampton, 14–15
University of Strathclyde, 15, 78, 79, 85–86, 135
User needs
computer-aided design and, 257–58, 269
as a design core stage, 105–6
Utterback, J. M., 239

-V-

Value analysis/value engineering (VA/VE), 147, 185
van den Kroonenberg, H. H., 503
Van Loon, P. P., 119, 123
VDI Guideline 2221, 376, 377, 499–500, 502
VDI Richtlinien, 422
Vigilant versus nonvigilant processing, 337–38, 511–13
Vinson, W. D., 336, 339
Voice of the customer (VOC)
research and, 464, 465
role of, 355–56
von Hippel, E., 496

-W-

Walker, Eric, 9, 454
Waterman, N. A., 422
Weisberg, R. W., 491, 492
Westlund, B., 180
Whipple, R., 496, 503
Wilkie, G. A. R., 281
William, J. R., 296
Wohl, J. G., 337, 338
Wood, W. M., 439
Wright, J., 257
Wyskida, R. M., 363

-X-

Xerox copier example, 188–94, 200, 202–3

-Y-

Yasdi, R. A., 281
Yoell, B., 118

-Z-

Zeigler, B. F., 280
Zero parts approach, 439